D0812359

METHODS OF BIOCHEMICAL ANALYSIS

Volume 16

Advisory Board

METHODS OF BIOCHEMICAL ANALYSIS

Edited by DAVID GLICK
Stanford University Medical School
Palo Alto, California

VOLUME 16

INTERSCIENCE PUBLISHERS a division of
John Wiley & Sons, New York • London • Sydney • Toronto

Library of Congress Catalog Card Number 54-7232
SBN 470 30750

PRINTED IN THE UNITED STATES OF AMERICA

PREFACE TO THE SERIES

Annual review volumes dealing with many different fields of science have proved their value repeatedly and are now widely used and well established. These reviews have been concerned primarily with the results of the developing fields, rather than with the techniques and methods employed, and they have served to keep the ever-expanding scene within the view of the investigator, the applier, the teacher, and the student.

It is particularly important that review services of this nature should now be extended to cover methods and techniques, because it is becoming increasingly difficult to keep abreast of the manifold experimental innovations and improvements which constitute the limiting factor in many cases for the growth of the experimental sciences. Concepts and vision of creative scientists far outrun that which can actually be attained in present practice. Therefore an emphasis on methodology and instrumentation is a fundamental need in order for material achievement to keep in sight of the advance of useful ideas.

The current volume is another in this series which is designed to try to meet the need in the field of biochemical analysis. The topics to be included are chemical, physical, microbiological, and if necessary, animal assays, as well as basic techniques and instrumentation for the determination of enzymes, vitamins, hormones, lipids, carbohydrates, proteins and their products, minerals, antimetabolites, etc.

Certain chapters will deal with well-established methods or techniques which have undergone sufficient improvement to merit recapitulation, reappraisal, and new recommendations. Other chapters will be concerned with essentially new approaches which bear promise of great usefulness. Relatively few subjects can be included in any single volume, but as they accumulate these volumes should comprise a self-modernizing encyclopedia of methods of biochemical analysis. By judicious selection of topics it is planned that most subjects of current importance will receive treatment in these volumes.

The general plan followed in the organization of the individual chapters is a discussion of the background and previous work, a critical evaluation of the various approaches, and a presentation of the procedural details of the method or methods recommended by the author. The presentation of the experimental details is to be given in a manner that will furnish the laboratory worker with the complete information required to carry out the analyses.

Within this comprehensive scheme the reader may note that the treatments vary widely with respect to taste, style, and point of view. It is the Editor's policy to encourage individual expression in these presentations because it is stifling to originality and justifiably annoying to many authors to submerge themselves in a standard mold. Scientific writing need not be as dull and uniform as it too often is. In certain technical details, a consistent pattern is followed for the sake of convenience, as in the form used for reference citations and indexing.

The success of the treatment of any topic will depend primarily on the experience, critical ability, and capacity to communicate of the author. Those invited to prepare the respective chapters are scientists who either have originated the methods they discuss or have had intimate personal experience with them.

It is the wish of the Advisory Board and the Editor to make this series of volumes as useful as possible and to this end suggestions will always be welcome.

David Glick

May 1968

CONTENTS

CONTENTS

METHODS OF BIOCHEMICAL ANALYSIS VOLUME 16

The Isotope Derivative Method in Biochemical Analysis

J. K. WHITEHEAD AND H. G. DEAN* *The Tobacco Research Council Laboratories, Harrogate, Yorkshire, England*

* Department of Pharmacology, School of Medicine, Leeds, Yorkshire, England.

I. INTRODUCTION

1. General

Isotopes are now common tools in biochemistry, and their use has greatly facilitated the recent advances in the understanding of biochemical mechanisms and metabolic pathways. Analytical techniques have made a similar advance, but this has been due to the further development of physiochemical methods rather than isotopic techniques. Thus the sensitivity of determination techniques increased from milligram to microgram amounts, whereas the most recently developed, isotopic analytical methods have a sensitivity in the nanogram range.

Physiochemical analytical techniques have many disadvantages, especially with material of biological origin. These arise mainly from the difficulty in isolating a material which has to be determined in a pure state, since the accuracy of the technique usually depends on the purity of the sample. Advances in chromatographic purification techniques during the past decade have greatly simplified these purifications, but they are still laborious and time consuming.

The greatest disadvantage associated with purification of material for analysis by physiochemical methods is the necessity for a quantitative, or accurately reproducible, recovery. This cannot be satisfactorily overcome in any analytical technique of this kind, with the possible exception of gas–liquid chromatography, where internal standards can be used. Even in this case the compound used is not the same as the unknown, although it may be very similar chemically. Thus the method assumes that the internal standard behaves in exactly the same way as the unknown throughout the initial isolation and purification or that the ratio of recovery of internal standard to recovery of unknown does not vary with the concentration of either.

Such losses can only be overcome in a reliable, reproducible, and accurate way by the use of isotopic tracer compounds. It is the use of such tracers which gives isotopic techniques, and particularly isotope-derivative techniques, their advantages. The simple use of tracers still does not overcome the necessity for the isolation of a chemically pure sample for assay by some physiochemical method. The application of a double isotope derivative technique, where one isotope gives a measure of the amount of the unknown while the second allows correction for any losses during the analytical procedure can overcome this, since such a method requires only the determination of radioactivity. Thus the criterion of chemical purity is replaced by one of radiochemical purity. The only labeled compound in the final sample assay must be the one of interest, although other unlabeled compounds may be present, providing that they do not interfere with the counting procedure. For this reason it is possible to add milligram amounts of carrier to a nanogram sample of doubly labeled derivative in order to facilitate the radiochemical purification without altering the accuracy or sensitivity of the method in any way.

Double isotope techniques have a further advantage over other analytical methods, because their sensitivity depends on the specific activity of the labeled compound available. Recent advances in the techniques of preparation, purification, and storage of these labeled materials have made them readily available at high specific activity. The sensitivity of analytical techniques has, therefore, been increased to the nanogram level which is now necessary in the advancing field of biochemical and physiological research. The study of hormones is a field in which this technique has found particular application.

Isotopes available from atomic reactors shortly after World War II were soon utilized in biochemical studies, and in the succeeding 20 years a large number of analytical methods were developed. These methods, although a considerable advance in themselves, had two disadvantages. First, the techniques depended on the availability of a labeled form of the compound to be determined, a difficulty which became more acute as the method was applied to complex biological molecules. Second, the methods were also limited by the amount of material which could be determined physiochemically or gravimetrically.

The first difficulty was overcome by the development of the isotope derivative technique of Keston, Udenfriend, and Cannan in 1946 (67). This group used *p*-iodobenzenesulfonyl chloride-^{131}I (pipsyl chloride) to react quantitatively with an unknown mixture of amino acids. A known excess of each of the unlabeled pipsyl derivatives was added to the mixture of labeled derivatives and pure specimens isolated. These samples were then assayed by counting and by a conventional analytical technique.

This procedure still involved the isolation of a pure sample large enough for conventional assay. Further development of the method by Keston, Udenfriend, and Levy in 1947 and 1950 (69,70) eliminated this; after the preparation of the ^{131}I-labeled pipsyl derivatives of all the amino acids in the unknown mixture, a known quantity of the ^{35}S-labeled derivative of the amino acid under assay was added. In this way they were able to add unknown amounts of unlabeled-carrier pipsyl derivative during the purification and to determine the weight of the amino acid in the original sample from the ratio of the ^{35}S/^{131}I counts in a radiochemically pure specimen of the derivative.

This double isotope derivative technique was further refined by Keston and Lospalluto in 1951 (66). Known samples of the amino acids labeled with ^{14}C were added before the esterification and a similar determination was carried out. This technique had the added advantage of compensating for any losses during the esterification, and still enabled the amino acids to be determined from the ^{14}C/^{131}I ratio.

Further developments of these methods pioneered by Keston and his co-workers have led to a large number of very refined analytical techniques for the determination of amino acids, proteins, peptides, and steroids in biological systems at levels down to 10^{-8} g. The most significant advances are the introduction of tritium-labeled compounds and reagents with specific activities in excess of 100 mc/mmole, the development of chromatographic separation techniques for the purification of derivatives, and the manufacture of liquid scintillation spectrometers for counting weak β emitters, including the simultaneous determination of two isotopes.

II. METHODS AND MATERIALS

1. Introduction

The two basic methods which make up isotopic derivative techniques are isotope derivative analysis, and double isotope derivative analysis. These techniques and their variations will be described in detail and the general mathematical principles will be derived for the calculation of results.

Each method involves the use of a specialized counting technique; in double isotope derivative analysis the counting of two isotopes in the same sample is required. With some of the instruments now available, simultaneous radioassay of certain pairs of isotopes is possible. Some of the counting systems will be described.

The labeled compounds, reagents, and derivatives* are often of a highly specialized nature, and therefore a full description of them, their availability, preparation, purification, storage, and standardization will be given.

The separation techniques used in isotope derivative methods are no less important than in any other form of analysis, although the ultimate criteria of radiochemical purity are quite different. These methods are, in effect, applications of more general methods, but as far as they have special applications relevant to these analyses, they will be described.

2. General Principles

A. ISOTOPE DILUTION ANALYSIS

Although this method of analysis is not a derivative technique, and not strictly relevant to this article, the principle of isotope dilution is very important to the understanding of derivative techniques.

Simple isotope dilution analysis is carried out by adding a known quantity of compound labeled with a radioactive isotope to the mixture under assay. A sample of the pure compound is then isolated

* Compound refers to the substance under analysis. Reagent describes the molecule which combines with the compound to form the derivative.

by conventional methods and assayed for radioactivity and mass. The material to be determined acts as a diluent for the labeled compound. Under normal laboratory conditions, the chemical procedure used in the isolation of the pure sample does not differentiate between identical chemical structures of different isotopic composition. The isolated material therefore shows the same isotopic composition as the whole of the substance under assay.

Calculation.

W_u = weight of the substance to be determined

W_0 = weight of labeled compound added

C_0 = counts/min of compound added

S_0 = specific activity (radioactivity/unit mass) of compound added

W_1 = weight of pure sample isolated

C_1 = counts/min of sample isolated

S_1 = specific activity of sample isolated

$$S_0 = C_0/W_0 \quad \text{and} \quad S_1 = C_1/W_1 = C_0/(W_0 + W_u)$$

$$W_u = (C_0W_1/C_1) - W_0 \quad \text{or} \quad (C_0W_1)/C_1 \quad \text{if} \quad W_u \gg W_0 \tag{1}$$

or

$$W_u = W_0[(S_0/S_1) - 1] \tag{2}$$

B. ISOTOPE DERIVATIVE ANALYSIS

a. Without Dilution. When a compound for analysis can be made to react quantitatively with a labeled reagent of known specific activity, the amount of that compound can be determined solely from the measurement of radioactivity in the isolated, purified derivative.

S_R = equivalent specific activity of the reagent; (counts/min/g-equiv); C_D = counts/min of the completely isolated derivative; and E = equivalent weight of the compound: Then by definition

$$S_R = (C_D/W_u)\,E$$

and

$$W_u = C_DE/S_R \tag{3}$$

b. With Reverse Isotope Dilution. Simple isotope derivative analysis without dilution is only of theoretical interest since it is almost impractical to quantitatively isolate the radiochemically pure material. This process can be simplified by the addition of a large known weight, W_{UL}, of the pure unlabeled derivative. The isolation of a sample of pure derivative is carried out and the specific activity of this material determined, S_1.

Calculation. In this case

$$S_R = (C_D E)/W_u$$

$$S_1 = C_1/W_1 = C_D E/(W_u + W_{UL})$$

and

$$W_u = W_{UL}[S_1/(S_R - S_1)]$$

or

$$W_u = (S_1/S_R)(W_{UL} + W_u)$$

$$W_u = (S_1/S_R)\, W_{UL} \quad \text{when} \quad W_{UL} \gg W_u \tag{4}$$

C. DOUBLE ISOTOPE DERIVATIVE METHOD

The sensitivity of the isotopic technique of analysis may be much increased by the combination of isotope dilution and isotope derivative methods. In this method two different radioactive isotopes are used, the first in a straightforward isotope derivative analysis, and the second in an isotope dilution as a check on the losses during the purification of the derivative. Thus the ratio of the counts, due to the second isotope in the final sample, to the amount of the second isotope added, is the factor which will correct the counts due to the first isotope. This will give the total activity in the whole sample due to the derivative and, knowing the specific activity of the reagent used, the amount of the unknown present.

This method overcomes the disadvantage of the single isotope method since the necessity for the determination of the mass of a pure sample of the derivative is eliminated. The second isotope can be introduced either before or after the derivative has been formed, that is either by the addition of a small amount of the labeled compound or of labeled derivative. The latter can be labeled either in the reagent residue or in compound.

It would appear from the general principles of this method that it can only be applied when the materials involved can be suitably prepared labeled with different isotopes. The early reagents, *p*-iodobenzenesulfonyl chloride (pipsyl chloride) and the same acid anhydride (pipsan), were specifically chosen because they could be labeled with two isotopes which could readily be counted in the same sample. Iodine-131 is a strong gamma emitter and sulfur-35 a weak beta emitter. However, with the advent of multichannel scintillation spectrometers for counting soft β's, tritium and carbon-14 were used, thus giving the method a much wider application.

Availability of labeled compounds and reagents in recent years has allowed for a number of possible variations within the isotope derivative method. These variations are outlined below.

a. Labeled Unknown Compound. When the compound is available in a labeled form, a known quantity (W_0 mg) of the labeled compound of known specific activity (S_0) is added to the mixture for determination. The compound is extracted and partially purified before the derivative is formed using a reagent labeled with the second isotope and of equivalent specific activity (S_r). The derivative is freed from all other labeled materials by chromatographic purification (a sample of unlabeled carrier derivative may be added to facilitate this), and then counts due to the two isotopes are determined. If C_1 equals the counts due to the first isotope and C_2 equals those due to the second isotope, the factor to correct for the losses of the reagent isotope during purification will be

$$S_0W_0/C_1 \tag{5}$$

If an equimolecular reaction takes place between the reagent and the compound, the equivalent specific activity of the derivative with respect to the second isotope will be the same as that of the reagent S_r, i.e.,

$$S_r = C_r/M_r$$

where C_r is the number of counts due to the second isotope, assuming complete reaction with all the compound, and M_r is the number of moles of reagent (and, in this case compound) in the derivative. If the reaction is not equimolecular, then

$$S_r = C_r/E_c \tag{6}$$

where E_c is the number of equivalents of the compound in the derivative.

E_c also equals the total weight of the compound present divided by its equivalent weight.

$$E_c = (W_u + W_0)/E \tag{7}$$

and $C_r = C_2$ multiplied by the correction factor (5)

$$C_r = C_2[(S_0W_0/C_1] \tag{8}$$

Substituting Eqs. (7) and (8) in Eq. (6), we have

$$S_r = [(C_2S_0W_0)/C_1][E/(W_u + W_0)]$$

$$\therefore\ S_rC_1(W_u + W_0) = C_2S_0W_0E$$

$$S_rC_1W_u + S_rC_1W_0 = C_2S_0W_0E$$

$$W_u = (C_2S_0W_0E - W_rC_1W_0)/S_rC_1$$

$$W_u = W_0\{[(C_2S_0E)/(S_rC_1)] - 1\} \tag{9}$$

When the weight of unknown is expressed as in Eq. (9), the similarity to simple isotope dilution is obvious.

This result can also be deduced from Eq. (2)

$$W_u = W_0[(S_0/S_1) - 1]$$

In this equation the only unknown besides W_u is S_1, but

$$S_1 = C_1/W_D \tag{10}$$

where W_D is the weight of the compound isolated as derivative. Assuming an equimolecular reaction $S_r = C_r/M_r$, but for the general case

$$S_r = C_2/E_D \tag{11}$$

where E_D is the number of equivalents of the compound in the derivative, and since $E_D = W_D/E$ by definition, substituting in Eq. (11) $S_r = C_2E/W_D$

$$W_D = C_2E/S_r$$

and substituting in Eq. (10), $S_1 = C_1S_r/C_2E$. Substituting this in Eq. (2), we get Eq. (9).

This method may be applied when the labeled unknown compound is present in a mixture as a result of some experiment where a tracer has been introduced at an early stage, but has become diluted with unlabeled material to give a total of W_x mg. Such a situation arises often in the determination of secretion rates of steroids. In this case S_0 is not known, but $S_0 = C_0/W_x$, $W_0 = W_x$, and $W_u = 0$. C_0 can be determined by counting an aliquot and, applying this to Eq. (8)

$$W_x = C_2C_0E/C_1S_r$$

b. Labeled Derivative. (*1*) *Labeled in Radical.* If the compound which is being determined is not available in a conveniently labeled form, it is still possible to carry out an accurate analysis, but the method does not compensate for any losses during steps up to and including formation of the derivative.

The unknown (W_u) is caused to react quantitatively with the reagent of equivalent specific activity S_r and before purification a known weight (W_0 mg) of the derivative of specific activity S_0 counts/mg is added to the crude reaction mixture. This derivative has been prepared from carrier compound and the same reagent as used to prepare the derivative, but labeled with a different isotope. As in the first method, a radiochemically pure sample of the derivative is then isolated and the counts C_1 due to the residue of W_0 added after the formation of the derivative and C_2 due to the reagent isotope, which reacted with the unknown W_u, are determined. As in Section II-1-C-a Eqs. (5), (6), and (8) apply, but Eq. (7) becomes

$$E_C = W_u/E \qquad (12)$$

Substituting Eqs. (12) and (8) into Eq. (6)

$$S_r = (C_2S_0W_0/C_1)(E/W_u)$$

$$W_u = (C_2S_0W_0E/S_rC_1)$$

(*2*) *Labeled in Compound.* If the unknown compound is available in a labeled form, it is preferable that the analysis be carried out as under Section II-2-C-a since this procedure compensates losses in the purification if the preparation of the derivative is not quantitative. However, occasion may arise where the compound is available in a labeled form, and where it has been necessary to prepare the derivative using unlabeled reagent. This situation may arise during the purification of a labeled steroid and in this case it may be more convenient to use this derivative in the analysis. In this case the method is the same as in Section II-2-C-b-(1), but the radioactive isotope is in a different part of the molecule. Again the unknown in the sample is given by

$$W_u = C_2 S_0 W_0 E / S_r C_1$$

3. Factors Affecting Accuracy

A. PURITY OF THE MATERIALS

In all analytical procedures involving radioisotopes, it is a basic essential that the labeled materials should be both chemically and radiochemically pure. Chemical impurities will decrease the specific activity and radiochemical impurities will increase it, both factors leading to errors in results for all the methods outlined above. The radiochemical and chemical purity of all radioactive materials should be determined before use. The radiochemical purity is usually determined by reverse isotope dilution technique.

In view of the sensitivity of isotope derivative analysis and the low level at which determinations are carried out, scrupulous care must be taken in the preparation and purification of the derivative. The necessity for purity of reagents and labeled compounds has already been stressed, but the purity of solvents and other reagents must also be considered. Such materials should be of analytical grade and further purified where required. For example, Kliman and Peterson, in their method for the assay of aldosterone (75), report the purification of dichloromethane, benzene, cyclohexane, and carbon tetrachloride by passage through a bed of silica gel. Pyridine was refluxed for 4–6 hr over barium oxide, fractionally distilled, and stored over calcium chloride. Acetic anhydride was refluxed over calcium carbide for 4–6 hr and distilled using a fractionating column after further reflux over chromium trioxide. Glacial acetic acid was similarly purified by reflux and distillation over chromium trioxide. Rigorous purification using similar methods is essential for all solvents and reagents if accurate and reproducible results are to be obtained.

B. COMPLETE EQUILIBRATION

It is a basic assumption in all isotope techniques involving dilution that a labeled molecule behaves in the same way as an unlabeled molecule with the same structure. It is therefore necessary that complete equilibrium be reached before any manipulation, purification, or sampling is carried out, so that the sample is homogeneous with respect to the ratio of labeled and unlabeled molecules throughout the analytical procedure.

C. PURITY OF FINAL SAMPLES

Where a result is calculated from the specific activity of a derivative or compound, it is essential that the isolated material be chemically and radiochemically pure, since no correction can be made for any impurities.

In double isotope techniques where the result is calculated from the activity due to the two radioactive isotopes present, the need for chemical purity is eliminated, but absolute radiochemical purity is essential. This is especially true when the level at which the analysis is carried out is very low. Extensive purification is therefore necessary to free the desired product from the excess of labeled reagent and from any other labeled compounds formed by reaction of the reagent with other substances which may be present.

D. UNSTABLE SUBSTANCES

Most analytical procedures become exceedingly difficult when unstable materials are involved; any method involving gravimetric or even physiochemical determinations is open to error. Double isotope techniques eliminate some of the difficulties with such compounds, since the final determination only involves a counting procedure; once a radiochemically pure sample is isolated, any decomposition, provided it does not involve loss of a radioactive atom, will not affect the result.

E. ADDITION OF TRACER COMPOUND IN DOUBLE ISOTOPE DERIVATIVE TECHNIQUES

In double isotope derivative analysis the tracer compound or derivative can be added at one of a number of stages in the analysis. For the highest accuracy it is necessary to add a labeled sample of the compound to the material for analysis before any manipulation is carried out. Correction will then be made for any loss caused by the preliminary purification procedure or the preparation of the derivative. This method has been called "the complete indicator technique" (18). The alternative procedure, which the same authors called "the partial indicator technique" involves the addition of a tracer after the formation of the derivative. This is necessary when it is not possible to prepare the unknown compound labeled with a suitable isotope and the tracer must be a derivative prepared using the reagent labeled with a second isotope.

The latter procedure, as the name implies, only partially corrects for losses during the analysis. Special precautions must therefore be adopted during the preliminary purification and the preparation of the derivative. In some cases these can be made quantitative; otherwise a correction factor must be applied. Either procedure, however, is open to error and partially defeats the end of the double isotope procedure.

4. Counting Methods

A. GENERAL

The counting of ionizing radiation is now a well-documented procedure. It is used in many branches of science, and many counting systems for all types of radiation are commercially available. Counting techniques will, therefore, only be described here insofar as they are specifically applicable to methods of isotope derivative analysis.

Simple isotope derivative analysis which involves only one isotope, therefore needs no further attention in this section. There are, however, a number of counting techniques specifically applicable to double isotope derivative analysis, since they permit the simultaneous determination of two isotopes and each of these merits description.

B. $\beta\gamma$ COUNTERS

Because of the large difference in penetrating power of these two types of radiation, it is possible to filter out the weaker β radiation while the γ is counted, and then determine $\beta + \gamma$ without any filter. Keston and his co-workers (69) reported the use of a 0.003-in. thick aluminum filter which removed 99.7% of counts due to sulfur-35, but allowed 56% of counts due to iodine-131 to pass to the counter. This was the method applied to counting of samples from double isotope derivative analysis at a time when the methods for the determination of mixed β-emitting isotopes had not been developed. Velick and Udenfriend (139) describe such a technique in their double isotope derivative method for a number of amino acids. Their samples were dried on to copper disks and counted under a Geiger tube in a system set up for automatic counting of a large number of samples and involving the use of an aluminum filter similar to the one described above.

This method has the advantage that the equipment required is simple and therefore much cheaper than that used for counting mixed weak β emitters. $\beta\gamma$ counting has been used successfully in the estimation of amino acids (68,70) and for the determination of steroid hormones (17,19,131).

An interesting recent development in the simultaneous counting of β- and γ-emitting isotopes has been reported by Bojesen et al. (18) in their method for the determination of estrone and estradiol. In this procedure ^{131}I is replaced by ^{125}I, which is a very weak γ emitter (see Section II-5-A-b). A methane flow counter with a thin plastic window is used to determine ^{35}S at 10% efficiency with a 2% breakthrough of ^{125}I in conjunction with low crystal–scintillation detector which will determine the ^{125}I (efficiency not quoted) with only a 0.007% breakthrough of ^{35}S.

Two disadvantages arise from the use of hard γ-emitting isotopes with short half-lives. Precautions are necessary against radiation hazards, and repeated standardization is required to correct for decay.

c. β COUNTERS

a. ^{14}C and Tritium Gas Counting. When the use of carbon-14 in conjunction with tritium was introduced into isotope derivative analysis by Avivi and his co-workers (3) in 1953, the only established method of determination of tritium was by gas counting. The purified derivatives were combusted to carbon dioxide and water in a tube furnace, introduced into high vacuum manifold where the products were separated, and the carbon dioxide counted. The tritiated water was then reduced over hot zinc and the hydrogen pumped into a second counter.

This method has the fundamental advantage of separation of the isotopes before counting. Although the counting apparatus required is not expensive compared to the highly sophisticated electronic equipment which has superseded it, the process of preparation of the samples is time consuming and can only be partially automated.

b. Liquid Scintillation Counting. In their report on the use of carbon-14 and tritium in microbiochemical analysis, Avivi and his co-workers (3) also describe a liquid scintillation counter for the determination of these two isotopes. This counting technique had then only recently been developed and the equipment available was not able to discriminate between ^{14}C and tritium, so that the former had to be determined using a thin end window counter and the latter found by difference.

In the past 10 years liquid scintillation counting has made very great advances and it is now generally possible to discriminate between two isotopes that differ in energy by a factor of 4 (see Table I). With the simpler single-channel or older double-channel equipment it is necessary to make two determinations, the first with the high volvate and discriminator settings adjusted to count ^{14}C with the exclusion of all tritium counts, and then at settings to count both isotopes. Such a procedure is described in detail for an early model Tri-Carb scintillation counter by Peterson (109). As a result of careful investigation of the optimum procedures for counting double labeled samples using a commercial dual-channel liquid scintillation spectrophotometer, Okita and his co-workers (107) concluded that, while both isotopes could be measured simultaneously at the same voltage, better separation was obtained by measuring them at different voltages in sequence.

c. Double-Channel Simultaneous-Scintillation Counting. Although the simultaneous measurement of ^{14}C and ^{3}H was first mentioned in 1954 by Werbin and LeRoy (141), it was not until after 1959 that the developments in electronic circuitry enabled counters to be developed which would discriminate between β radiations of different energies in the same sample simultaneously. This is possible by dividing the output signal from the photomultiplier and feeding it into two distinct amplifiers. Each of these can be attentuated separately; thus by reducing the gain on the ^{14}C channel and having the tritium channel at full gain, it is possible to obtain results better than those from two separate counters. This equipment is now standard in most multichannel scintillation counters, and many are in fact designed specifically for this purpose. This technique has been fully reviewed by Kabara and his co-workers (65).

These scintillation counters, although very expensive in comparison to the other systems outlined above, have many advantages. The counting efficiency for both isotopes is high and, because the scintillation system is a solution, usually in toluene, the sample preparation is minimal. Moreover, in methods where a large number of determinations must be carried out, the time-saving factor is most important.

D. AUTOMATIC SYSTEMS

Many of the procedures in routine biochemical analyses tend to be both exceedingly tedious and time consuming. It is therefore highly desirable that automatic equipment be used wherever possible. Many of the processes in isotope derivative analysis must be carried out by hand, but counting of samples by methods other than gas counting can be readily automated. Automatic sample changers of various commercially available units can handle from 20 to 200 samples, which are transferred to the counting chamber and automatically counted to a preset time or count and the results recorded and/or calculated. Such highly automated systems are essential in the laboratory where double isotope methods are routine.

E. OTHER EQUIPMENT

Labeled materials on chromatograms can be located by radioautography, but this process may require days or weeks if low activities of weak β emitters are being determined. Rapid location of active spots can be accomplished by use of a chromatogram scanner. These scanners may be simple devices in which the chromatogram is passed manually under a thin end window Geiger counter connected to a simple ratemeter or sophisticated automated machines using recording equipment.

5. Labeled Materials

A. GENERAL

a. Availability. Many labeled compounds and reagents required for the established double isotope derivative techniques are now available from a number of commercial sources (126). It should be remembered, however, that a labeled compound becomes commercially available only when a new isotopic technique has created a demand for it. Thus the development of a new double derivative analysis may involve the synthesis of new labeled compounds.

b. Isotopes Used in Derivative Analyses. The characteristics of the isotopes generally used in derivative analysis are shown in Table I. The properties required of an isotope for this method are:

1. Ease of incorporation into organic compounds and/or reagents

2. Availability at high specific activity and possibility of incorporation into compounds at that activity

3. Suitable half-life

4. Easy handling technique

5. Radiation of type and energy which can be determined in presence of another isotope giving the same type of radiation with a different energy

TABLE I

Isotopes Used in Derivative Analysis

Isotope	Principal particle	Energy, MeV	$t^{1/2}$
^{3}H	β	0.018 (100%)	12.26 yr
^{14}C	β	0.155 (100%)	5760 yr
^{35}S	β	0.167 (100%)	87.2 days
^{125}I	γ	0.035 (7%)	60 days
^{131}I	β	0.61 (87%)	8.04 days
	γ	0.36 (80%)	

Although the early work on derivative analysis was carried out using ^{35}S and ^{131}I, the obvious isotopes for general use are ^{14}C and ^{3}H since they can, theoretically, be incorporated into all organic compounds. When ^{35}S and ^{131}I were first used, ^{14}C and ^{3}H were not readily available; this necessitated the development of special reagents such as *p*-iodobenzene sulfonyl chloride, which could be labeled with the former isotopes.

Both ^{14}C and ^{3}H are now available at very high specific activities. With tritium, the only limit to the amount of isotope which can be incorporated is the self-decomposition of the labeled compound resulting from radiation damage. These isotopes are cheap and have half-lives which are long enough to eliminate decay correction during a determination. They are both weak β emitters and thus precautions against radiation hazards are at a minimum. In this respect it may be mentioned that Andersen and his co-workers (2) introduced ^{125}I in place of ^{131}I. The former isotope, although a γ emitter, has a radiation energy so weak as to obviate the heavy precautions required in the use of the latter isotope. The β radiation of ^{14}C and ^{3}H is sufficiently different in energy to allow for simultaneous measurement in scintillation counters. ^{35}S however, has a β radiation so similar in energy to that of ^{14}C that it is not possible to use this pair of isotopes together in analytical derivation techniques.

c. Isotopic Synthesis. Labeled materials are much more unstable than their nonradioactive counterparts and their susceptibility to self-decomposition increases with the specific activity of the compound. Commercially available materials usually have a given purity and rate of decomposition, but for accurate analytical work reagents and compounds need elaborate purification and standardization before use.

Custom synthesis facilities are available through some commercial concerns, but this procedure is exceedingly expensive and often gives no guarantee of purity. In the authors' experience the estimated cost of the custom synthesis of an organic biochemical would convert and equip a laboratory for the synthesis of simple labeled compounds. Materials synthesized as required have the advantage of being freshly prepared (most important with isotopes of short half-life such as iodine-131) and of guaranteed purity.

Many concise texts are available on organic synthesis with isotopes (28,48,102), but what may not be realized by the biochemist or analyst is that the apparatus, technique, and facilities required for such work on a small scale are little more than those used for micro or semimicro organic work, with the possible addition of some high-vacuum apparatus.

d. Storage of Labeled Compounds. Much published literature has been devoted to the discussion of storage conditions for labeled compounds (e.g., 8,47–50,117,123). The review of Evans and Stanford (49) is particularly valuable since it deals specifically with tritium-labeled amino acids and steroids of high specific activity.

In general, however, the following conditions will minimize decomposition of small samples during laboratory storage.

1. Store samples in clean, neutral, and chemically inert containers.

2. If possible, dissolve the samples in benzene or a mixed solvent consisting largely of benzene. Concentration should be low and the container free from oxygen.

3. If benzene is not suitable, alcohol or water should be used; otherwise the sample may be dried on filter paper.

4. Solids should be stored in small quantities to reduce self-absorption; dry and free from oxygen.

5. Always store samples at the lowest possible temperature.

B. REAGENTS

a. Required Properties. Before a full description of the common reagents used in derivative analysis is given, the required properties of such reagents will be outlined.

a. The reagent should be available at a high specific activity (about 100 mc/mmole).

b. It should be easy to store, handle, and dilute, a characteristic which is better in solids than liquids.

c. It should react quantitatively with the compound in a crude form. Failing this, the yield of derivative should be high.

d. The isotope in the reagent should be such that a tracer compound or derivative can be easily prepared at high specific activity ($>100\ \mu c/\mu g$) using a second isotope. If such conditions are attained, the weight of the added tracer can be kept small in comparison with the unknown, thus increasing the accuracy.

e. The reagent should be selective for the compound under assay and not react with metabolites and related compounds.

f. The derivative obtained should be readily extractable from the reaction mixture and easily purified by chromatography. It is an additional advantage if derivative material can be located by its ultraviolet absorption and/or fluorescence.

g. The reagent isotope used along with a second isotope must be such that the counting procedure is as simple and yet as accurate as possible.

h. Reagents should give a low blank value.

b. Pipsyl Chloride (*p*-Iodobenzenesulfonyl Chloride). (*1*) *Pipsyl Chloride-^{131}I.* This reagent was the first to be used in derivative analysis (67) and is now commercially available from a number of sources. Its synthesis was described shortly after the publication of the analytical method and by the same group of workers (68).

^{131}I, free from tellurium (84) and in the form of the iodide ion, is made up to 25 mg by the addition of potassium iodide. After the pH has been adjusted to be above 7, the solution is evaporated to 0.2 ml, and a few crystals of sodium sulfate are added. An equal volume of concentrated hydrochloric acid is then added to the cooled solution, followed by 25 mg of 4-diazobenzene sulfonic acid. This mixture is allowed to react in the cold until the evolution of nitrogen ceases, when it is warmed to complete the reaction. The cooling is repeated and a further 15 mg each of 4-diazobenzenesulfonic acid and potassium iodide is added; these are allowed to react as before. The resulting solution is saturated with sodium chloride and cooled when the sodium salt of the 4-iodobenzenesulfonic acid separates. This is separated by centrifugation, washed with saturated brine, and the crystallization of the original solution is repeated several times with the addition of 40 mg of sodium *p*-iodobenzenesulfonate and a little water after each crystallization.

The sodium salt is dried, dissolved in 1 ml of phosphorus oxychloride containing an excess of phosphorus pentachloride, and warmed. This mixture is then transferred to a separating funnel containing 1 ml of water and 50 ml of benzene. Two hundred milligrams of carrier pipsyl chloride is added, the benzene layer is washed several times with cold water and dried over sodium sulfate. After the benzene has been evaporated, the product is distilled at 150°C and 3.5 mm onto a cold finger. This material is crystallized from benzene–light petroleum and dried at 70°C for 1 hr, mp 86°C.

Increased specific activity can, of course, be attained by separate preparations of the acid chloride from each batch of the acid and without the addition of carrier sodium salt.

Pipsyl chloride-^{131}I has only been used for the determination of amino acids in protein hydrolysates; the original method developed by Keston and his co-workers has now been superseded by methods using ^{3}H and ^{14}C, particularly ^{3}H acetic anhydride (143).

Before the consideration of the characteristics of this reagent, it is necessary to point out the disadvantages arising from the use of ^{131}I. It is a gamma emitter of short half-life. Special radiation protection is therefore necessary when the isotope is handled, and in any work correction must always be made for the decay.

Although ^{131}I is available at very high specific activity, pipsyl chloride-^{131}I can only be obtained from commercial sources at low specific activities. Storage, handling, and dilution of the compound are complicated by the half-life of the isotope, the type of radiation, and the reactivity of the reagent toward water. It will, however, react quantitatively with the amino groups of amino acids and can be counted very simply in the presence of β emitter by use of filters in conjunction with a Geiger tube, as described in Section II-4-B. It is therefore very suitable for use with ^{14}C- or ^{35}S-labeled tracer compounds. Unfortunately it is not highly selective and will react with many compounds which occur in samples of biological origin. Extraction and purification of the derivatives are satisfactory, but the derivatives can only be located on chromatograms by scanning or radioautography.

(*2*) *Pipsyl Chloride-^{35}S.* The reagent used in conjunction with ^{131}I-labeled pipsyl chloride in the later work of Keston and his group (70) was pipsyl chloride-^{35}S. It was used to prepare standard derivatives of amino acids for use as tracers in the double isotope procedure which had been developed.

^{35}S is an excellent isotope for use in analytical procedures, but suffers from the disadvantage that it emits a β particle of very similar energy to ^{14}C (see Table I), thus preventing these two materials from being simultaneously determined.

The synthesis of ^{35}S pipsyl chloride was also described by Keston and his co-workers (70).

Sulfuric acid-^{35}S [ref. 70 quotes A.E.C. catalog item S44 (quantity not specified)] is evaporated to 0.5 ml, mixed with 40 mg of carrier sulfuric acid, and evaporated until just fuming. When this solution has been cooled, 0.7 ml of glacial acetic acid, 0.3 ml of acetic anhydride, and 80 mg of acetanilide are added. This mixture is heated at 60°C for 2 hr and then kept at room temperature overnight. The product is diluted with 10 ml of water and 2 ml of concentrated HCl and hydrolyzed by heating for 2 hr on a water bath. The resulting solution of sulfanilic acid is cooled and diazotized using a sevenfold excess of sodium nitrite. After 10 min the excess nitrous acid is destroyed with ammonium sulfamate and an excess of potassium iodide is added to the mixture. Finally the product is crystallized in the presence of carrier *p*-iodobenzene sulfonic acid by the addition of sodium chloride. The acid chloride is prepared as described above for pipsyl chloride-^{131}I.

Pipsyl chloride-^{35}S is available commercially at up to 120 mc/mmole (27) and could be synthesized at activities considerably higher than this since carrier-free sulfate-^{35}S is available. The ^{35}S reagent only differs from the ^{131}I-labeled compound in that it has a better storage life because of the difference in half-life and, being a medium energy β emitter, can be determined using a liquid scintillation counter.

c. Pipsan (*p*-Iodobenzenesulfonyl Anhydride). In the application of the method developed by Keston and his co-workers (68,70) for the analysis of mixtures of amino acids to the determination of steroid hormones, pipsyl chloride was found to be an unsuitable reagent because it led to the formation of a derivative chlorinated in the side chain (20). Use of the corresponding sulfonic acid anhydrides was found to be quite satisfactory.

The acid anhydrides of ^{131}I- and ^{35}S-*p*-iodobenzene sulfonic acid can be prepared by refluxing the acid with thionyl chloride [see Sections II-5-B-b-(1) and II-5-B-b-(2)] (19), or alternatively by the methods of Christensen (33,34).

A much simpler two-step method for the preparation of pipsan-^{131}I has been described in detail by Christensen (33). This involves the direct sulfonation of iodobenzene with oleum and has been designed to work on 0.1-g quantities with an activity of 50 mc with a yield of 25%.

Benzene diazonium chloride prepared from aniline (1.5 mmole) is transferred to the reaction vessel in a shielded cell, cooled, and ^{131}I (50 mc) as a solution in 1 ml of 0.01*N* $Na_2S_2O_3$ is added, followed by 1.5 mmole of carrier KI in 2 ml of water. The mixture is warmed to 20°C and, when the iodination is complete, is made alkaline and steam distilled. (The steam distillation unit is incorporated into the apparatus.) The iodobenzene-^{131}I is centrifuged, separated, and dried in a desiccator over silica gel for 12–16 hr. The dry product is mixed with 4 parts oleum containing 15% SO_3. After 3–4 hr the precipitate of pipsan is separated by centrifugation, washed with acetic anhydride, ether, and finally pentane, and dried in a stream of nitrogen. The crude product is purified by reprecipitation from dry chloroform with pentane.

Christensen describes in detail two pieces of apparatus which are convenient for use in this synthesis. The first is a combined reaction vessel and steam distillation unit and the second a rotating pipet stand which is used for all manipulations after the steam distillation. His method is most suitable for regular preparation of large batches of high activity pipsan-^{131}I.

d. Tosan-^{35}S (*p*-Toluenesulfonylanhydride-^{35}S). With the development of double isotope derivative analysis based on tritiated steroids, an attempt was made to overcome the disadvantages of ^{35}S-pipsan by the use of ^{35}S-tosan. A recent synthesis of this material has been described by Thunberg (135).

The ^{35}S as sulfate, 100 μl of concentrated sulfuric acid, and 0.5 ml of water are shaken together and heated in an oil bath at 110°C until all the water is removed. The reactants continue to be maintained at 110°C and continuously shaken under reflux, while 500 μl of toluene followed by 50-μl portions of thionyl chloride are added. Further 50-μl portions of thionyl chloride are added until the evolution of gases ceases and the toluene phase disappears. The reaction is stopped by the addition of 0.5 ml of water; the product with the excess toluene is then distilled *in vacuo*, and the temperature is finally raised to 130°C to produce the anhydrous acid. This acid is converted to the acid chloride by heating with 150 μl of thionyl chloride at 80°C for 30 min. The excess thionyl chloride is removed by distillation *in vacuo* at 40°C followed by codistillation with hexane. The product is dissolved in benzene, washed with water to free it from sulfuric acid, dried over sodium sulfate, and the solvent evaporated. Pure *p*-toluene sulfonyl chloride is obtained by repeated crystallization from 3 ml of heptane at −20°C to mp 69–69.5°C.

The acid chloride is hydrolyzed with 1 ml of water at 100°C for 30 min, the water distilled *in vacuo,* and the anhydride prepared, either by heating with thionyl chloride at 30°C for 2 hr, or by reaction with carbodiimide. It is purified by crystallization from benzene/hexane.

A further refinement of this procedure (136) simplifies the carbodiimide preparation of tosan-^{35}S, enabling the preparation of 10–50 mg batches of tosan from the acid monohydrate on a silic acid column. This procedure only involves mixing the reagents on the column in ether, evaporation of the solvent, development of the column, and evaporation of the solvent fraction containing the tosan-^{35}S. It was found to be more convenient to store the acid monohydrate and prepare small amounts of the anhydride in this way; yields were reported 90% mp 129–129.5°C.

e. Acetic Anhydride-^{3}H and -^{14}C. It was pointed out by Avivi and his collaborators (3) that some of the disadvantages in the procedures of Keston and his co-workers (68,70) using pipsyl chloride-^{35}S and -^{131}I could be overcome by the use of ^{14}C- and ^{3}H-labeled acetic anhydride. These reagents are more generally applicable than pipsyl chloride since they will react with alcohols and amines. Avivi, Simpson, Tait, and Whitehead described the synthesis of these materials, but since 1953 their use in derivative analysis has created a demand which is now readily met commercially. The self-decomposition of high-activity tritiated acetic anhydride has, however, caused difficulties in the double isotope derivative assays of some steroids, especially when the steroid is present in the sample at a very low level. Henderson and his co-workers (58) report that these difficulties are caused by nonvolatile, high specific-activity impurities in the 20% benzene solution of the reagent as supplied commercially. They describe a method of purification and an investigation of the rate of buildup of the impurities under storage conditions. As a result of their investigation it is recommended that solutions of tritiated acetic anhydride be redistilled at intervals of one to two weeks when used for investigations requiring the maximum sensitivity, and that solutions which have been stored for several months be redistilled before use. The procedure they recommend is a rapid vacuum transfer from one flask to another through a U-tube. Nonvolatile impurities are reduced tenfold by this procedure.

^{14}C-Acetic anhydride is commercially available at high specific activities and the tritiated reagent is listed at activities not less than 3000 mc/mmole; the latter is therefore most useful in derivative analysis. Both isotopes have long half-lives and the reagents can therefore be stored for long periods without appreciable loss of activity; however, if so stored they require, as explained above, purification before use. Although the volatility of the material simplifies purification, it does present handling, dilution, and standardization problems.

When these reagents have been used in derivative analysis, usually the ^{14}C-acetic anhydride has been employed for the preparation of the tracer compounds and the tritiated material to form the derivative with the unknown. Since tritium can be determined in the presence of most other β-emitting isotopes, tritiated acetic anhydride is potentially a very versatile reagent for derivative analysis. The ^{14}C-labeled material also has a wide range of use since the weak ^{14}C β particles can be detected in the presence of all other β emitters except ^{35}S and in the presence of γ-emitting isotopes. The acetyl derivatives of amino acids and steroid acetates are readily extracted and purified, but the latter often need further chemical modification during the chromatographic procedure to attain the required purity (see Section II-6-C-a). Counting of both these isotopes individually and simultaneously is now a very well-documented, exact, and often completely automated procedure, making this aspect of their use very simple.

f. Thiosemicarbazide-^{35}S. Bush (25) and Pearlman and Cereco (108) suggested the use of thiosemicarbazide as a reagent for steroids containing reactive carboxyl groups because of the satisfactory chromatographic properties of the derivatives with this reagent; Tait and his co-workers (133) developed its use as a labeled reagent.

Thiosemicarbazide-^{35}S is commercially available at reasonable cost at high specific activity (1200 mc/mmole). It can also be synthesized easily from elemental ^{35}S (available up to 1500 mc/mm) via potassium thiocyanate-^{35}S (35). The latter is commercially available but only at a specific activity of up to 100 mc/mmole.

Potassium thiocyanate is prepared from elemental ^{35}S (129) following the procedure for the ^{14}C-labeled compound. Sulfur (40 mg) and potassium cyanide

(100 mg) are fused together in a Pyrex test tube at 310°C under nitrogen for 15 min and allowed to cool. The crude compound is used to prepare thiosemicarbazide-^{35}S following the procedure of Bambas (5). It is treated with warm water (0.5 ml) and a solution of hydrazine sulfate [(0.3 ml) of a solution containing hydrazine hydrate (900 mg) and hydrazine sulfate (2 g) with water to 3 ml] added. This mixture is allowed to stand at room temperature for 30 min. It is then filtered and the filtrate heated under reflux in an oil bath at 150–179°C until the solution temperature reaches 130°C, which is maintained for 2 hr. The mixture is then transferred quickly to a beaker and the solvent evaporated under nitrogen. Clark and Roth (35) report 96 mg, 60 ml, 179–181°C after crystallization from water.

The reagent can be stored without deterioration in solution in methanol for short periods, but long-term storage is better in water at −30°C or at room temperature *in vacuo* over phosphoric oxide (105). Unlike the pipsyl or acetic anhydride reagents, it is not affected by water. Thiosemicarbazide will react quantitatively with keto-steroids in crude extracts from plasma. Steroid thiosemicarbazones are readily extractable from the reaction mixture with methylene dichloride after the addition of water, a procedure which leaves the excess reagent in the aqueous phase. The derivative can be chromatographed on paper or on silica-gel plates. The partially purified derivatives can frequently be acetylated with acetic anhydride at 65°C for 15 hr, giving the acetyl derivatives. These semicarbazone acetates have different chromatographic characteristics facilitating further purification. For example, testosterone 3-thiosemicarbazone-2,4-diacetate can be separated from testosterone 3-monothiosemicarbazone (113). Detection of the steroid derivative is facilitated by their ultraviolet absorption characteristics. The differential rates of hydrolysis of the various thiosemicarbazones under controlled conditions (3 keto $\gg$ 20 keto $>$ Δ^4-3-keto $>$ 17-keto) (57,127,133) can be exploited in further purification of derivatives (e.g., 113,114). Since tritiated steroids with specific activity of 100 $\mu c/\mu g$ can now be prepared (80), a negligible amount of tracer compound need be added to plasma during the analytical procedure. The $^3H/^{35}S$ ratios are as easily measured by liquid scintillation techniques as $^3H/^{14}C$ ratios.

Unfortunately, the thiosemicarbazides tend to absorb strongly on paper and silica gel, necessitating high carrier levels and very careful elution. The use of glass-fiber paper has improved recovery after chromatography (86). High blank values are sometimes obtained using this reagent; this is probably due in part to inadequate separation of excess reagent before addition of the carrier. Care must therefore be taken to ensure that the reagent is completely removed, since it cannot be destroyed with water.

g. Sodium Borotritide. The application of double isotope derivative methods to steroids containing no acetylatable hydroxyl group was only possible by the development of special reagents like thiosemicarbazide-^{35}S described above. Before this occurred a most ingenious method was developed by Woolever and Goldfien (149) using sodium borotritide for the reduction of progesterone, followed by partial reoxidation using manganese dioxide. The tritiated reagent is readily available commercially at very high specific activity and has been used in the analysis without purification. However, it is not stable in solution, even when stored at -15°C. The steroid must be partially purified before reduction, necessitating the addition of a ^{14}C tracer to the sample at the beginning of the analysis. The reduction is complete in a relatively short time and is quantitative. Sodium borotritide has the advantage of reducing only aldehydes and ketones and therefore has a degree of specificity comparable to thiosemicarbazide-^{35}S. Partial reoxidization of the tritiated material removes labile tritium. This operation enables the derivative to be extracted and purified more easily. Location and counting is also simplified.

h. 2,4-Dinitrofluorobenzene-^{3}H and -^{14}C (DNFB). The application of this reagent to derivative analysis was first described by Beale and Whitehead (10) for the determination of amino acids in protein hydrolysates and for the estimation of *N*-terminal amino acids of peptides. These authors (10) describe the preparation of both the ^{14}C- and ^{3}H-reagents, the former by nitration of chlorobenzene-^{14}C followed by halogen exchange with potassium fluoride, and the latter from bromobenzene by a similar method. Both of these preparations are limited by both the specific activity and cost of the commercially available starting material and by the comparatively large scale on which the reactions must be carried out.

DNFB-^{3}H and -^{14}C reagents are now available commercially, but the specific activity of the ^{14}C material is rather low in comparison to the ^{3}H product. A synthesis of tritiated material at extremely high specific activity (19 c/mmole) has recently been described (59).

Ten microliters (0.09 mmole) of 1-bromo-2-fluorobenzene, 5.5 mg of 5% palladium on charcoal catalyst, and 300 μl of 3% potassium hydroxide in methanol in a 2-ml conical flask on a vacuum manifold are shaken at room temperature for 3 hr with 2370 μl (STP) of 98% tritium gas. During this time, 1720 μl (0.077μM) are taken up. The product is then extracted with carbon tetrachloride, washed with water, and chromatographed on 1 g of alumina (grade 1), using carbon tetrachloride. One milliliter of 96% sulfuric acid, 0.01 ml of 30% oleum, and 1.0 ml of 98% nitric acid are added to the purified carbon tetrachloride solution in a 10-ml conical flask, and the mixture stirred for 15 min at 0°C and for 45 min at room temperature. The nitration mixture is decomposed on ice, the carbon tetrachloride evaporated, and the product distilled *in vacuo* (150°C) using a water pump. Hesselbo reports a yield of 50% based on tritium at a specific activity of 19 c/mmole.

DNFB has all the characteristics of a good reagent for derivative analysis. It is highly selective toward primary amines and has the added advantage that, since derivatives are colored, they are self-indicating on paper chromatograms. Although no synthesis of high activity ^{14}C-labeled material has been reported, the ^{3}H reagent can be used in conjunction with the high specific activity ^{14}C-amino acids, giving a very accurate method of analysis.

i. Phenylisothiocyanate-^{35}S. Edman's original technique for the determination of *N*-terminal amino acids using phenylisothiocyanate (44) was carried out on a microscale by Cherbuliez and his co-workers using the ^{35}S-labeled reagent (30–32).

Phenylisothiocyanate ^{35}S is available commercially at moderate specific activities (27). It has been synthesized from hydrogen sulfide-^{35}S via ammonium hydrogen sulfide-^{35}S and phenyldithiocarbamate-^{35}S by Wieland, Merz, and Rennecke (145), but the theoretical radiochemical yield from this method is only 12½%. More recently Moye (98) has described the synthesis of this material from carbon disulfide-^{35}S in 75% yield. Since carbon disulfide-^{35}S is available commercially at a cost which is only slightly less than that of phenylisothiocyanate-^{35}S, the synthesis of the latter only becomes economical if carried out from elemental sulfur (90,103). Such a preparation would enable specific activities of 250 mc/mmole to be attained, but the method has considerable practical difficulties. For these reasons the details of the synthesis of these materials will not be included.

j. Standardization of Reagents. Since the result of any isotope derivative analysis is directly dependent on the specific activity of the labeled reagent, this must be determined very accurately for each batch of reagent used.

The method for such a determination always follows the same pattern, although the precise conditions vary from one reagent to another. In general, a known weight of the labeled reagent, which has been diluted with carrier material to the approximate specific activity required for the determination, is allowed to react with an excess of a suitable compound to form a stable derivative. This derivative is then rigorously purified and assayed, first by counting and then by a physical method to determine the concentration of the solution counted. The latter can be carried out in a number of ways: (*1*) gravimetrically, if the solution has been made up from pure solid derivative, (*2*) fluorescence estimation, (*3*) colorimetric estimation, and (*4*) ultraviolet absorption estimation.

Thus, if the solution of the pure derivative gives C cpm per milliliter and contains x mg per milliliter, then the specific activity of the reagent $= CE/x$ counts per minute per mmole where E = milliequivalent weight.

C. OTHER LABELED COMPOUNDS AND DERIVATIVES

a. Label in Reagent Residue. Keston and his co-workers developed a procedure for eliminating the quantitative isolation of the labeled derivative in their first method for the analysis of amino acids by the introduction of a second isotope. They reported that the label could be introduced in two ways; the simplest of these is the addition of a small sample of the pipsyl derivative, which had been prepared from the amino acid and pipsyl chloride-^{35}S, after the pipsyl-^{131}I derivative of the compound had been formed. The second method is the addition of a sample of ^{14}C-labeled compound before the derivative is formed. Similarly, if a steroid analysis is to be carried out using tritiated acetic anhydride, the tracer derivative can be prepared by acetylating the same steroid with acetic anhydride ^{14}C (e.g., ref. 75). A sample of this material, after purification, can then be added to the reaction mixture after acetylation with the tritiated reagent. This method of preparation of a labeled derivative is theoretically applicable to all double derivative analyses, where the reagent is available and labeled with two different isotopes, which can be counted in the same sample. The advantage obtained by the simplicity of this preparation of the labeled derivative is often outweighed by the disadvantage of adding the tracer late in the analytical procedure (see Section II-3-E).

The reagents used in the preparation of such derivatives have been fully described in the preceding section.

b. Label in the Original Compound. The highest accuracy in derivative analysis is obtained when a tracer is added to the sample for analysis before any manipulations are carried out. For example the second method of Keston and his group of workers (66) was to add a small sample of the amino acid or acids under investigation labeled with ^{14}C to the crude mixture at the start of the procedure. Such conditions necessitate the use of samples of the compound under analysis labeled with an isotope different from the one in the reagent. Reagents may be developed so that ^{131}I or ^{35}S can be used, but in general, compounds contain carbon or hydrogen and nitrogen, making ^{14}C and tritium the only universally applicable isotopes.

(*1*) *^{14}C Compounds.* Most compounds used as tracers in derivative analysis are commercially available labeled with ^{14}C. These are usually prepared by synthesis from $^{14}CO_2$ with an absolute limit of specific activity of about 63 mc/mmole per ^{14}C atom introduced into the molecule. In practice, compounds are available at 20–40 mc/mmole and, because of the method of preparation, are exceedingly expensive, especially steroids.

Amino acids are, however, prepared by biosynthetic methods. These compounds are also expensive but are available at specific activities of 100–300 mc/mmole, making them very suitable for use in protein and peptide analysis by isotopic derivative techniques.

(*2*) *Tritiated Compounds* (48). The sensitivity of any derivative procedure is limited mainly by the specific activity of both the reagent and the compound. High specific activity ^{14}C-labeled amino acids are ideal for use as tracers. The ^{14}C steroids which are available are usually at lower specific activity, and reduce the sensitivity of the method for the assay of these compounds.

This disadvantage can be overcome by the use of tritiated tracers as proposed and described by Tait and his collaborators (133). The specific activity at 100% isotope enrichment for this isotope is of the order of 29 c/milliatom and in most organic compounds a large number of hydrogens can be replaced by tritium using exchange techniques. The possible specific activity is therefore very high indeed.

Tritium labeling can be carried out in two ways, either chemically or by exchange. The chemical introduction of tritium usually involves catalytic reduction of an unsaturated compound with tritium gas (56); Gut and Hayano (56) describe some general methods for the stereoselective introduction of tritium into steroids, and Laumas and Gut (80) report a synthesis of high specific-activity aldosterone by catalytic reduction of 1-dehydroaldosterone-21-acetate in dioxan with tritium over 5% palladium on charcoal. These chemical methods generally lead to much purer products than the use of exchange reactions. The former methods are quick, clean, and controlled, whereas the latter are slow and lead to much radiation damage of the product. (The exchange method using alumina referred to below is an exception to this.)

Gaseous exchange is carried out by the Wilzbach technique (147). The following procedure has been reported for C-21-hydroxy-steroids (109). The steroid C-21-acetate is deposited in a thin film inside a flask and exposed to about 5 c of carrier-free tritium gas for 10 days at 27°C and 0.39 atm. The time of exposure, which must be determined for each compound tritiated, is arrived at by a consideration of the amount of radiation damage which can be tolerated to obtain a product of suitable specific activity. Favorable conditions will lead to a final pure product containing 0.1–1% of the tritium gas used. Most of the tritium is incorporated in labile positions of the molecule. A large proportion of this unwanted activity can be removed by equilibration of the crude material with ethanol and removal of the solvent by distillation. The steroid acetate is now subjected to an extensive purification to separate it from its own radiation-decomposition products, all of which will be extensively labeled. The dry, crude compound, dissolved in 20% aqueous ethanol, is extracted into dichloromethane recovered from the solvent and purified by column or paper chromatography. The steroid acetate, which is less polar than the impurities, is eluted and hydrolyzed enzymatically with acetylcholine esterase. The free steroid is then extracted with dichloromethane and further purified by repeated paper chromatography. The purity of the final product can be determined by reverse dilution or, with the addition of some ^{14}C-labeled material, by the procedure described in Section II-6-A.

Compounds which are stable in aqueous media at temperatures up to 120°C can be labeled by exchange with tritiated water in the presence of a platinum catalyst. About 1 g of the compound and 200 mg of platinum catalyst are stirred with tritiated water, or 70% acetic acid, containing 100 c of tritium for about 20 hr under reflux. The catalyst and solvent are then removed and the product freed from labile tritium by washing with water (134).

This method generally gives products containing fewer impurities than those obtained by gaseous exchange, and purification is therefore usually possible by chromatography and crystallization alone.

Activities from 40–400 mc/g can be obtained with chemical recoveries of 20–80%. Tritium-labeled steroids are now usually prepared by this method, rather than the Wilzbach method.

Klein and Knight (73) report that steroids can be labeled with tritium by chromatography on basic alumina which has been previously treated with tritiated water. Their results indicate that it is possible to obtain labeled steroids of activities of 5–10 mc/mmole, in which the tritium is not labile under ordinary conditions, by a simple procedure and with remarkable economy. These activities are too low to be of much use in derivative analysis. Further development of the method of exchange may yield steroids of higher specific activities.

When a tracer of the required specific activity cannot be obtained by the introduction of the label into one part of the molecule as described above, it may be possible to overcome this difficulty by reacting a labeled compound with the reagent labeled with the same isotope. Such conditions might possibly occur for a ^{14}C compound where it is difficult to incorporate more than one labeled atom per molecule, thus limiting the maximum specific activity.

6. Separation Techniques

A. GENERAL

Since radiochemical purity of the double-labeled derivative is absolutely essential in this method of analysis, purification must be carried out with maximum efficiency. Only by the development of special chromatographic techniques has the required purity been obtained. The method very conveniently provides its own check of the purity of the final sample of the isolated derivative. The final purification step will usually be some form of chromatographic separation from which the product can be obtained as a number of fractions. If the derivative is pure, the ratio of the two isotopes in each fraction will be constant.

A similar check can be made if the final purification is carried out using paper chromatography. The paper spot containing the derivative is cut into a number of strips, each of which is counted before or after elution, and radiochemical purity is determined by the degree of constancy of isotope ratio for each strip. The reliability of this method is shown by its use in the positive identification of

γ-amino butyric acid in mouse brain (137). The compound isolated from the brain tissue was converted into the ^{131}I-pipsyl derivative, which was mixed with a synthetic sample of the ^{35}S-pipsyl derivative of γ-amino butyric acid. When the mixture was applied to a paper chromatogram with the ^{131}I-pipsyl derivative of the unknown and the pure ^{35}S-pipsyl-γ-amino butyric acid, the three spots were found to run with the same R_f and, when the spot containing the mixture was divided into four samples and counted, the same ratios for ^{131}I/^{35}S (0.433, 0.449, 0.439, and 0.433) were found. This identified the unknown with the amino acid in the ^{35}S-pipsyl-γ-amino butyric acid which was added.

B. CHROMATOGRAPHY

a. General. The purification of the doubly labeled derivative in this method of analysis is dependent on an extensive use of chromatographic procedures. These techniques are extremely well known, but certain systems and methods merit special mention.

b. Column Chromatography. Although Avivi and his co-workers (3) used partition columns in the purification of hydrocortisone acetate and the attempted purification of aldosterone acetate, this method found no further application in double derivative analysis until the recent work of Dray (40), who in his method for the estimation of testosterone used partition columns on Celite for the preliminary purification of the steroid before acetylation. The same method, in conjunction with thin-layer chromatography, was used for the purification of the acetate and its phenylhydrazide.

c. Paper Chromatography. Several detailed reviews and studies on the paper chromatography of steroids have been published (24,26,150,151)and these provide much information on the methods and systems available for particular compounds. No such detailed reviews have been published on the paper chromatography of amino acid derivatives, but Whitehead has described the separation of *N*-acetyl amino acids (142). This work describes a number of methods employing two-dimensional ionophoresis or one-dimensional ionophoresis followed by paper chromatography in the second dimension.

The use of glass-fiber paper for the chromatography of testosterone thiosemicarbazones diacetate has already been described [Section II-5-B-b-(6)]. The overall recovery during the analytical procedure was increased from 7% using thin-layer and conventional paper chromatography (86) to 22% by the use of thin-layer chromatography in conjunction with chromatography on glass-fiber paper. It therefore appears that glass-fiber paper could find further application in the chromatography of derivatives which bind to paper, particularly steroid thiosemicarbazones.

d. Thin-Layer Chromatography. It is well known that this method has all the characteristics of paper chromatography in the separation of mixtures, and that it does so very much more quickly. As a general technique it has been used considerably in the purification of labeled steroid derivatives, particularly in the past few years. Moreover, it has led to the development of existing derivative methods by modification of the chromatographic purification procedures. Hudson and his co-workers (64) described a method for the determination of plasma testosterone using ^{3}H-acetic anhydride in 1963 in which paper chromatographic purification was employed. In 1964, Burger, Kent, and Kellie (23) reported a similar determination using thin-layer chromatography.

Peterson, in association with a number of other workers, describes the use of marker compounds in the location of labeled derivatives on chromatograms [aldosterone acetate (75) and digitoxin acetate (88)]. These markers run with the same or very similar R_f as the derivative in the particular chromatographic system used. As they are not radioactive, they can be applied in such quantities to enable them to be located easily under UV light, thus enabling the active derivatives to be pinpointed for counting.

e. Ion-Exchange Chromatography. Levy (83) has described the use of Dowex 1 × 2 Cl^- ion-exchange columns for the separation of ^{131}I-pipsylated peptides (see Section III-1-B-c). Mechanic and his co-workers (94) also used Dowex 1 for the separation of pipsyl derivatives of amino acids.

f. Gas–Liquid Chromatography. In a recent study of plasma ketosteroids and testosterone in man, Kirschner and his co-workers (71) used a double derivative technique involving ^{14}C-steroids and ^{3}H-acetic anhydride as reagent. Their initial purification of the derivatives were by standard TLC and paper systems but the final purification was by GLC.

A dual-column gas chromatograph was used with an effluent fraction collector. The detector was of the argon ionization type, and the effluent corresponding to each peak was absorbed on *p*-terphenyl coated with silicone. Recovery of radioactivity from the column was 50–70%. The *p*-terphenyl was counted by standard procedure following dissolution of the crystals in 5 ml of a toluene PPO, POPOP, scintillator.

Kliman (74) has recently published a comparison of the determination of urinary aldosterone by double isotope dilution using paper and gas–liquid chromatographic separation techniques. The purification of the derivative by paper chromatography was accomplished following the method which involved the use of at least four solvent systems. The preliminary purification of the samples for gas–liquid chromatography was carried out on one paper system and two thin-layer systems. The GLC separation was effected on a 6 ft × ½ in. coiled-glass column packed with 2% SE30 on 30–60 mesh Chromosorb-W. The other conditions were: flash heater, 280°C; column, 245°C; cell, 260°C; argon carrier gas 20 psi, flow rate 50 ml/min.

The gas chromatograph was coupled to an automatic fraction collector in which the column effluent corresponding to the aldosterone peak was condensed on *p*-terphenyl crystals which were then dissolved in a liquid scintillator and counted (see Table III, no. 20).

g. Location of Materials. The use of a second isotope in this method of analysis makes it possible to add very large amounts of carrier derivative to the isotopic material to facilitate chromatographic separation of the desired material from other labeled materials. This also helps in the location of fluorescent (113) and colored (10) materials on chromatograms when the quantities of labeled material alone are far too small to be visible. These procedures are far superior to radioautographic methods (17) which, while quite satisfactory for high energy isotopes, are usually unsuitable for ^{14}C and ^{3}H. The availability of commercial radiochromatogram scanners, which are sensitive to β radiation of very low energy, has much-simplified location procedures.

C. SPECIAL TECHNIQUES

a. Chemical Modification of the Derivative. (*1*) *General.* In order to attain the necessary purity of a derivative in the most efficient way, it is often necessary to modify the molecule chemically. The introduction of a new physical property will often shorten the number of chromatographic steps necessary to obtain the radiochemically pure derivative.

(2) Second Derivative. Steroid thiosemicarbazones can be acetylated to the 2,4-diacetyl derivatives (133). These compounds are more polar than the nonacetylated thiosemicarbazones, which in turn chromatograph more slowly than the parent steroids. Thus, in the estimation of progesterone using thiosemicarbazide-^{35}S, Riondel and his co-workers report the acetylation of progesterone 3-thiosemicarbazone to the 2,4-diacetyl derivative before the final purification of the doubly labeled derivative by paper chromatography. A similar procedure involving acetylation of a thiosemicarbazone was used by the same workers in their method for the determination of testosterone in human peripheral blood.

A number of second derivatives have been used to facilitate the purification of steroid acetates by chromatography. Hudson and his co-workers prepared the thiosemicarbazide (64) and oxime (42) of testosterone acetate whereas Dray (40) prepared the phenyl hydrazone. Kirschner and his group (71) prepared the 1,1-*N*-dimethylhydrazone of the same compound and the acetates of dehydroepiandrosterone, androsterone, and etiocholanone for purification by GLC.

(3) Oxidation. Kliman and Peterson (75) purified aldosterone diacetate by chromatography on two systems and then oxidized the derivative with chromic acid in acetic acid before the final chromatography. The residue from the chromatogram is dried in a tube and 0.5% solution of chromium trioxide in glacial acetic acid (0.1 ml) added. The residue is well wetted with the solution and then left at room temperature for 10 min. The oxidation is stopped by shaking the mixture with 1 ml of 20% ethanol and 10 ml of dichloromethane. The aqueous phase is discarded and the dichloromethane evaporated in air at 30–40°C.

An oxidation of this type is now general in all methods for the estimation of aldosterone and corticosteroids as their acetates by double isotope derivative methods, although one group of workers (60) found it unnecessary in their procedure for the determination of cortisone and cortisol in plasma.

(4) Partial Hydrolysis. In an investigation of the characteristics of steroid thiosemicarbazones, Southern (127) determined the rates of hydrolysis of various keto steroid derivatives. His results showed that the rates were in the order 3-keto > 20-keto > 21-hydroxy-20-ketone > 4-ene-3-keto > 17-hydroxy-20-keto > 17,21-dihydroxy-20-keto > 17-keto.

These characteristics can be used to eliminate certain steroidal impurities by controlled hydrolysis with pyruvic acid. For example, saturated 3-ketone can be hydrolyzed when estimating 4-ene-3-ketones, and most other steroid thiosemicarbazones can be hydrolyzed and eliminated when 17-ketones are being estimated.

This method can also be used to obtain a derivative with different chromatographic properties. Thus progesterone-3,20-bis-thiosemicarbazone can be chromatographed, hydrolyzed to the 3-monothiosemicarbazone in 60% yield, and rechromatographed giving a hundredfold radiochemical purification (114,133). A similar hydrolysis procedure was used by Riondel and his associates (113) in the purification of testosterone thiosemicarbazone, where the derivatives of impurities were hydrolyzed in preference to the testosterone derivative.

D. ISOTOPE FRACTIONATION

It has been stated above that the criterion for radiochemical purity of a doubly labeled derivative is the constancy of the isotope ratio in a number of chromatographic fractions of the same compound. While this statement still stands in the general case, a number of exceptions have been reported.

Klein (72) reported that a number of examples of isotope fractionation with inorganic ions (128), organic acids (22), and amino acids on ion-exchange columns occur in the literature. Partial resolution has also been observed in countercurrent extraction of arabinose-^{14}C (91) and on partition column chromatography of aldosterone diacetate (see below) labeled with ^{14}C and ^{3}H (29). Klein goes on to report an investigation of isotope fractionation during the chromatography of cholesterol acetate-^{14}C and -^{3}H. His results show premature elution of the ^{14}C-acetate peak by the increase in $^{3}H/^{14}C$ ratio across the band. A statistical investigation showed these results to be significant, and a full mathematical study of the results was reported.

The presence of isotope fractionation during an isotope dilution analysis has been shown by Laragh and his co-workers (79). They detected and investigated this effect during an isotope-derivative analysis used in the determination of aldosterone section rates in man. In their studies, using both laboratory-prepared standards and clinical material, they found isotopic fractionation when aldosterone mono- or diacetates were chromatographed on paper or a Celite partition column.

The error introduced by isotope effects is usually small and has little or no effect on practical results; however, its recognition is necessary so that proper correction can be applied where required and the accuracy of the method maintained. Laragh and his co-workers therefore propose that if partial isolation of one radiolabel is indicated by a progressive increase in one isotope across a chromatograph peak, particularly during the estimation of aldosterone or tetrahydroaldosterone, the following steps should be taken:

1. The entire area under each radioactive peak should be taken into account in the calculation of the isotope ratio.

2. If possible, the proportion of each isotope should be the same.

3. Possible fractionation in other stages of the purification should not be overlooked.

4. When peaks are truncated, this should be done in a symmetrical manner about the maximum, in that it should be collected so that it is divided into at least 8 and not more than 10 fractions, so that

5. The error function of the isotope ratio can be determined by probit analysis.

Klein (72) suggests that these isotope effects can also serve useful ends since they provide an opportunity for the study of intramolecular contributions to the adsorption or partition process involved in chromatography as related to the stereochemistry of the molecule. They also offer a precise method of identification of substances which have steric arrangements which minimize these efforts.

III. APPLICATION

1. Amino Acids, Proteins, and Peptides

A. AMINO ACID COMPOSITION ANALYSIS

a. Pipsyl Method. (*1*) *Simple Isotope Derivative Method with Reverse Isotope Dilution.* The use of an isotopically labeled reagent in an analytical procedure was first reported by Keston, Udenfriend, and Cannan (67) in 1946. These authors used pipsyl chloride-^{131}I in an analysis of β-lactoglobulin for glycine, isoleucine, and alanine. The same group of workers extended the procedure to the analysis of the crystalline proteins β-lactoglobulin, human hemoglobin, aldolase, and phosphoglyceraldehyde dehydrogenase for glycine, alanine, and proline (68).

The derivatives of the amino acids corresponding to 0.3–2.0 mg of protein are prepared with 9 mg of pipsyl chloride-^{131}I in the presence of 15 mg of sodium bicarbonate by shaking at 100°C for 10 min. After the products have been extracted with ether, the aqueous phase is evaporated and the residue reacted with a further 9 mg of pipsyl chloride-^{131}I. The residue from the aqueous phase from this reaction is then reacted with a further 12 mg of pipsyl chloride-^{131}I, combined with the other ether extracts, and added to the carrier samples of the pipsyl derivatives of the amino acids for analysis (200 mg of each in ammoniacal solution). The ether is evaporated and the solution made up to 150 ml with 0.2*M* hydrochloric acid. This solution is then subjected to a ten-plate countercurrent distribution between chloroform and 0.2*M* hydrochloric acid. The pipsyl derivatives of the amino acids are obtained from the appropriate tubes, as follows. Pipsyl glycine is obtained from the aqueous phase of tubes 3, 4, and 5 and by extraction of the organic phase from tubes 2, 3, and 4 with alkali, all of which is then acidified and extracted with *n*-butanol. The chloroform phase from tube 8 and both phases from tubes 9 and 10 are extracted with alkali; the aqueous phase is acidified and extracted five times with chloroform–carbon tetrachloride, (60:40 v/v). The first three extracts are pooled, extracted with alkali, and then treated with carrier dipipsyl alanine and potassium chloride to remove any labeled dipipsyl derivatives from the solution of the pipsyl proline. The aqueous phase from the chloroform–carbon tetrachloride extraction is then extracted with *n*-butanol to remove pipsyl alanine. The solutions containing each of the required derivatives are extracted with alkali, acidified, and the solids purified to constant specific activity by reprecipitation with acid and treatment with activated charcoal.

(2) Isotope Derivative without Dilution Using Paper Chromatography. Keston, Udenfriend, and Levy (69) used paper chromatography to separate the pipsyl derivatives of glutamic acid, serine, glycine, and alanine virtually quantitatively from the hydrolysate of silk protein.

The pipsyl derivatives of 1 mg of amino acid from the hydrolysate are prepared as described above [Section III-1-A-a-(1)], and an aliquot of the ether extract equivalent to 1.7 μg of protein is chromatographed on Whatman No. 1 paper using *n*-pentanol saturated with 2*N* ammonia.

(3) Double Isotope Derivative Analysis. The same group of workers improved the method by the addition of a small sample of the derivative labeled with a second isotope (^{35}S) to act as a loss indicator. This eliminated the necessity for quantitative recovery from each band of the paper chromatogram or the isolation of a pure

solid sample of the derivative after dilution. This method was fully described later (70) for the determination of glutamic acid, aspartic acid, hydroxyproline, threonine, and serine in 0.2–1.0 mg of hydrolyzed β-lactoglobulin, bovine serum albumin, and human serum albumin.

The amino acids from 0.2 to 1.0 mg of protein are converted into the ^{131}I-pipsyl derivatives as described above [Section III-1-A-a-(1)]. The extracts containing these derivatives are added to a mixture containing accurately known amounts (0.3–0.9 μmole of each) of the ammonium salts of ^{35}S-pipsyl derivatives of glutamic acid, aspartic acid, serine, threonine, and hydroxyproline. The *p*-iodobenzenesulfonic acid-^{131}I arising from the excess of labeled reagent is removed by a four-plate countercurrent extraction between ether and 0.2*M* hydrochloric acid, the ether phase from which is separated, evaporated to dryness, and the residue dissolved in 1 ml of 2*M* sodium hydroxide. ^{131}I-pipsyl glycine is removed from this solution by repeated precipitation with acid after the addition of 50-mg batches of carrier material. The residual pipsyl glycine is removed with other unwanted ^{131}I-pipsyl derivatives by chloroform extraction, and the aqueous phase extracted with ether. After this ether has been evaporated, the residue is dissolved in alcoholic ammonia and chromatographed on Whatman No. 1 paper. For aspartic and glutamic acid development is with isopropanyl–*n*-butanol 35/50 saturated with 1*M* ammonia; for the remainder, *n*-butanol saturated with ammonia is satisfactory. Radioautograms of the chromatograms are prepared and, using these as a guide, slots are cut into strips which are eluted and counted.

A slight modification of this method was reported by Keston and Lospalluto (66). In a determination of glycine in a mixture of amino acids they added ^{14}C-glycine to the unknown mixture before the preparation of the derivative with pipsyl chloride, thus increasing the accuracy of the method (see Section II-3-E).

The method described above for the estimation of aspartic and glutamic acids, serine, threonine, and hydroxyproline was modified by Velick and Udenfriend (139) to include glycine and alanine. These workers also developed fractionation procedures which made the determination of proline, methionine, phenylalanine, and valine possible by isotope derivative methods.

The pipsyl-^{131}I derivatives of the amino acids resulting from the hydrolysis of 1 mg of protein or less were prepared, separated from excess reagent, and added to the ^{35}S-derivatives as described above [Section III-1-A-a-(3)]. The mixture of derivatives is subjected to the same counter-current partition between chloroform and 0.2*M* hydrochloric acid, but the fractions are divided into three groups which are further purified and then chromatographically separated (see Fig. 1).

1. Protein hydrolysate treated with pipsyl chloride and extracted with ether
   (*a*) Aqueous phase, Group 0, contains *p*-iodophenylsulfonic acid, pipsyl arginine, and other material
   (*b*) Ether extract evaporated to dryness with a few drops of 0.1*N* ammonia and then diluted to volume with water
2. To aliquots of (1, *b*) appropriate ^{35}S-labeled indicators are added
   (*a*) Traces of *p*-iodophenylsulfonic acid washed out by 0.2*N* HCl
   (*b*) Washed aliquots containing indicators distributed in 10 plate counter-current partition between $CHCl_3$ and 0.2*N* HCl

3. Plates 1–3, Group I
   (*a*) Pipsylglycine contaminant removed by carrier precipitation
   (*b*) Pipsylaspartic and glutamic acids isolated on butanol–isopropanol–ammonia chromatograms
   (*c*) Pipsylserine, threonine, hydroxyproline isolated on butanol–ammonia chromatograms

Plates 9–10, Group III
Group distributed in 15 plate counter-current partition between CCl_4 and 0.2*N* HCl

Plates 4–8, Group II
(*a*) Pipsylserine contaminant removed by solvent partition
(*b*) Pipsylglycine and alanine isolated on butanol–ammonia chromatograms

Plates 1–5, Group III, *a*
Pipsylphenylalanine removed on hydrochloric acid chromatogram and valine–methionine area eluted and divided into two portions

Plates 6–9, Group III, *b*
(*a*) Phenylalanine and proline derivatives first separated on 0.1 *N* HCl chromatogram
(*b*) Bands eluted and rechromatographed with butanol–ammonia for final purification

Pipsylmethionine removed by oxidation beyond sulfone state and pipsylvaline isolated on butanol–ammonia chromatogram

Pipsylproline removed on a butanol–ammonia chromatogram; valine–methionine area eluted and methionine oxidized to sulfone in which form methionine is determined from butanol–ammonia chromatogram

Fig. 1. Scheme for the analysis of amino acids by the isotope derivative method. Reproduced from Velick and Udenfriend (139) with their permission.

(*4*) *Specific Applications.* Pipsyl chloride-^{35}S has been used to determine the amino acid content of two peptides containing hydroxyproline found in the urine of patients with rheumatoid arthritis (94). The procedure for the pipsylation of the amino acids was essentially as described by Keston, Udenfriend, and Cannan (67) and reported in Section III-1-A-a-(1), except pipsyl chloride-^{35}S was used in the presence of sodium borate. The extracted derivatives with 2.0 μmoles of the inactive pipsyl derivatives of hydroxyproline, proline, glutamic acid, and aspartic acid were separated on ion exchange column using Dowex 1 and the peaks assayed by spectrophotometry and counting.

b. Acetic Anhydride Method. (*1*) *General Application.* The double isotope derivative method was also applied to the determination of amino acids at the submicrogram level by Whitehead (143). Tritiated *N*-acetyl derivatives of the amino acid components in hydrolysates of crystalline insulin and chymotrypsinogen were prepared using tritiated acetic anhydride and ^{14}C-labeled acetylamino acids, prepared using added ^{14}C-acetic anhydride, were added. The derivative of each amino acid was separated by paper chromatography. In the original work, the two isotopes were separated and determined by gas counting, as described in Section II-4-C-a.

An alkaline solution of the hydrolysate from 5–20 μg of the protein is treated with excess tritiated acetic anhydride, previously standardized as hydrocortisone acetate, in toluene. Samples of the ^{14}C-labeled acetyl derivatives (approximately 50 μg containing 6000–8000 cpm) are added to the reaction mixture and the toluene removed by evaporation under high vacuum. The residual solution is passed through an acid Zeo-Karb 225 column to remove sodium ions, and the solution taken to dryness. The residue is dissolved in 0.1 ml of 1:1 v/v ethanol–2*N* aqueous ammonia solution and the *N*-acetyl amino acids are separated by paper chromatography or electrophoresis on Whatman No. 2 paper (see Section II-6-B-c). The spots on the chromatograms are shown up by treating the dry papers with bromcresol green in acetone (100 mg/liter) made alkaline with 1 drop of morpholine; the exceptions are histidine and arginine, where the reagent is used in the neutral form.

The use of a scanning technique to locate the individual derivatives and the development of methods for the elution of the *N*-acetylamino acids from the paper would enable scintillation counting to be applied in the determination of the ^{14}C and ^{3}H activities following this method of derivative analysis (see also Section II-4-C).

(*2*) *Special Applications.* The procedure described above for the estimation of amino acids using tritiated acetic anhydride as the labeled reagent was extended to the determination of thyroxine by Whitehead and Beale (144). All the free amino acids, including thyroxine and its analogs, are acetylated in a sample of serum with tritiated acetic anhydride of high specific activity. A known amount of ^{14}C-labeled thyroxine acetate is added to the mixture after acetylation and the derivative purified by chromatography.

A 5-ml sample of serum is covered with 5 ml of ether and acetylated in alkaline solution by five successive additions of the reagent (1 mg) in toluene and 0.2 ml of 0.1*N* sodium hydroxide solution. To this mixture is added about 100 μg of standard *N*-acetyl thyroxine-^{14}C containing approximately 10,000 cpm. The solution is then reduced to 2 ml by distillation *in vacuo* and the serum proteins precipitated at pH 6 with ethanol. The solid is separated after centrifugation, and after the residue has been washed with aqueous ethanol the derivative is extracted from the solution with chloroform and back-extracted into water. Fatty acids, precipitated from the solution with calcium acetate (prepared by neutralizing a suspension of 7 g of calcium hydroxide in 50 ml of water with 10 ml of glacial acetic acid and making up to 200 ml with water) are separated by centrifugation. The supernatant is passed through an acid Zeokarb 225 column, the eluates from which are adjusted to pH 9 and taken to dryness *in vacuo*. This residue, in water (4 ml) and acetic acid (1 ml, 10% v/v) is extracted thoroughly with *n*-butanol–chloroform (30% v/v half saturated with water) and the extract evaporated *in vacuo*.

The separated material in ethanol–2*N* ammonia (0.2 ml, 1:1 v/v) is applied to Whatman No. 2 paper, run to the origin with acetone–2*N* aqueous ammonia (1:1 v/v) and the chromatogram is developed using *n*-butanol–methanol–2*N* ammonium hydroxide, 4:1:5 (v/v) and ethyl acetate–methanol–2*N* ammonium hydroxide 4:1.5:1.9 (v/v) in two dimensions. The derivative was shown up by treatment with aqueous sodium carbonate (5% v/v) and then diazotized sulfanilic acid, cut out, and counted.

Whitehead and Beale (144) also describe the determination of thyroxine using ^{131}I-thyroxine and tritiated acetic anhydride, the former replacing the *N*-acetyl thyroxine-^{14}C.

Standardized ^{131}I-thyroxine (approximately 0.01 mg and 30,000 cpm) is added to the serum, which is then acetylated and the derivative purified as described above. The ^{131}I-activity is determined in a well scintillation counter and the tritium by gas counting (Section II-4-C-a). With this method it is also possible to extract the thyroxine from the serum after addition of the ^{131}I-labeled compound and partially purify it before acetylation.

This procedure is quoted (144) as estimating the readily available thyroxine in serum with an accuracy of 2.3% and has been used to determine levels of from 3.0 to 6.5 μg/100 ml of serum.

It has also been applied to the determination of 3-monoidotyrosine and 3,5-diodotyrosine by Beale and Whitehead (9).

c. 2,4-Dinitrofluorobenzene (DNFB) Method. (*1*) *Procedure.* A method for the estimation of the amino acids arising from protein hydrolysates, at a level of 10^{-5} μmole of each acid, has also been described by Beale and Whitehead (10). This procedure is an application of the reagent first described by Sanger (120) for the determination of amino acid groups of peptides to microscale amino acid composition analysis. The mixture of amino acids is treated with tritiated DNFB in alkaline solution, and known amounts of the ^{14}C-labeled 2,4-dinitrophenyl amino acids are added. The derivatives are separated from the reaction mixture by solvent extraction, purified by paper chromatography, and the tritium and ^{14}C determined by gas counting. This method, carried out on microgram quantities of protein hydrolysate, gives results in agreement with those obtained using tritiated acetic acid on milligram quantities of hydrolysate.

The hydrolysate of 0.9 μg of protein in water (1 ml) is diluted with ethanol (1 ml) and reacted with standardized ^{3}H-DNFB (5 μl of a solution of 0.79 g of ^{3}H-DNFB in 1 g of nitrobenzene) in the presence of sodium bicarbonate (1 mg) at 40°C for 2 hr. When the reaction is complete, known amounts of the ^{14}C-DNP derivatives of the amino acids (approximately 6000 cpm, 10 μg) are added and the ethanol removed *in vacuo*. The aqueous residue is extracted with ether and chromatographed in two dimensions on Whatman No. 2 paper, first using toluene, pyridine, 2-chlorethanol, 0.8*N* aqueous ammonia (30:9:18:18) by an ascending technique, and then by using a phosphate system (NaH_2PO_4:Na_2HPO_4, 2:1, 1.5*M*, pH 6) by a descending technique.

(*2*) *Special Applications.* A modification of the method described above was applied to the microdetermination of β-aminoisobutyric acid in blood, urine, and tissue by Gerber and Remy-Defraigne (55). They reacted the amino acids in urine, deproteinized blood or tissue homogenate with tritiated FDNB in alkaline medium, extracted the derivatives, and separated the 2,4-dinitrophenyl-β-aminoisobutyric acid by thin-layer chromatography after carrier 2,4-dinitrophenyl-β-aminoisobutyric acid had been added to all the material at the origin of the chromatogram.

Ten microliters of urine, 5 μl of a solution of ^{3}H-DNFB in benzene containing 400 μc/ml (specific activity 156 mc/mmole), 5 μl of 0.5% solution of carrier DNFB and 50 μl of a pH 8.8 buffer are heated together at 40°C for 1 hr. The mixture is washed with ether, acidified with 30 μl of 1*N* hydrochloric acid, and extracted into 200 μl of chloroform. Fifty microliters of this solution is applied to a Kieselgel H plate which has been activated at 110°C for 15 hr, 5 μl of a solution of carrier derivative containing 40 μmole/ml is applied on top of the active spot, and the plate developed using benzene–pyridine–acetic acid 80:20:2 (v/v). The required spot is scraped from the plate and counted in a scintillation counter.

Twenty microliters of deproteinized blood or tissue homogenate are treated with 5 μl of 5% solution of DNFB, 20 μl of ^{3}H-DNFB, and 50 μl of pH 8.8 buffer as described above to form the derivative which is also purified in the same way.

The accuracy of this method quoted as 10–15% could be improved by the use of a second isotope. This would also eliminate the use of internal standards for calibration, and ensure reproducibility of results.

d. Continuous Isotope Derivative Procedure. The main disadvantage of the methods described above for the determination of amino acids in mixtures by isotope derivative techniques is that each method involves lengthy, complex preparation and purification steps. In an attempt to overcome these difficulties, Blaedel and Evenson (15) have reported the development of a continuous procedure for isotope derivative analysis of amino acids with direct automatic readout of results. The method is based on the formation of ^{58}Co(III) complex with the amino acid to be determined which is then separated, assayed chemically, and counted. The method has only been investigated for alanine. (See Fig. 2.)

The solution containing the unknown amino acid and a solution containing an excess of ^{58}Co(II) are pumped together using peristaltic pumps. The resulting stream is passed through a graphite-filled electrolysis column to oxidize the Co(II) to Co(III), which is then bound quantitatively to the amino acid in a complex. This material is diluted with a known stream of the carrier complex and the desired material separated from excess ^{58}Co and other complexes by ion-exchange chromatography. (This procedure has not been fully investigated.) The amount of complex in the purified sample is determined chemically and counted. Provided the mass of the carrier complex is very much greater than the labeled complex, then the concentration of the unknown is directly proportional to the ratio of these two results (see Section II-2-B-b).

B. END GROUP AND SEQUENCE ANALYSIS OF PROTEINS AND PEPTIDES

a. Introduction. Over the past twenty years nonisotopic methods have been developed for the determination of terminal amino acids in

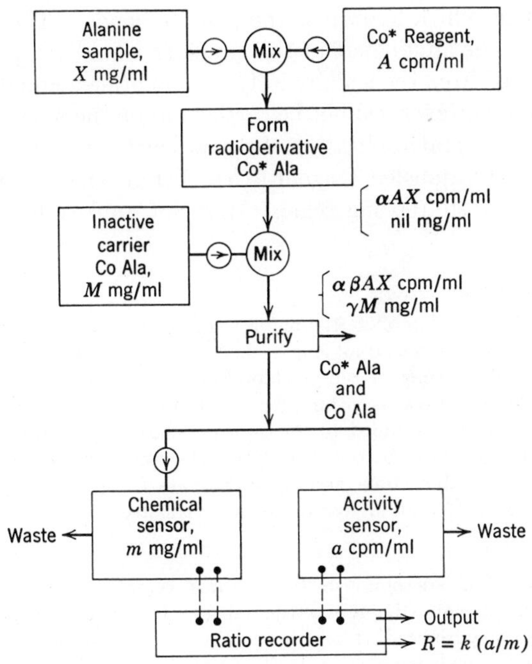

Fig. 2. Block diagram of continuous isotope derivative procedure. Reproduced from Blaedel and Evenson (15) with permission.

protein and for sequential degradation of peptides; they are well documented in the literature and have been extensively reviewed (e.g., 39, 52, 111). These methods of analysis are carried out on milligram quantities of the protein, but the application of isotopic reagents to the same procedures has made the same analyses possible at the microgram level. The first double isotope derivative method to be applied to the analysis of peptide end groups was, however, not an application of this kind but a technique developed primarily as an isotope derivative method for amino acid composition analysis which was applied to this problem.

b. Pipsyl Method for Amino End Group Analysis. Udenfriend and Velick described the application of the pipsyl method (138) for amino acid composition analysis to the determination of amino end groups of proteins. The free amino groups of the proteins react with pipsyl chloride-^{131}I and the sulfonamide derivative so formed is stable under the conditions which hydrolyze the peptide bonds. The ^{131}I-labeled derivative of the amino acid at the amino end (and ε-*N*-pipsyllysine) can be isolated from the acid hydrolysate, identified, and determined. This method, however, cannot be carried out on the microgram scale, but the most general application of the two methods of reaction of the protein with the labeled reagent (the "heterogeneous" method) is described on a 20–100 mg scale. It is not applicable to sequence determinations.

Heterogeneous Reaction

To a stirred suspension of 20–100 mg of the protein, 3 ml of water, and 90 mg of sodium bicarbonate in a 50-ml long-necked Erlenmeyer flask are added 6 ml of absolute ethanol; 400 mg of pipsyl chloride-^{131}I in 1.5 ml of benzene is then added, and the mixture shaken for 2 hr. The product is extracted with ether and 20 ml of acetone is added to the aqueous phase to precipitate the pipsyl protein, which is separated by centrifugation and washed twice with acetone, once with water, twice with acetone, and finally with ether. The solid is dried in air, dissolved in 1 ml of 1*N* sodium hydroxide, and estimated.

Homogeneous Reaction

For proteins which are soluble in 60% aqueous acetone containing tetramethylammonium bicarbonate, the above reaction may be carried out in this system and the product hydrolyzed without insolation although the product obtained is less pure than by the heterogeneous method above.

Hydrolysis

A known amount of the ^{35}S-pipsylamino acid is added to the pipsylated protein and the mixture heated with 6*N* hydrochloric acid at 115°C in a sealed tube until the protein is hydrolyzed.

Determination and Identification of Derivatives

This is carried out as described by the same workers for the application of this technique to composition analysis [Section III-1-A-a-(3)]. The procedure can be simplified since the fractionation only involves a small number of derivatives, arising from the *N*-terminal amino acids and ε-*N*-pipsyllysine. The ten-plate chloroform acid partition is carried out and each of the three groups is checked for activity by extracting into alkali and counting in a liquid counter. The distribution of activity will indicate which group may be discarded; otherwise the full separation must be carried out. The derivatives other than ε-*N*-pipsyllysine are identified by paper chromatography, as described in Section III-1-A-a-(3), and are determined by counting.

Identification and Determination of Lysine

ε-*N*-Pipsyllysine occurs in Group 0 and α-ε-*N*-dipipsyllysine in Group III of the 10 plate chloroform–acid extraction. (The latter had not been detected by Udenfriend and Velick.) ε-*N*-Pipsyllysine is of importance since it will give an indication of the completeness of the formation of the derivative. The ^{35}S-labeled compound is prepared from the amino copper complex of lysine in the usual way. It is isolated as the complex salt, free from copper by treatment with hydrogen sulfide or $2N$ hydrochloric acid, purified by crystallization and chromatography (butanol–ammonia system if required). Group 0 solution is evaporated to dryness and the residue is dissolved in $2N$ ammonia in 50% ethanol and centrifuged. When an aliquot of the supernatant is chromatographed on the butanol–ammonia system, the R_f of the lysine derivative is 0.45 and the sulfonic acid from the reagent is 0.65. Analyses for lysine by this method do not give such reproducible results as can be obtained for other amino acids.

Udenfriend and Velick applied this method to a number of proteins and found one phenylalanine and one glycine amino end group in bovine insulin, two N-terminal valine residues in horse hemoglobin, and two N-terminal proline groups in rabbit muscle aldolase. Their values for insulin have since been shown to be correct, based on the formula of Ryle and his co-workers (118), although at the time they disagreed with the quantitative results of Sanger (120).

c. Pipsyl Method in Sequence Analysis. Levy (83) has described the application of the pipsyl method to the determination of N-terminal amino acid in peptides and proteins. The preparation of the pipsyl proteins is not general and conditions vary from one protein to another.

Pipsyl peptides have been prepared in 85–95% yields as follows:

To pipsyl chloride-^{131}I (15 mg) in a Folin-Wu tube is added 0.1 ml of a solution of the peptide in water and Na_2HPO_4 (28 mg) in water (0.5 ml). The mixture is heated to 90°C and then shaken until the pipsyl chloride-^{131}I disappears. $3N$ hydrochloric acid (1 ml) and a glass bead are added to the cooled mixture and this is then extracted with 3 × 3-ml portions of ethyl acetate. The aqueous residue may be further pipsylated.

Levy also describes the purification of pipsyl peptides on ion-exchange columns.

Dowex 1 × 2Cl^- resin is set up in 0.9-cm diameter columns in 50% alcoholic $0.001N$ hydrochloric acid. The column is compacted by forcing a further 50–100 ml of the solvent through as rapidly as possible and the pipsyl peptide allowed to flow through under gravity using 20% alcoholic $0.001N$ hydrochloric acid solution.

This method has special applicability to the determination of peptides resulting from protein hydrolysis during sequence studies. For the determination of C-terminal amino acids, this method can be applied in conjunction with the hydrazinolysis procedure of

Akabori and his co-workers (1). Investigations of silk proteins by this method have been carried out by Levy and Slobodian (85,125).

d. 2,4-Dinitrofluorobenzene Method of Amino End-Group Analysis. The method first described by Sanger (120) and subsequently considerably developed by many other workers was applied by Beale and Whitehead (10) to the determination of amino end groups on microgram samples of insulin using a method similar to that described in Section III-1-A-c.

One milliliter of ethanol is added to 1 ml of an aqueous suspension containing 1.8 μg of protein. To this is added 1 mg of sodium bicarbonate and 5 μl of the standard ^{3}H-DNFB solution and the reaction carried out as in Section III-1-A-c. When the reaction is complete, known activities of ^{14}C-DNP's of the unknown amino acids are added and thoroughly mixed. The ethanol is removed *in vacuo* and after the aqueous phase has been extracted with 3 $\times$ 1-ml portions of ether, it is refluxed with 1 ml of concentrated hydrochloric acid for 8 hr. The hydrolysate is extracted with 2 $\times$ 1-ml portions of ether, which is evaporated, and the residue chromatographed using the toluene and phosphate systems as in Section III-1-A-c above. The spots are then cut out and counted.

The results obtained by this method for insulin and ribonuclease were in complete agreement with those calculated from the formulas of Ryle and his co-workers (118) and Redfield and Anfisen, respectively (112).

e. Hydrazinolysis Method for Carboxyl End Group. Beale and Whitehead (10) also used their modification of the DNFB method for the analysis of C-terminal groups of proteins using the hydrazinolysis technique of Niu and Fraenkel-Conrat (104). When proteins are heated with anhydrous hydrazine, all but the C-terminal amino acids can be separated by treatment with benzaldehyde. The remaining C-terminal amino acids can be determined as the DNP derivatives as described above [Section III-1-A-c-(1)].

Three to four micrograms of protein and 5 μl of hydrazine (98%, distilled from barium oxide) are heated at 100°C for 4 hr in a sealed tube. The tube is then cooled, opened, excess hydrazine is removed *in vacuo,* and the residue dissolved in 0.5 ml of water. This solution is shaken with 0.1 ml of benzaldehyde and the aqueous phase removed. The benzaldehyde is washed with 3 $\times$ 0.1 ml of water and the amino acid composition of all the aqueous phase determined by the ^{3}H-DNFB technique described in Section III-1-A-c-(1).

This method gave results for the C-terminal amino acids of insulin and ribonuclease in agreement with the formulas (see Section III-1-B-b), although hydrolysis of C-terminal asparagine gave a significant value for C-terminal aspartic acid in insulin.

f. Phenylisothiocyanate-^{35}S Method for Stepwise Degradation from the Amino End. This method of end-group analysis was first described by Edman (43,44) and has since been considerably developed. Two reactions are involved, first the preparation of the *N*-phenylthiocarbamyl derivative of the peptide (PTC peptide) by the action of the reagent on the peptide in an alkaline medium, and then cleavage of the derivative of this group and the *N*-terminal amino acid from the peptide to produce the phenylhydantoin (PTH).

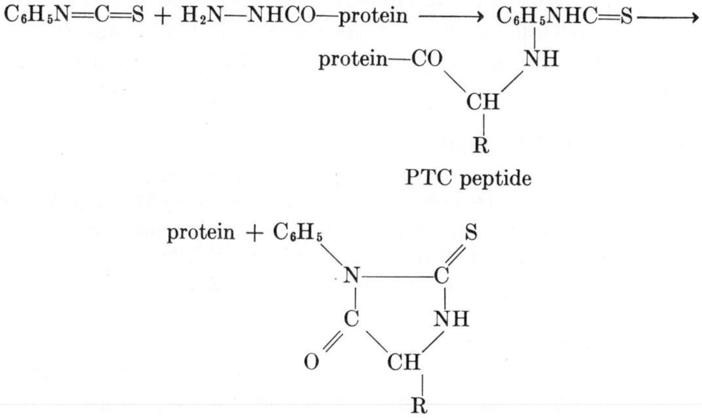

$$C_6H_5N{=}C{=}S + H_2N{-}NHCO{-}\text{protein} \longrightarrow C_6H_5NHC{=}S{-}NH{-}CH(R){-}CO{-}\text{protein} \longrightarrow$$

PTC peptide

$$\text{protein} + C_6H_5{-}N{<}\,[\,N{-}C({=}S){-}NH{-}CH(R){-}C({=}O){-}N\,]$$

PTH of *N*-terminal amino acid

Because of the difficulty of carrying out the second reaction to give a high yield of either the shortened peptide or the PTH, sequence determinations have often been carried out by the subtractive procedure. This method relies on the comparison of quantitative amino acid analysis carried out on the peptide before and after the removal of the *N*-terminal amino acid as its PTH. This method has serious limitations since yields of the shortened peptide are not only small but the amount available is considerably reduced at each stage by material used in the constituent amino acid analysis. Since most proteins and peptides are only available in a pure form in very small quantities, this presents a serious limitation to the length of chain which can be degraded.

Aqueous hydrolysis techniques, which have been developed by Fraenkel-Conrat (51) and Dahlerup-Petersen and his group (38) favor a high rate of hydrolysis and cyclization with a minimum of damage to the more labile PTH's and the main peptide chain. In this way the shortened peptide can be separated from the PTH and the second *N*-terminal residue determined. Thus the whole chain can be sequentially degraded and each amino acid characterized in turn as its PTH.

The use of phenylisothiocyanate-^{35}S has been applied to modifications of these techniques. The labeled reagent has not only simplified the detection and determination of the PTH's, but has enabled the procedure to be carried out on much smaller samples of protein.

Two procedures using this reagent have been reported. The first, due to Cherbuliez and his co-workers (30–32), describes a micro method for amino end-group assay and the stepwise degradation of polypeptides.

Synthesis of the PTC-Peptide

To not more than 0.02 μmole of the peptide in 5 μl of a 0.05*M* sodium diethylbarbiturate buffer in a microtube is added 0.2 μmoles of PTC-^{35}S in *N*, *N*-dimethylfomamide. The tube is stoppered and shaken with a vibrator while maintained at 40°C for 2 hr. The cooled solution is diluted with 20 μl of the buffer, extracted with 5 × 20 μl of benzene containing 10% pyridine, and the aqueous phase is evaporated *in vacuo*.

Cyclization of the PTC-Peptide

The residue from this evaporation is dissolved in 20 μl of a mixture of 4 parts concentrated hydrochloric acid and 1 part glacial acetic acid and, after 2 hr at room temperature, the acids are evaporated in a vacuum desiccator over potassium hydroxide.

Isolation of the PTH

The solid residue from the acid hydrolysis is dissolved in 15 μl of 0.05*N* sodium hydroxide and 5 μl of the buffer and extracted with 3 × 20-μl portions of benzene. The aqueous phase is evaporated and treated with PTC-^{35}S for the isolation of the second amino acid PTH while the benzene solution is chromatographed for the identification of the first PTH.

Chromatography of the PTH

Silica gel, 250 μ thick on 250 × 50-mm plates, is activated at 140°C for 2 hr and cooled in a vacuum desiccator. Spots are applied 15 mm from the edge of the plate and 10 mm apart. Development is with heptane–pyridine–ethyl acetate 5:3:2 (v/v) for 30 min, giving a run of 100 mm. After the plates have been dried for 10 min at room temperature and 5 min at 105°C the PTH's show as white spots on a brown background when they are sprayed with 5% starch solution followed by a solution of iodine and sodium azide. Radioactivity is located with a chromatogram scanner and the PTH's are identified by comparison of the R_f's of the located radioactive material with known synthetic PTH's.

The original method of Cherbuliez and his group of workers was applied to a number of di- and tripeptides with some success, but the results are confused by the occurrence of radioactive artifacts on the paper chromatograms. These were eliminated by the use of the reagent in *N*,*N*-dimethylformamide, instead of alcohol or pyridine, and by the thin-layer chromatographic system as described above. In this way 0.005 μmole of di-, tri-, and a pentapeptide have been successfully determined (32), but fuller use could have been made of the labeled reagent if it had been standardized and the yields of each PTH calculated.

Laver (82) has reported a different modification of the original method which, although theoretically applicable to stepwise degradation, has only been applied to amino end-group analysis. He has, however, applied the reagent on a quantitative basis to the estimation of *N*-terminal groups from 10^{-4} μmoles of a number of known proteins and of influenza B virus (Lee strain) (81).

Standardization of PTC-^{35}S

The labeled reagent is diluted with carrier PTC, dissolved in light petroleum bp 80–100°C and a slight excess of aniline is added. The crystals of the ^{35}S-*sym*-diphenylthiorea which separate on standing overnight at 20°C are recrystallized from aqueous ethanol, dried (mp 151–153°C), and counted at infinite thinness with an end-window Geiger Muller Counter.

Preparation, Purification, and Determination of ^{35}S-PTH's

To 1–2 mg of protein in 0.1 ml of water and 0.1 ml of a *N*-allylpiperidine–pyridine buffer (prepared from 1 g, *N*-allylpiperidine in 39 ml of pyridine adjusted to pH 9 with 1*N* acetic acid) is added 5 μl of PTC-^{35}S and the mixture is heated at 40°C for 1 hr with shaking. The excess of reagent and by-products is removed by 3 $\times$ 3-ml extractions with acetone. After the residue has been allowed to stand at 20°C for 5 min in 0.3 ml formic acid and 0.2 ml of 6*N* hydrochloric acid, 3 ml of acetone is added and the protein removed by centrifugation. The acetone supernatant is evaporated in a stream of air, the residue reextracted with 2 ml of acetone to remove all traces of protein, and the acetone again evaporated. The solid which remains is dissolved in 1 drop of acetone, 0.2 ml of 0.1*N* hydrochloric acid is added, and the solution is left at room temperature for 15 hr. The PTH which is formed is separated by evaporation of the solvent with a stream of air.

The mixture of PTH's can be separated by partition chromatography on Whatman No. 1 paper using a number of systems (45) of which Laver used three: xylene–formamile, *n*-butyl acetate (water saturated)–propionic acid–formamide, and *n*-heptane–ethylene chloride–formic acid (75% 1:2:2).

For the third system the material is dissolved in glacial acetic acid, the carrier PTH's are added, and the material applied to the paper. For the first two, however, the PTH's are applied in ethylene chloride and, since the PTH's of aspartic acid, glutamic acid, arginine, and histidine have a low solubility in this solvent, it is better that these be added as carrier only when using the third system.

The dry chromatograms are scanned for radioactivity or each spot is cut out, eluted immediately after drying with hot glacial acetic acid, and counted in a scintillation counter.

The results obtained by this method with crystalline horse hemoglobin bovine serum, albumin, and insulin were in agreement with other published results. The quantitative studies indicated that yields of PTH's were between 30 and 40%.

2. Steroids

A. INTRODUCTION

Since the double isotope derivative procedure of Keston and his group (68,70) gave accurate analytical results for amino acids it was applied to steroid hormones (20). The esterification of corticosteroids with pipsyl chloride proved unsatisfactory due to low yields of the esters, and unwanted side reactions, but the use of pipsan (*p*-iodobenzenesulfonyl anhydride) was reported satisfactory for cortisol, corticosterone, and 11-desoxycorticosterone. Svendsen (131), however, found the original pipsyl reagent of the Keston group satisfactory for the esterification of estrone and 17β-estradiol, but used the more stable ^{35}S isotope in the derivate-forming reagent and the ^{131}I as the tracer, thus greatly increasing the sensitivity. These two pipsyl reagents have led to the development of a number of procedures for the determination of submicrogram quantities of steroids.

Labeled acetic anhydride was described as a reagent for identification of steroid hormones by Simpson and Tait in 1953 (122). Tritiated acetic anhydride was later proposed by Avivi and his co-workers (3) as a reagent for double isotope-derivative analysis, which had all the advantages of pipsyl chloride and less of the disadvantages. It was applied to the analysis of steroids before its use in amino acid analysis and first proved satisfactory for the determination of the readily acetylatable hydroxyl groups in crude aldosterone. Satisfactory results were obtained for the estimation of hydrocortisone in blood by a double isotope derivative procedure. Although the same method was applied to the estimation of aldosterone in blood, the chromatographic procedure was not concise enough for the isolation of a radiochemically pure derivative. Berliner (12,13) used the

^{14}C-labeled reagent for detection and determination of cortisol in plasma by a simple isotope derivative procedure and for the detection of less than 0.1-μg quantities of cortisone, tetrahydrocortisol, tetrahydrocortisone, and pregnandiol in plasma. These derivatives were determined directly from the paper chromatograms using a strip counter (14). Hollander and Vinecour (61) used the isotope derivative method, again with ^{14}C-acetic anhydride, for the determination of cortisone at the 0.2-μg level, and investigated the separation and determination of hydrocortisone, coricosterone, androsterone, and dehydroespiandrosterone.

Acetic anhydride was finally used successfully in a double isotope derivative procedure for aldosterone by Kliman and Peterson (75). This was the first application of labeled acetic anhydride to this procedure since its suggestion by Avivi and his co-workers and has provided the basis of a method for the determination of many steroids using ^{14}C- and ^{3}H-acetic anhydride.

The only other reagent which has found wide use in the determination of steroids by double derivative analysis is thiosemicarbazide-^{35}S. The unlabeled reagent was suggested by Pearlman and Cerceo (108) and Bush (25) for the characterization and determination of steroids because steroid thiosemicarbazones have good chromatographic properties in the systems usually used for the parent steroids. In 1962 Riondel and his co-workers developed double isotope derivative methods using thiosemicarbazide-^{35}S and high activity-tritiated steroids (115). This method has recently found wide application in the determination of ketosteroids.

A number of special reagents have been devised to overcome difficulties experienced in particular steroid analyses. Before the development of thiosemicarbazide-^{35}S as a reagent for ketosteroids, progesterone, which will not react with acetic anhydride, was estimated using sodium borotritide in a double isotope derivative analysis (149). This reagent was used to reduce the keto group which was then partially reoxidized to incorporate the tritium in a nonlabile position. Bromine-82 generated from potassium bromide-^{82}Br has been used for the determination of estrogens by an isotope derivative procedure with reversed dilution (124). Another interesting procedure is based on the use of ^{14}C-labeled trazolinium blue (93) for the semiquantitative determination of corticosteroids or total steroids in blood by a simple derivative procedure.

B. DETERMINATIONS USING SULFONYLATING AGENTS

All the published methods for the determination of steroid hormones using sulfonylating agents are outlined in Table II. These methods have all been progressively developed by a group of Scandinavian workers over the past 10 years and the procedures which they have arrived at for the estimation of aldosterone, corticosteroids, and estrogens have recently been fully described in review (18). This work is of particular value, since it is the result of continuous application of the methods to routine analysis for the compounds in physiological samples. Each of these methods and the significant stages in its development will be described with the details required to supplement the information in Table II.

a. Corticosteroids. The method described by Buus et al. (18) is based on the procedure of Bojesen (17) with modifications and additions. The main developments are the use of ^{125}I in place of ^{131}I for the tracer derivative, the preparation of the ^{35}S-derivative at -20°C, and the application of the method to a number of corticosteroids.

Procedure. *Extraction and Preliminary Purification*

Heparinized plasma (10 ml) is extracted twice with chloroform (15 ml) by stirring and centrifugation. After the chloroform has been evaporated at <45°C under nitrogen at reduced pressure, the residue in aqueous methanol (1 ml 70%) is washed twice with toluene hexane, 20:80 (0.5 ml). The aqueous phase is then evaporated under nitrogen, dissolved in ethyl acetate (1 ml), washed twice with water (0.5 ml), and dried over sodium sulfate.

Esterification

The dry residue from the evaporation of the ethyl acetate is dissolved in 7% pyridine–chloroform (25 μl). This solution is cooled to -20°C and shaken while dry chloroform (0.1 ml) containing pipsan-^{35}S (3 mg) and then 1% pyridine–chloroform (25 μl) are added. The tube is shaken for 1 min, allowed to stand at room temperature for 4 min, and then the reaction is stopped by the addition of 1 drop of water. The tracer ^{125}I-pipsyl derivatives are then added and the mixture washed twice with methanol–water, 25–75 (0.2 ml) and finally evaporated to dryness under nitrogen.

Chromatography

Purification is carried out on the systems shown in Table II. Transfers and elution are carried out with chloroform–methanol, 90:10.

b. Aldosterone. In the same review (18) Bojesen et al. describe a method for the determination of aldosterone which is based on the procedure of Bojesen and Degn (19). In this case, however, the latter procedure has been considerably modified so that the tracer compound can be added to the plasma sample and can thus give an indication of all losses. This has been accomplished by the use of very high specific activity ^{3}H-aldosterone in conjunction with high specific activity tosan-^{35}S as the labeled reagent. The latter is preferable to the pipsan-^{35}S which was used by Bojesen and Degn (19).

Procedure. *Extraction and Preliminary Purification*

1,2-^{3}H-aldosterone (1000 cpm) is added to plasma (10 ml) and the mixture extracted with chloroform (3 × 15 ml). The extracts are washed with 0.1*N* sodium hydroxide (1 ml) and water (2-ml portions) until neutral, dried over sodium sulfate, and evaporated to dryness under nitrogen. The residue, in 70% methanol–water, is washed with 10% toluene–hexane (2 × 0.5 ml) and taken to dryness at 40°C.

For samples >10 ml, the residue from the chloroform extraction is washed with 70% aqueous methanol (3 × 1 ml) and this solution chromatographed (system *i*, Table II, no. 9). The aldosterone is recovered by chloroform extraction and this solution is washed with water (3 × 0.5 ml), dried over sodium sulfate, and evaporated.

Esterification

To the residue from either preliminary purification is added 10% pyridine in acetonitrile (10 μl) and tosan in acetonitrile (10 μl) of a solution containing 7 mg in 100 μl). After 10 min at room temperature the reaction is stopped by the addition of water (25 μl). Five minutes later benzene (100 μl) and carrier aodosterone tosylester (20 μg) in 10% acetone–benzene (25 μl) are added. This solution is washed with 0.1*N* sulfuric acid (100 μl) and water (3 × 150 μl), dried over anhydrous sodium sulfate, and evaporated.

Purification

Purification is carried out as outlined in Table II. The residue from the second chromatogram is dissolved in 50 μl of glacial acetic acid oxidized with fresh 0.5% chromium trioxide in glacial acetic acid (10 μl) for 30 min at room temperature. The product is extracted with chloroform, washed with water, and the solution is evaporated to dryness. After the third chromatogram the residue is mixed with a solution of 2,4-dinitrophenylhydrazine in acetic acid (50 μl of a solution containing 25 mg in 15 ml), left for 30 min at room temperature, and the mixture evaporated to dryness for the final chromatography.

c. Estrone and Estradiol. In the same publication Bojesen and Svendsen describe a method for the determination of estrone and estradiol based on the original work of Svendsen (131). This procedure has since been applied to the determination of the two estrogens and estradiol precursors in plasma by Andersen and his co-workers (2).

TABLE II

No.	Steroid, source, and tracer	Reagent, reaction standardization	Isolation and/or prelim purification
1	17-Hydroxycorticosterone. Human and dog peripheral plasma. ^{131}I-pipsyl derivative.	^{35}S-Pipsan in 7% pyridine in chloroform, standardization with glycine.	$CHCl_3$ extraction: solid in 70% MeOH, washed with 20% toluene in *n*-hexane. To dryness, in ethyl acetate washed twice with water and dried.
2	17-Hydroxycorticosterone. Human plasma only (referred to as hydrocortisone).	^{35}S-Pipsan in 7% pyridine in chloroform, standardization with glycine.	As above.
3	Aldosterone. 10–30 ml of plasma from heparinized blood. ^{131}I-pipsyl derivative 60–80 mc/mmole.	^{35}S-Pipsan in 7% pyridine in chloroform, standardization with glycine.	Chloroform extraction, residue in aq. $CHCl_3$/MeOH, washed 20% toluene in hexane. Residue in chloroform washed dilute $NaHCO_3$ and H_2O.
4	Estrone (A). 17-β-Estradiol (B). 10 ml of heparinized plasma. ^{131}I-pipsyl derivative. 60 mc/mmole.	^{35}S-Pipsyl chloride 60 mc/mmole in acetone aq. NaOH. Finally taken into chloroform. Standardized with glycine.	Chloroform extraction, purified by partition, extracted alkali, washed in chloroform with $0.1M$ Na_2HPO_4 and dried.
5	17-β-Estradiol and estrone as above.	As above. Specific activity 100 mc/mmole.	Chloroform extraction purified by partition, extracted alkali, washed in chloroform with $0.1M$ Na_2HPO_4 and dried.
6	Estrone, 17-β-estradiol, and precursors, fetal and cord plasma. ^{125}I-pipsyl derivatives.	^{35}S-Pipsyl chloride as No. 5 above.	Separation from precursors by $CHCl_3$, extraction, otherwise as No. 5 above.
7	Estrone and 17-β-estradiol heparinized plasma. ^{125}I-pipsyl derivatives, 50 mc/mmole.	^{35}S-Pipsyl chloride as No. 5 above.	Solvent extraction and partition after Nos. 4–6 above.

Sulfonylating Agents

Purification of derivative	Special purification procedures, etc.	Level of estimation and accuracy/error	Authors and remarks	No.
Wash H_2O–50% MeOH. Chromatographed Whatman No. 1 paper. (*i*) 70% MeOH-H_2O, 40% toluene–ligroin, bp 68–75°. (*ii*) 90% MeOH-H_2O, 40% toluene–ligroin, bp 68–75°.	Radioautography.	0.2–10 μg/5 ml plasma; SD ± 6%.	Bojesen (17).	1
As above.	Duplicates. Radioautography.	4–18 μg/per cent; 0–15 μg % SE of mean ±0.45 μg, % maximum technical error ± 1.1 μg %.	Engell et al. (46).	2
Two times reverse-phase paper chromatography; ethyloleate; acetic acid/H_2O 50/50 at 4°C. 1x formamide; ligroin/toluene, 20/40 at 4°C. Oxidize and repeat formamide system.	CrO_3 oxidation. Radioautography and use of selective filters in counting.	1–10 ng; 80% recovery.	Bojesen and Degn (19).	3
Paper chromatography 2 × ligroin: water 1:1. A in ligroin: MeOH-H_2O (10:1:9) B in ligroin- (10:9:1)	Use of scanner and radioautography.	Sensitivity 2 ng: recoveries estrone, 65 ± 9%; 17-β-estradiol, 64 + 6%.	Svendsen (131).	4
1 × ligroin: H_2O 1:1 then both 2× with system for B above, No. 4, but see special procedure, right.	Use of scanner and radioautography. Between chromatography 2 and 3 preparation of *p*-carboxyphenyl hydrazone of estrone and oxidation of estralaliol to estrone with CrO_3.	1 ng.	Svendsen and Sorensen (132).	5
As above.	As above.	Detect at 2 ng.	Andersen et al. (2).	6
		1 ng.	Bojesen et al. in rev. of 4–6 (18).	7

(continued)

TABLE II

No.	Steroid, source, and tracer	Reagent, reaction standardization	Isolation and/or prelim purification
8	Cortisol (a), cortisone (b), deoxycortisol (c), corticosterone (d), heparinized plasma. ^{125}I-pipsyl derivatives. 60 mc/mmole.	^{35}S-Pipsan 22 mc/mmole at −20° 7% pyridine in chloroform.	Chloroform extraction and as No. 1 above.
9	Aldosterone, plasma. ^{3}H aldosterone 30–50 c/mmole.	^{35}S-Tosan. 150 mc/mmole added in acetonitrile react in any chloroform containing 10% pyridine.	Chloroform extraction addition of ^{3}H-aldosterone as 3 above (large samples) chromatographed on Kieselgel/formamide plate with 70% chloroform/benzene.

The most important difference from the original procedure (131) is the use of ^{125}I instead of ^{131}I as the isotope for labeling the tracer derivatives. The advantages of the latter isotope over the former are described in Section II-5-A-b.

Procedure. *Extraction and Preliminary Purification*

Heparinized plasma (10 ml) is extracted with chloroform (3 × 1.5 ml) by stirring with a glass rod. The chloroform is dried over sodium sulfate and reduced to 0.5 ml by evaporation under nitrogen at 50°C *in vacuo,* diluted with 1:1 chloroform–methanol (3–4 ml), and transferred to a test tube. This solution is taken to dryness, dissolved in 50% aqueous ethanol, washed with pentane (3 × 0.5 ml) and the solvent again evaporated. The residue is dissolved in 5:1 (v/v) carbon tetrachloride–chloroform, washed with water (2 × 0.5 ml), and extracted with 1*N* sodium hydroxide (3 × 0.5 ml). The alkaline solution is acidified with 10*N* hydrochloric acid (0.15 ml) and extracted with chloroform (3 × 0.5 ml), which is then washed with 1*M* disodium-hydrogen phosphate solution (0.5 ml) and water (0.5 ml), evaporated, and the residue dried *in vacuo* at 50°C.

Esterification

The dry samples in 4 mm × 75-mm test tubes are dissolved in acetone (25 μl) and water (25 μl) and shaken while ^{35}S-pipsyl chloride (0.2 mg) in acetone (25 μl)

(continued)

Purification of derivative	Special purification procedures, etc.	Level of estimation and accuracy/error	Authors and remarks	No.
Paper chromatography. Following No. 1. Whatman No. 2 paper.		56% recovery for cortisol a ± 0.03 μg b ± 5 ng c ± 5 ng d ± 10 ng standard deviations	After Bojesen (17). Buus et al. in rev. (18).	8
(*i*) Paper chromatogram reverse phase as No. 3 but using acetic acid and water 30:70 (v/v). (*ii*) TLC Kieselgel/formamide developed with toluene. (*iii*) TLC Kieselgel/formamide, Petroleum b.p. 115–125°/toluene 15:85. (*iv*) TLC Kieselgel activated 1 hr 110° chloroform, ethyl acetate 95:5.	CO_3 oxidation after second chromatogram. 2,4-DNP prepared in glacial acetic acid after third chromatogram. PPO (2,5-diphenyl-oxazole) 10 mg/ml added to stationary phase in reverse phase chromatography to facilitate detection of derivative under UV light. A similar material (Leucht pigment Z.S. Super, Ridol de Haem Germany) was added to the thin-layer plates.	50 pg.	Bojesen et al. in rev. (18) after Bojesen and Degn 1961 (19).	9

is added. The sides of the tube are finally washed with acetone (25 μl). After 30 min the ^{125}I-labeled tracers (0.2 ml of a solution in chloroform) are added and the mixture washed first with 50% aqueous methanol (2 × 0.25 ml) and then with water (0.25 ml).

Purification

The derivatives are purified on the chromatographic systems as outlined in the Table II. The estradiol ester from the second chromatogram is applied to the starting line of the third paper and oxidized with 5% aqueous chromic acid solution (20–25 μl), which is left in contact with the spot for 15 min before the paper is developed. The estrone ester is converted to the *p*-carboxyphenylhydrazone on the portion of paper cut from the second chromatogram by the addition of *p*-carboxyphenylhydrazine hydrochloride [50 μl of a solution of 5 mg in methanol (4 ml)]. After 15 min the hydrazone is eluted with chloroform onto the starting line of the third paper.

C. DETERMINATIONS USING ACETIC ANHYDRIDE

a. General. Although the first reagents to be used in double isotope derivative analysis for steroids were the sulfonylating compounds which had been applied to the determination of amino acids by Keston and his co-workers (68,70), the most widely used labeled reagent for the estimation of steroids is acetic anhydride. The use of this reagent by Avivi and his group of workers in 1954 (3) for the estimation of hydrocortisone and in an attempt to estimate aldosterone was not further developed until 1960, when Kliman and Peterson (75) published a detailed method for the determination of aldosterone in urine and plasma. In the interim, however, a number of workers (e.g., 13,61) did use the labeled reagent in simple isotope derivative procedures which claimed to be quantitative but lacked the conciseness of the double isotope method.

In a review of this type it is not possible to describe all the methods which utilize acetic anhydride as a labeled reagent or even to report the most important of them, although Table III gives most of the relevant details. However, since all of the double isotope derivative methods are either developments of the Kliman and Peterson method or are very similar in procedure and detail, that one method will be described here.

b. Procedure for the Determination of Aldosterone (75)

Extraction

Concentrated hydrochloric acid (0.1 ml) is added to urine (10 ml) (fresh or preserved by freezing) and the pH adjusted to 1.0. After 24 hr at room temperature this solution is extracted with dichloromethane (60–70 ml), the extract washed with 0.1N sodium hydroxide (6 ml), 0.1M acetic acid (6 ml), and finally with water (6 ml), before it is evaporated to dryness.

Plasma is extracted with dichloromethane following a similar procedure.

Acetylation

The residue from the extract is transferred to a tapered tube with ethanol and dichloromethane and the solvent evaporated. The solid, concentrated at the tip of the tube, is dried *in vacuo* over anhydrous calcium chloride. This material is acetylated by the addition of anhydrous pyridine (0.025 ml) and ^{3}H-acetic anhydride (0.03 ml) and left for 24 hr at 37°C. A known amount of aldosterone diacetate-^{14}C (approximately 1000 cpm in 0.1 ml of ethanol) is then added followed by water (0.5 ml) and carbon tetrachloride (0.5 ml).

Purification

The products of the acetylation are extracted into the carbon tetrachloride, which is then washed with water (0.4 ml) and evaporated. The chromatographic purification is as outlined in Table III. Between the second and third chromatograms the derivative is oxidized with chromic acid as described in Section II-6-C-a-(3).

The other methods using this reagent differ mainly in their chromatographic and other special procedures; the important ones are described in Section II-6-C and Table III.

The procedure described by van der Molen and his co-workers (97) for the determination of progesterone by formation of the acetate of an enzymic reduction product is most elegant. Unfortunately this method is not very concise and is described with a large number of other methods of determination which are all carried out on the same sample.

c. Procedure for the Determination of Digitoxin. Lukas and Peterson (88) have recently reported a method for the determination of submicrogram quantities of digitoxin in the plasma, urine, and stools of patients maintained on the drug. Their method, which uses the tritiated compound and acetic anhydride-^{3}H as the reagent, has increased the sensitivity of the determination of digitoxin from 2 μg to 0.01 μg.

Digitoxin-^{3}H was prepared by the Wilzbach technique (147) [see Sect. II-5-C-b-(2)]. The labeled compound was purified by partition between dichloromethane and water followed by paper chromatography on Whatman No. 1 paper at 26°C with benzene, methanol, water (4:2:1) by the descending technique. The tritiated digitoxin was located by running markers of unlabeled compound on the same chromatogram and dipping the strips with the markers only in a freshly prepared mixture of methanolic solutions of *m*-dinitrobenzene (1% w/v, 3 vol) and benzyltrimethyl ammonium hydroxide (40% w/v, 1 vol.). Digitoxin gives blue spots under this treatment. The located tritiated material on the untreated chromatogram was eluted with methanol and rechromatographed by downward development with cyclohexane–dioxan–methanol–water (4:4:2:1 v/v) for 3 days. A further separation was carried out using the same solvents in the proportions 4:3:2:1 (v/v) for 4 days. The methanol eluate from the final chromatogram was evaporated in a stream of air at 37°C and the residue taken up in 40 ml of dichloromethane. This solution was washed with water (2–3 ml), the organic phase evaporated, and the residue taken up into ethanol to make a stock solution which was stored at −20°C. The acetic anhydride-^{14}C was standardized as cortisol acetate (75) (see Sections III-2-C and II-5-B) and then used to standardize the tritiated digitoxin.

D. DETERMINATIONS USING THIOSEMICARBAZIDE-^{35}S

a. General. The procedures described by Riondel and his co-workers for the determination of progesterone and testosterone (113,114) and of Horton and Tait for the determination of androstenedione (62,63) using thiosemicarbazide-^{35}S all follow a very similar pattern. The methods for the testosterone and androstenedione analysis seem to be reapplications of the procedure for the determination of progesterone rather than developments from it. All the methods using this reagent are outlined in Table IV and only the determination of testosterone will be described further.

TABLE III

No.	Steroid, source, and tracer	Reagent, standardization, reaction	Preliminary purification, etc.	Purification of derivative	Special procedures
					Simple Isotope
1	"Acetylable" steroids. Blood, urine. No tracer.	Acetic anhydride-^{14}C.	Chloroform extraction.	Paper chromatography. Hexane/formamide or benzene/formamide (151).	—
2	3-20 or 21-hydroxy steroids. Plasma. Unlabeled carrier.	Acetic anhydride-^{14}C, 2 mc/mmole in pyridine.	Chloroform extraction. Hydrolysis, β-glucuronidase, second chloroform extraction, partition.	Paper chromatography (151).	Oxidation in identification (150).
3	Androsterone, dehydroepiandrosterone, cortisone, hydrocortisone, corticosterone in mixture and from plasma.	Acetic anhydride-^{14}C benzene/pyridine.	Chloroform extraction. Solvent partition.	(150).	—
					Double Isotope
4	Aldosterone, hydrocortisone in human peripheral blood. Tracer aldosterone-^{14}C. Acetate hydrocortisone-^{14}C 1.5 mc/mmole.	Acetic anhydride-^{3}H in pyridine.	Partition column chromatography.	2 × partition columns (3 for aldosterone).	—
5	Aldosterone. Biological extracts, urine and plasma. Tracer aldosterone diacetate-^{14}C.	^{3}H-acetic anhydride 100 mc/mmole, standardized as cortisol acetate. Reaction in pyridine, 24 hr at 37°C.	Urine first hydrolyzed at pH 1. Dichloromethane extraction; washed in acid, alkali, and water.	3 × paper chromatography after Bush (24). Whatman No. 1 paper.	CrO_3 oxidation after second paper chromatography.

Acetic Anhydride

Level of estimation	Chromatography system			References	No.
Derivative Procedures					
Quantitative recovery 0.1 μg.				12 (no details of quantitative method)	1
Cortisol. 0.86 μg/10 ml plasma (SD 0.042); others identified only.				13	2
Satisfactory results only from mixtures, not from plasma.				61	3
Derivative Procedures					
2.5 ± 0.5 μg per 100 ml blood for hydrocortisone; final sample of aldosterone not sufficiently pure for result.				3	4
	1	2	3		
Samples containing 0.1 μg at ±10%.	Andrenosterone as standard. Cyclohexane/benzene/methanol/water, 4:2:4:1.	21-deoxycortisone as standard. Cyclohexane/dioxan/methanol/water 4:4:2:1.	Standard as in 2. Cyclohexane/benzene/methanol/water 4:3:4:1.	75	5

(*continued*)

TABLE III

No.	Steroid, source, and tracer	Reagent, standardization, reaction	Preliminary purification, etc.	Purification of derivative	Special procedures
6	Cortisol. Plasma, plasma ultra filtrate. Cortisol-21-acetate-^{14}C or cortisol-4-^{14}C.	^{3}H-acetic anhydride in pyridine/benzene about 100 mc/mmole, standardized as cortisol-21-acetate-^{3}H.	Dichloromethane extraction and washed in alkali, acid, and water.	3 × paper chromatography with addition of carrier cortisol acetate. Whatman No. 1 paper.	CrO_3 oxidation to cortisone acetate.
7	Testosterone. Plasma, tracer Testosterone-^{14}C 22.2 mc/mmole.	^{3}H-acetic anhydride, 100 mc/mmole, standardized as cortisol acetate-^{3}H reaction in pyridine.	Dichloromethane extraction; washed alkali; water dried.	Extracted carbon tetrachloride; washed; chromatographed Whatman No. 1 paper at 20° with addition of carrier.	Formation of thiosemicarbazide after third paper chromatography.
8	Testosterone. Adrenal venous plasma. Testosterone-^{14}C 21.5 mc/mmole.	^{3}H-acetic anhydride, 14.2 mc/mmole standardized as cortisol acetate, after TLC and paper chromatography. Reaction in pyridine.	Dichloromethane extraction; washed alkali, water dried.	Extracted carbon tetrachloride; washed; chromatographed 2 × thin layer (2 dimensions) 3 × paper.	—
9	Aldosterone, corticosterone cortisol. Incubation media rat/beef adrenals. Tracers aldosterone di-^{14}C-acetate. Corticosterone-4-^{14}C 50 μc/mg, cortisol-4-^{14}C 70 μc/mg.	^{3}H-acetic anhydride, 40 mc/mmole, standardized cortisol. Reaction in pyridine.	Dichloromethane extract from blood or incubation media. Extract washed with alkali and water.	4 × paper chromatographies for each steroid involving five systems on Whatman No. 2 paper.	Aldosterone and corticosterone derivatives oxidized CrO_3 in AcOH after 2nd chromatography, and cortisol after 3rd chromatography.

(continued)

Level of estimation	Chromatography system 1	Chromatography system 2	Chromatography system 3	References	No.
60–65% recovery up to acetylation, often not reproducible, overall recovery 40–60%.	Cyclohexane/dioxan/methanol/water 100:100:50:25.	Cyclohexane/benzene/methanol/water 50:100:100:20.	Carbon tetrachloride/methanol/water 100:100:25.	Method of Peterson, Schedl et al. (121).	6
Mean recovery = 17%; sensitivity = 0.05 μg/100 ml of plasma; accuracy 0.75 μg/100 ml ±10%, 0.14 μg/100 ml ±16%.	(*i*) Heptane/phenyl-cellusolve. (*ii*) Hexane/formamide. (*iii*) Dekalin/nitromethane/methanol 100:50:50.	(*iv*) Petroleum ether (80–100)/methanol/water 100:90:10. (*v*) isooctane/*t*-butanol/methanol/water 500:100:350:50.		64	7
Sensitivity 0.08–1.0 μg/100 ml. Accuracy in range 0.55–1.45 μg/100 ml of plasma ±1.8 SEM.	Thin layer silica gel. (*i*) 1st dimension 25% (v/v) acetone/hexane, 2nd dimension 2% (v/v) ethanol benzene. (*ii*) 1st dimension 15% (v/v) ethanol in petroleum ether 2nd dimension 25% ethyl acetate/hexane.	Paper systems Whatman No. 2. (*i*) Methyl cyclohexane/methanol/water 2:1:1; reverse phase. (*ii*) (After Bush A) Petroleum ether/acetic acid/water 5:4:1. (*iii*) Cyclohexane/nitromethane/methanol 2:1:1.		64 23	8
0.1 μg corticosterone 0.1 μg aldosterone.	Chromatography: 1. Isooctane/*t*-butanol/methanol/water 500:225:225:50. 2. Skellysolve C/benzene/methanol/water 667:333:800:200. 3. Isooctane/toluene/methanol/water 250:250:350:150. 4. Cyclohexane/benzene/methanol/water 600:450:600:150. 5. Carbon tetrachloride/methanol/water 800:800:200. Aldosterone, 1,2,3,2.	Corticosterone, 1,2,1,4.	Cortisol 1,4,5,5.	130	9

(continued)

TABLE III

No.	Steroid, source, and tracer	Reagent, standardization, reaction	Preliminary purification, etc.	Purification of derivative	Special procedures
10	Cortisone and cortisol, from plasma. Cortisol-4-^{14}C, 66 μc/mg, cortisone-4-^{14}C, 12.5 μc/mg.	^{3}H-acetic anhydride, 40 mc/mmole standardized with cortisol as in No. 5.	Dichloromethane extract; alkali and water wash.	4 × paper chromatographies on Whatman No. 2 paper.	Oxidation between third and fourth chromatographies found unnecessary.
11	Corticosteroid sulfatesin urine of human neonate as in No. 10 after hydrolysis, but including tetrahydracortisone.	^{3}H-acetic anhydride, 40 mc/mmole standardized with cortisol as in No. 5.	Dichloromethane extract; alkali and water wash.	Preliminary column and four paper chromatographies.	Oxidation between second and third chromatographies.
12	Aldosterone in urine. Tracer 1,2-^{3}H-aldosterone 100 mc/mg	Acetic anhydride 1-^{14}C 2 mc/mmole standardized as No. 5, derivative prepared as No. 5.	As No. 5, followed by 70% aqueous methanol/toluene–petroleum ether 50:50 (v/v) partition methanol phase evaporated water, extracted dichloromethane; TLC purification silica gel.	TLC silica gel; 1 × 1 dimension; 1 × 2 dimensions; 1 × paper chromatogram on Whatman No. 1 paper; 1 × TLC silica gel 1 dimension.	
13	Testosterone (I), dehydroepiandrosterone (II), androsterone (III), etiocholanolone (IV), in plasma. Tracers ^{14}C-steroids specific activity 22, 50, 23, and 21 mc/mmole, respectively.	Acetic anhydride-1-^{3}H; (100 mc/mmole) standardized with testosterone-^{14}C and dehydroepiandrosterone-^{14}C. Derivative prepared in benzene/pyridine.	40 ml of plasma contaning 1 ml 1N NaOH, extracted ether/chloroform (3:1) and washed water. TLC followed by paper to separate four compounds. Each one further TLC.	Methylene chloride extract. Washed, in alkali and water; dried; TLC on silica gel G; paper on Whatman No. 1; GLC on 25% SE 30 on Gas Chrom P 60–80 mesh.	1,1-N-dimethyl hydrazones prepared and used for GLC.

(*continued*)

Level of estimation	Chromatography system	References	No.
Sensitive to 0.05 μg of each steroid.	Chromatography systems: Light petroleum/benzene/methanol/water 666:333:800:200 Cyclohexane/benzene/methanol/water 4:3:4:1. Isooctane/*t*-butanol/methanol/water 500:225:225:50. Carbon tetrachloride/methanol/water 100:100:25.	60	10
Nonquantitative.	13 paper chromatography systems used. Procedure basically as Nos. 9 and 10.	41	11
Overall recovery 20%; blank values 1 μg.	Chromatography systems: 1. Chloroform/ethanol 99:1 (v/v). 2. Ethylacetate/ethylenedichloride/water 90:10:1 (v/v). 3. Chloroform/methanol/water 94:6:0.5 (v/v). 4. Chloroform/acetone 90:10 (v/v). Before acetylation: same plate 2× system 1, then 2 and 3. After acetylation: TLC 4:2 dimensions 2 and 4: paper–cyclohexane/benzene/methanol/water 6:2:5:1 (v/v) TLC 4.	11	12
Theoretical error 8%; coefficient of variation maximum 20% for lowest levels. Sensitivity 0.1 μg per 100 ml of plasma at ±17% for II, III, and IV. 0.05 μg per 100 ml of plasma, ±20% for I.	Thin-layer systems: 1. Benzene/ethyl acetate 60:40. 2. Benzene/ethyl acetate 80:20 Paper systems: 3. Ligroin/methanol/water 100:90:10 4. Ligroin/methanol/water 100:70:30 5. Decalin/nitromethane/methanol 100:50:50 Preliminary separation by TLC System 1, giving II and III, and I and IV; II and III separated on system 3; I and IV on system 4; individuals then all on system 1. Chromatography of acetates: I, system 2 and GLC; II, III, and IV, system 4, then 2, and GLC.	71	13

(*continued*)

TABLE III

No.	Steroid, source, and tracer	Reagent, standardization, reaction	Preliminary purification, etc.	Purification of derivative	Special procedures
14	Progesterone in plasma. Tracer progesterone-4-^{14}C specific activity 24.1 mc/mmole.	Acetic anhydride-^{3}H; specific activity 57.8 mc/mmole reaction in benzene/pyridine.	10 ml of plasma. Ether extraction solvent partition petroleum ether, 70% aqueous MeOH; water, ether. 2 × TLC (alumina containing phosphor) GLC, IR (KBr disk) KBr in water, benzene extract, enzyme reduction.	Reverse phase system 2. TLC developed three times with toluene, then system 3, and finally system 2.	Enzyme from rat ovary used to convert progesterone to 20α-hydroxypregn-4-en-3-one.
15	Testosterone in human plasma. Testosterone-4-^{14}C specific activity 45.2 mc/mmole. Purified partition column V.	Acetic anhydride-^{3}H; specific activity 50 mc/mmole standardized as in No. 5.	Plasma from 30 ml of blood added to testosterone-^{14}C, extracted 3 × 60 ml of dichloromethane, washed Na_2CO_3 solution, H_2O, taken to dryness, and purified on a partition Celite column with hexane/benzene/MeOH/H_2O 3:7:7:3.	Residue from column dried and acetylated in pyridine at 25°C for 18 hr. Product taken to dryness repeatedly from benzene and MeOH, 2 × TLC 2 dimensions Systems I and II. 3 × column chromatography, systems III–V.	Formation of phenyl hydrazone after second column chromatography. Dry residue from column treated with 1 ml of 1% phenylhydrazine for 1 hr at 60°C or 24 hr at 25°C. Product taken to dryness and dissolved in dichloromethane.
16	Testosterone (*i*) and androst-4-ene-3,11-dione (*ii*) in human plasma. 1,2-^{3}H-testosterone 130 μc/μg. ^{3}H-androst-4-ene-3, 17-dione prepared from 1,2-^{3}H-testosterone by chromic acid oxidation.	^{14}C-Acetic anhydride 20 mc/mmole as 5% solution in benzene. 30 μl of this solution and pyridine (10 μl) used for acetylation at 37°C for 24 hr.	To plasma (10–20 ml) 3*N* NaOH (0.4/10 ml) is added. Extracted. Ethyl ether washed in 0.1*N* HCl, water, and evaporated. Chromatographed system I, Eluate	Reaction stopped with water (0.4 ml). Testosterone acetate (20 μg) in ethanol (20 ml) added. After 1 hr extracted carbon tetrachloride (2 ml), washed 2 × 0.1*N*	Androst-4-ene-3, 17-dione reduced with $NaBH_4$ in methanol (0.1 ml of solution containing 20 mg/100 ml) at 4°C for 10 min, water (1 ml) added, extracted ether, and taken to dryness.

No. 16 is continued on pp. 70 and 71.

(continued)

Level of estimation	Chromatography system	References	No.
High blank value due to ^3H impurity.	Chromatography systems: 1. Cyclohexane/ethyl acetate 4:1 (v/v) 2. Decalin/methanol/water 10:10:1 3. Ethyl acetate/toluene 5:95	97	14
Determined at 0.79 μg ±0.08 (1σ) per 100 ml of plasma; sensitivity = 0.05 μg per 100 ml.	Thin-layer systems: I Silica gel, 2 dimens. (*i*) benzene/ethyl acetate, 3:1 (*ii*) hexane/ethyl acetate, 1:1 II alumina, 2 dimens. (*i*) dichloromethane (*ii*) benzene/ether 1:1 Partition cols (all Celite) III cyclohexane/nitromethane/MeOH 2:1:1 IV hexane formamide V hexane/benzene 3:7 formamide.	40	15
Recovery: (*i*) 30–40%. (*ii*) 10–25%. Blank: (*i*) 8.7 mμg in 10 ml. (*ii*) 24.5 mμg in 100 ml. Results: (*i*) males, 551 ± 151 mμg/100 ml. females, 47 ± 14.8 mμg/100 ml.	I. Residue in benzene (25 ml) applied to a florisil column. Developed 1% methanol in benzene (25 ml), 8% methanol in benzene (25 ml). II. Whatman 3-mm paper, 7 hr; hexane/methanol/water (100:90:10). III. Whatman No. 2 paper, 4 hr; benzene/heptane/methanol/water (33.3:3:66.7:80:20) at 37°. IV. Thin-layer silica-gel; 2 dimensions. (*i*) Hexane/acetone (225:25). (*ii*) Benzene/dichloromethane/methanol (160:160:4). V. Whatman No. 2 paper, 4 hr, ligroin (bp 66–75°) methanol (60:40) at 4°C.	116	16

(continued)

TABLE III

No.	Steroid, source, and tracer	Reagent, standardization, reaction	Preliminary purification, etc.	Purification of derivative	Special procedures
16 *(continued)*			to system II. Both compounds eluted and dried. Androst-4-ene-3,17-dione reduced to testosterone and chromatographed; system III and both samples acetylated.	NaOH, three times in water and evaporated to dryness. Chromatography system IV, eluted methanol. Chromatographed system V. Reduced and chromatographed system II using 10% ethanolic phosphomolybdic acid to locate the derivative.	Testosterone acetate is reduced to 3β-hydroxyandrost-4-ene-17β-acetate with $NaBH_4$ in methanol (0.1 ml of a solution containing 6 mg/ml), reaction as androst-4-ene-3,17-dione.
17	Digitoxin in plasma, urine, and stool. Digitoxin-^3H 0.76 mc/mg (33% of ^3H activity in genin).	Acetic anhydride-^{14}C 10 mc/mmole. Standardized as 5 above. Reaction in pyridine/benzene at 56°C for 4 days.	Plasma and urine extracted, dichloromethane washed 0.1*N* NaOH, 0.1*M* acetic acid. Ethanol (5 ml) added. Solution evaporated. Residue in ethanol (4 ml) reevaporated. Dissolved in ethanol (0.5 ml), water (2 ml), and extracted cyclohexane (10 ml). Aqueous phase extracted dichloromethane and organic phase evaporated. Residue in methanol–dichloromethane chromagraphed system I.	Reaction terminated with 50% aqueous ethanol (0.5 ml) washed isooctane (2 ml), extracted carbon tetrachloride (3 ml), washed water (0.5 ml), and evaporated. Chromatographed systems II–V.	Stools, as plasma but first homogenized. Extract washed two times 0.1*N* NaOH and 0.1*M* acetic acid. Dried extract in 25% ethanol (3 ml) washed cyclohexane (10 ml) until colorless. Dichloromethane extract of aqueous ethanol washed with water until clear. Residue chromatographed systems I and VI before acetylation. Whole blood diluted water (1 vol), then as plasma but 2 × washing with 0.1*N* NaOH and 0.1*N* acetic acid.

(*continued*)

Level of estimation	Chromatography system	References	No.
(*ii*) 99 ± 19.8 in males, 180 ± 56.7 in females.			
0.01 μg detected accuracy 101 ± 3% (mean ± standard error) for 0.01–0.2 μg in plasma. Coefficient of variation 4% for 0.2 μg in 10 ml of plasma.	Aldosterone diacetate, Rhodamine B and F11 as markers in II–V. Whatman No. 1 paper. I. Cyclohexane/dioxane/methanol/water (4:4:2:1) for 20 hr, Δ^1-cortisone as marker. II. Cyclohexane/benzene/methanol/water (100:25:100:15) for 24 hr. III. Cyclohexane/dioxane/methanol/water (10:2:10:1) IV. Isooctane/*t*-butanol/methanol/water (4:2:4:1) for 62 hr. V. Mesitylene/methanol/water (3:2:1 reverse phase) for 20 hr. VI. Benzene/methanol/water (4:2:1).	88	17

(*continued*)

TABLE III

No.	Steroid, source, and tracer	Reagent, standardization, reaction	Preliminary purification, etc.	Purification of derivative	Special procedures
18	Testosterone in sulfate fraction of plasma. 7α-^{3}H-testosterone, 10 c/mmole.	As No. 16 above.	Plasma washed ether (3 × 2 vol) proteins ppted. ethanol; separated, washed, and ethanol evaporated*; residue in water (100 ml) at pH 11 extracted butanol, washed water, and butanol evaporated. Chromatographed system I. Radioactive fractions taken to dryness. Chromatographed system II and eluate solvolyzed method I. Residue chromatographed systems III and IV. Eluate taken to dryness. Alternative: As above to *; residue in water (25 ml) with ketodase (18,750 units) and 2*M* acetate buffer (2.5 ml), incubated at 37°C for 4 hr and then washed three times with ether. Solvolysis of aqueous phase by method II; residue as above.	Acetylation and purification as No. 16 above. 3β-hydroxyandrost-4-ene-17β-acetate oxidized with chromic acid to testosterone acetate and chromatographed on system V.	Solvolysis method I. Dry residue moistened with water (2–4 drops); dissolved in ethanol (4 ml). Equilibrated ethyl acetate (20 ml) added, kept at 37° for 18 hr, washed 10% $NaHCO_3$, twice with water, and taken to dryness. Solvolysis Method II. Aqueous solution extracted ethyl acetate, the organic phase kept at 37°C for 18 hr, washed as method I; dried Na_2SO_4 taken to dryness and residue in hexane extracted twice with 20% aqueous methanol. Methanol phase evaporated to dryness.

(continued)

Level of estimation	Chromatography system	References	No.
Recovery 9% (using alternative method). Blank 13–27 mμg per 100 ml. Mean and standard deviation 139.5 ± 4.8 mμg/100 ml of plasma.	System I. Alumina (activated 600° for 1 hr) (10 g) in 1-cm diameter column set up in anhydrous *n*-butanol. Residue applied in anhydrous *n*-butanol and developed 100 ml of 2% aqueous *n*-butanol. 50 ml of 6% aqueous *n*-butanol 100 ml of 10% aqueous *n*-butanol 200 ml of saturated 0.1*N* NH_4OH in *n*-butanol System II. Whatman 3 mm paper. Butyl acetate, *n*-butanol, 10% formic acid (640:160:100) at 37° for 12 hr. System III. Florisil column. Residue applied in 15 ml of chloroform, developed with 25 ml of 2% ethanol in chloroform. System IV. Whatman No. 2 paper. Hexane/methanol/water (100:90:1) for 7 hr. System V. Eastman chromatogram silica-gel sheet type K 301 R in two dimensions. (*i*) benzene/ethyl acetate (4:1). (*ii*) cyclohexane/ethyl acetate (1.5:1).	119	18

(continued)

TABLE III

No.	Steroid, source, and tracer	Reagent, standardization, reaction	Preliminary purification, etc.	Purification of derivative	Special procedures
19	Testosterone (*i*) and androstenedione (*ii*) in human peripheral plasma. Testosterone-4-^{14}C 50 mc/mmole, androstenedione-4-^{14}C 50 mc/mmole prepared from testosterone-4-^{14}C by oxidation with 0.2% chromic acid. Both purified by thin-layer chromatography.	^{3}H-acetic anhydride 100 mc/mmole standardized as No. 13 above.	To plasma (10 ml) 1*N* NaOH (0.5 ml) is added. Extracted ether/chloroform (3:1), taken to dryness, and chromatographed system I (13 above). (*i*) rechromatographed System II; (*ii*) reduced enzymatically to (*i*); acetylated.	20% ethanol (1 ml), testosterone acetate (50 μg) added, extracted dichloromethane; washed 2 × water; chromatographed systems IV, V; residue converted to *O*-methyloxime and chromatographed in system VI, Isomers eluted and chromatographed system VII in ethanol (11 μl) and counted.	Isatin and 1,4-diamino-anthraquinone in ethanol as dye markers in chromatography. Androstenedione in ethanol (0.03 ml) reacted with 17β-hydroxysteroid dehydrogenase (0.05 mg) in 0.05*M* potassium phosphate buffer (2.5 ml pH 5.8) containing NADH for 30 min. Extracted dichloromethane, washed two times water, evaporated, and chromatographed system III. *O*-Methyloxime prepared in pyridine (2 drops) with methoxyhydroxylamine hydrochloride (2–3 crystals) at 26°C for 1 hr.
20	Aldosterone in urine, and aldosterone secretion rate. 4-^{14}C-aldosterone, 40 mc/mmole.	Acetic anhydride-^{3}H 100 mc/mmole standardized as 5 above. Derivative prepared in pyridine (0.02 ml) with 20% acetic anhydride in benzene (0.02 ml) at 37°C for 18 hr; stopped 20% ethanol (1 ml).	Secretion rate studies: Urine (250 ml) hydrolyzed for 24 hr at pH 1 with HCl. Extracted methylene chloride (800 ml) washed 0.05*N* NaOH (3 × 200 ml) and 0.1*N* acetic acid (200 ml), then taken to	Either: chromatograpny system II, oxidation, and chromatography systems III and IV, or gas chromatography system VI.	Oxidation with 0.2 ml of a solution of 5.00 g CrO_3 in 1000 ml of acetic acid containing 12.5 ml of water, for 5 min. Stopped 20% ethanol (2 ml) extracted methylene chloride (10 ml) washed water (2 ml).

No. 20 continued on pp. 76 and 77.

(continued)

Level of estimation	Chromatography system	References	No.
Theoretical error 7%. Coefficients of variation 7–16%. Sensitivity for 0.007 μg of (*i*) or (*ii*) theoretical precision of 35 and 22%, respectively. Blank value 0.4 mμg for (*i*) and 0.1 mμg for (*ii*).	No. 13 above. System I. Isatin R_f 0.32; (*i*) R_f 0.23; (*ii*) R_f 0.38. System II. 1,4-Diaminoanthraquinone R_f 0.42; (*i*) 0.50 TLC alumina; benzene/ethyl acetate (3:2). System III. Isatin R_f 0.20 (*i*) R_f 0.20 TLC silica gel Benzene/methanol (9:1). System IV. (*i*) Acetate R_f 0.29 TLC silica gel; benzene, ethyl acetate (4:1). System V. (*i*) Acetate R_f 0.31 glass fiber paper; propylene glycol, ligroin; eluates partitioned between water (1 ml) and chloroform (10 ml) to remove propylene glycol. System VI. (*i*) Acetate *O*-methyl oximes R_f's 0.30 and 0.40 TLC silica-gel; benzene, ethyl acetate (9:1). System VII. Gas-liquid chromatography on 1% XE-60 on diataport S; 200° retention time (*i*) acetate *O*-methyloxime 0.61 relative to (*i*).	6	19
Paper chromatography 52.0 ± 3.3 μg per 24 hr. SD ± 6.3%. TLC and GLC. 54.5 ± 3.8 μg/24 hr SD ± 7.0%.	System I. Paper; toluene/ethyl acetate/methanol/water (90:10:50:50). System II. Paper; cyclohexane/benzene/methanol/water (100:40:100:20). System III. Paper; cyclohexane/dioxane/methanol/water (100:80:100:40). System IV. Paper; cyclohexane/benzene/methanol/water (100:50:100:25). System V. Paper; cyclohexane/dioxane/methanol/water (100:60:100:25). System VI. GLC; 6 ft × ⅛ in. glass column, 2% S.E.-30 on 30–60 mesh Chromosorb W; injected in chloroform; column at 245°; argon 20 psi 50 ml/min. System VII Silica-gel 0.4 mm; methanol/chloroform (5:95). System VIII, as VII, hexane/ethyl acetate (1:3).	74	20

(continued)

TABLE III

No.	Steroid, source, and tracer	Reagent, standardization, reaction	Preliminary purification, etc.	Purification of derivative	Special procedures
20 (*continued*)			dryness at 40°C *in vacuo*. Chromatographed system 1; eluted and acetylated; acetic anhydride specific activity 1 mc/mmole.		
			Urine content studies: As above, using urine (15 ml), but without chromatography system 1. Acetylation with acetic anhydride-^{3}H specific activity 100 mc/mmole.	Either: chromatography systems II and V, then oxidation and chromatography system IV and possibly III, or chromatography systems II, VII, and VIII and then gas chromatography system VI.	
21	Aldosterone in peripheral plasma. 1,2,3-^{3}H-aldosterone. 90 mc/mg.	Acetic anhydride-^{14}C 50 mc/mmole 20% (v/v) in benzene. Standardized with cortisol. Reaction in pyridine (0.02 ml) with 0.02 ml of reagent; 24 hr at 18–24°C stopped 50% ethanol (1 ml).	Plasma (30 ml) and 1N NaOH (1.5 ml.) extracted dichloromethane (240 ml); extract washed 0.01N NaOH (24 ml) and 0.1M acetic acid (24 ml). Ethanol (5 ml) added, and evaporated to dryness. Chromatographed by system I; acetylated.	Product washed isooctane (3 ml), extracted carbon tetrachloride (5 ml), washed water (1 ml), and evaporated to dryness. Chromatographed on systems II, III, and IV. Benzylhydrazone formed. Chromatographed system V and counted.	To dried eluate from chromatogram is added benzylhydrazine (0.3 ml of a solution containing 0.5 mg/ml of methanol) and the solvent evaporated.
22	Aldosterone, (*i*) cortisol (*ii*) and corticosterone (*iii*); adrenal vein blood of sheep.	Acetic anhydride-^{3}H 100 mc/mmole redistilled; 10% solution in benzene.	Plasma extracted dichloromethane (5–10 vol). Extract washed 0.05N	Reaction mixture extracted CCl_4 (5 ml). Extract washed 20% aqueous EtOH	Oxidation with CrO_3 in acetic acid as No. 5 above.

No. 22 continued on pp. 78 and 79.

(*continued*)

Level of estimation	Chromatography system	References	No.
Recovery 5–20%. For samples containing 150, 10, 6.6, and 2 mμg % coefficient of variation of 5, 8, 14, and 50%, respectively. Normal males, 2–15 mμg % mean 6.6 mμg ± 1.20 SD.	System I. Paper, cyclohexane/dioxane/water (100:100:25). System II. Paper, cyclohexane/benzene/methanol/water (100:40:100:20). System III. Paper, cyclohexane/dioxane/methanol/water (100:75:100:25). System IV. Reversed phase paper, mesitylene/methanol/water (3:2:1). System V. Paper, isooctane/*t*-butanol/methanol/water (100:60:100:30).	110	21
Data from six year's application of the method.	All systems Whatman No. 1 paper. System I. cyclohexane/benzene/methanol/water (100:40:100:20); gives (*i*) + (*iii*) and (*ii*). System II. [(*i*) + (*iii*)] cyclohexane/dioxane/methanol/water (100:75:50:25); gives (*i*) and (*iii*).	36	22

(*continued*)

TABLE III

No.	Steroid, source, and tracer	Reagent, standardization, reaction	Preliminary purification, etc.	Purification of derivative	Special procedures
22 (*continued*)					
	4-^{14}C-aldosterone 10–139 μc/mg; 4-^{14}C-cortisol 42–130 μc/mg; 4-^{14}C-corticosterone 44–145 μc/mg.	Standardized as No. 5. Reaction in pyridine (20 μl) with reagent solution (25 μl) 18 hr at 37°C stopped 20% ethanol in distilled water.	NaOH (1/10 vol), 0.1*N* acetic acid (1/10 vol), and evaporated. Residue in 20% aqueous ethanol (5 ml) washed cyclohexane, (10 ml) extracted dichloromethane (30 ml).	(0.5 ml). Chromatographed systems I–VI with oxidation of (*i*) and (*ii*) between II and IV, and III and VI, respectively.	
23	Aldosterone and cortisol in whole blood, after Kliman and Peterson, No. 5 above. Aldosterone-^{14}C and cortisol-^{14}C.	As No. 5 above.	As No. 5 above, but including one chromatography.	As No. 5 above, but the oxidation product is chromatographed in three different systems.	As No. 5 above.
24	Aldosterone in urine as No. 5 above, but in order to determine *in vivo* stability of 1,2-^{3}H-aldosterone.	Acetic anhydride 1-^{14}C 1.25 mc/mmole as No. 5 above.	Urine (200 ml) washed dichloromethane (1000 ml) and hydrolyzed at pH 1 for 24 hr. Extracted dichloromethane (2 × 500 ml) at 4°C. Extracts washed 0.1*N* NaOH (100 ml), 0.1*N* acetic acid (100 ml), and water (100 ml), and taken to dryness. Residue in 20% aqueous ethanol (5 ml) washed cyclohexane (25 ml), extracted dichloromethane (25 ml) and this extract evaporated.	Chromatography (systems I–IV) and oxidation applied at varying stages in procedure.	As No. 5 above.

(continued)

Level of estimation	Chromatography system	References	No.
	System III. (*ii*) cyclohexane/dioxane/methanol/water (100:100:50:25). System IV. (*i*) cyclohexane/benzene/methanol/water (100:70:100:25). System V. (*iii*) mesitylene/methanol/water (reverse phase). System VI. (*ii*) cyclohexane/benzene/methanol/water (100:70:100:25).		
	Six in all, no details.	54	23
Stability of tritium shown and variations of chromatographic procedure discussed.	System I. 3 mm paper; toluene/ethyl acetate/methanol/water (450:50:250:250). System II. 3 mm paper; benzene/methanol/water (500:250:250). System III. No. 2 paper; cyclohexane/dioxane/methanol/water (360:360:180:90). System IV. No. 2 paper; cyclohexane/benzene/methanol/water (300:160:330:80). All systems at 37°.	78	24
		See also 77,148	

TABLE IV

No.	Steroid, source, and tracer	Reagent, standardization, reaction	Preliminary purification, etc.	Purification of derivative	Special procedures
1	Testosterone. Peripheral blood. 1,2-^{3}H-testosterone specific activity 137 mc/mg.	^{35}S-Thiosemicarbazide. 50–120 mc/mmole standardized with testosterone. Reaction MeOH/acetic acid at 65°.	20–25 ml heparinized plasma made alkaline; extracted ether, washed water, 10% acetic acid, water; taken to dryness.	Extracted methylene chloride, washed (addition of carrier thiosemicarbazide), taken to dryness. TLC system 1.* Paper chromatography system 2. TLC system 3. TLC system 4. Paper chromatography system 5.	Use of carriers and special elution solvents after chromatography. Pyruvic acid hydrolysis after second chromatography. Acetylation after third chromatogram; acetic anhydride in benzene/pyridine.
2	Progesterone. Peripheral blood. 1,2-^{3}H-progesterone 87 mc/mg.	As above.	As above.	As testosterone (1 above) up to *, then TLC system 3. TLC in system 4. Paper chromatogram system 5.	As above.
3	Androsteneidone. Peripheral blood. 1,2 -^{3}H-androstenedione 137 mc/mg.	As above.	As above.	Can be as testosterone (1 above) up to *, but alternatives systems 1 and 2. TLC system 3. Paper chromatography systems 2 and 4.	As above up to end of acid hydrolysis. Complete removal of propylene glycol after paper chromatography.
4	Testosterone. 1,2-^{3}H-testosterone 152 mc/mg.	Thiosemicarbazide-^{35}S 195 mc/mmole. Derivative prepared in methanol/acetic acid at 65° 5 hr. Product to dryness and extracted with chloroform.	Alkali plasma extracted ether, extracts washed water, dilute acetic acid, and water. Purified TLC system (*i*) acetylated.	Chromatography on glass-fiber paper (*iii*) two-dimensional TLC system (*ii*) acetylation. Paper chromatography (system *iv*). Paper chromatography (system *v*).	Acetylation acetic anhydride in pyridine before preparation of thiosemicarbazide; special elution techniques and addition of carrier at each stage during purification. Further acetylation after second chromatography of derivative.

Thiosemicarbazide-^{35}S

Level of estimation, etc.	Chromatographic systems, etc.	Reference	No.
Mean overall recovery 7%. Coefficient of variation = 5.5%. Mean percentage divergence 5%.	1. Silica Gel G. 2 dimensions; benzene-acetone-methanol, 89:1:10 (v/v). 2. Propylene glycol/toluene–acetone, 200:1 at 31°C. 3. Silica gel G. 2 dimensions benzene–acetone 75:25 (v/v). 4. Silica gel G. 2 dimensions (*i*) methylene chloride/ethyl acetate 70:30 (v/v) (*ii*) methylene chloride/methanol 95:5 (v/v). 5. Benzene/Skellysolve C 1:2 (v/v), methanol/water 4:1 (v/v).	113	1
Mean overall recovery 3%. Blank values 2 ng. Coefficient of variation 7.5%.	3. Silica Gel G. 2 dimensions; (*i*) Benzene/acetone/methanol 96:1:3 (v/v). (*ii*) dichloromethane/acetone/methanol 98:1:1 (v/v). 4. Silical Gel G. 2 dimensions; (*i*) Benzene/methanol 90:10 (v/v). (*ii*) Methylenedichloride/methanol 95:5 (v/v). 5. Propylene glycol–toluene.	114	2
Overall recovery after fourth chromatography 4.5%; after fifth, 1.2%. Coefficient of variation 4.3%.	1. Silica Gel G. 2 dimensions, both with benzene/methanol/acetone 92:5:3. 2. Paper chromatogram propylene glycol–toluene at 33°C. 3. Silica Gel G. 2 dimensions, both with benzene/methanol/acetone 89:10:1. 4. Paper chromatography. cyclohexane/dioxane/methanol/water 10:10:4:2.	63 62	3
Overall recovery 22%. Sensitivity 20 ng/100 ml using 10 ml of plasma. Blank 10 ± 0.9 ng/100 ml.	(*i*) Silica Gel benzene–ethyl acetate, 3:2 (v/v). (*ii*) Silica Gel 2 dimensions. Methylene chloride–methanol 100:3 and benzene/ethyl acetate 4:1 (v/v). (*iii*) 30% propylene glycol in methanol/ligroin/toluene 2:1 v/v. (*iv*) 50% propylene glycol in methanol/toluene. (*v*) 30% propylene glycol in methanol ligroin/toluene 3:2 (v/v).	86 After ref. 113	4

b. Procedure for Estimation of Testosterone (113)

Addition of Tracer and Extraction

Heparinized plasma (10 ml) is added to the residue from the evaporation of a solution of 1,2-^{3}H-testosterone (0.075 μc, 0.5 mμg) or the solution itself is shaken with heparinized plasma (10 ml) for 30 min; in either case, 3N sodium hydroxide (0.4 ml) is added and the mixture extracted with ether (3 $\times$ 10 ml). The extracts are washed with water (3 ml), 10% acetic acid (3 ml), water (3 ml), and taken to dryness.

Reaction

To the above residue from the extract is added ^{35}S-thiosemicarbazide (1 mg) in methanol (0.5 ml) and glacial acetic acid (0.05 ml). After the mixture has been heated at 65°C for 5 hr in a closed tube and cooled, 10% pyruvic acid solution (0.1 ml) and sodium carbonate solution (1 ml of solution containing 5.9 g per 100 ml) are added. This solution is then extracted with methylene dichloride (2 $\times$ 6 ml) and the extract washed first with water (1 ml), then with a solution of thiosemicarbazide (10 mg) in water (1 ml), and finally with water (1 ml). Carrier testosterone thiosemicarbazone (20 μg) and thiosemicarbazide (50 μg) in methylene dichloride–methanol (3:7 v/v) (0.2 ml) are then added to the extract which is washed once more with water (1 ml) and taken to dryness.

Purification of the Derivative

The chromatographic purification is carried out as outlined in Table IV. The material from the second chromatogram is dried, dissolved in methanol (0.4 ml), and 10% aqueous pyruvic acid solution (0.2 ml) added. After the mixture has been heated at 45°C for 45 min, sodium carbonate solution (1 ml of a solution containing 1.52 g per 100 ml of water) is added, the hydrolyzed solution is extracted with methylene dichloride (2 $\times$ 6 ml), washed with water (2 $\times$ 1.5 ml), and taken to dryness.

The testosterone thiosemicarbazone from the third chromatogram is converted to the diacetate before rechromatography. The dried residue is dissolved in pyridine (0.3 ml), acetic anhydride (0.15 ml) is added, and the mixture maintained at 65°C overnight. Methanol is then added and the solution taken to dryness.

The modification of this procedure by Lim and Brooks (86) merits special mention since their work increases the recovery threefold and introduces some unusual techniques.

The testosterone is partially purified by thin-layer chromatography and then acetylated with acetic anhydride in pyridine. This testosterone acetate is reacted with the ^{35}S-thiosemicarbazide by the usual procedure and the thiosemicarbazide purified by chromatography without pyruvic acid treatment but including further acetylation of the derivative. The paper chromatography uses Zaffaroni-type (151) systems, but with glass-fiber paper which facilitates the elution of the spots, thus giving the higher recoveries.

E. DETERMINATIONS USING OTHER LABELED REAGENTS

Of the reagents mentioned above, both the sulfonylating compounds and acetic anhydride have readily found wide general application. The development of thiosemicarbazide has provided alternative and often superior methods for the analysis of steroids for which some method had already been described. There are three exceptions to this pattern; the first concerns a compound which would not react with either of the first two types of reagent, namely progesterone (149); the second is a reagent with limited application but which was successfully applied to the analysis of estrogens (124), namely elemental bromine-82; and the third is the development of an isotopic version of the tetrazolium blue colorimetric method using ^{14}C-labeled dye (93).

a. Sodium Borotritide Method for Progesterone. The difficulties in the development of a suitable double isotope derivative analysis for progesterone were overcome in 1963 by Woolever and Goldfien (149) when they described an ingenious method using sodium borotritide as the reagent and progesterone-^{14}C as the tracer. The procedure for this determination is outlined in Table V.

b. Determination of Esterogens Using Bromine-82. Slaunwhite and Neely (124) have described a method for the determination of estrone, estriol, and estradiol as their 2,4-dibromo derivatives by a simple isotope derivative procedure following reaction with bromine-82. This isotope, although commercially available, has a half-life of 36 hr; it also has the disadvantage of being a hard gamma emitter. The analytical procedure is simple in comparison to the double isotope methods, each determination taking only 4–5 hr. Recoveries of about 90% are reported with a sensitivity of 1 mμg and an accuracy of $\pm 3\%$. The procedure is outlined in Table V.

c. Estimation of Corticosteroids Using ^{14}C-Tetrazolium Salts. Marton and his group of workers have prepared ^{14}C-labeled tetrazolium salts (92) and an abstract of their method for the estimation of corticosteroids has been published (93). The method is described as semiquantitative and satisfactory for 0.01 μg of individual cortico-

TABLE V

No.	Steroid, source, and tracer	Preliminary purification, etc.	Reagent, standardization, reaction	Purification of derivative
1	Plasma progesterone. Progesterone-4-^{14}C.	Extraction 3 × 2 vol. 1:1 (v/v) Ethyl acetate/chloroform. Residue chromatographed on paper (system I).	Sodium borotritide. Standardized during analytical procedure with pure progesterone and progesterone-4-^{14}C. Reaction in isopropanol ethanol, 4:1, 16 hr at −15°C, stopped glacial acetic acid. Take to dryness; water and chloroform added.	Partition chromatography system II. Paper chromatography system III.
2	Estrone, estradiol, and estriol. (Simple derivative procedure with reverse dilution.) Urine and plasma.	Not described.	^{82}Br from $K^{82}Br$ at approximately 0.2 mc/mg (24 mc/mmole). Reaction in glacial acetic acid under reflux for 10 min, then neutralize.	Chloroform extraction. Wash 2 × 2 ml 3% (v/v) acetic acid evaporate at 50°. Residue in benzene, with carriers, on alumina column. Fractions in 95% ethanol assayed by adsorption at 291 mμ. 50 mg of carrier added to each peak and crystallized to constant specific activity.
3	Corticosteroids. (Simple derivative procedure.) Plasma.	Paper chromatography to separate individual corticosteroids. (No details given.)	Tetrazolium salts-^{14}C, 2 mc/mmole TTC or BT. Derivatives prepared on paper from preliminary purification, or Steroids eluted (EtOH) and treated for 20 min at pH 8.9. Stopped at pH 6 with alcoholic formic acid. Solvent evaporated; residue washed with water.	
4	Estrone and estradiol in plasma of laying hens. ^{14}C-labeled methyl ethers.	Hydrolysis with HCl and separation of estrone and estradiol from estriol following Svendsen. (Nos. 4 and 5, Table II.)	Dimethyl sulfate-^{3}H. Excess decomposed alkali. Extracted hexane.	Thin-layer chromatography on silica gel systems I–III. Column chromatography system IV.

Determinations Using Other Reagents

Special procedures	Level of estimation, etc.	Chromatographic systems	Ref.	No.
Specific activity of progesterone-^{14}C determined at each set of determinations. After reaction, solution taken to dryness, oxidized MnO_2 (20 mg) in chloroform (0.3 ml) 2 hr with shaking. Extracted three times 3 ml of chloroform; filtered.	0.1 μg standard deviation; 0.007 recovery 22.5 ± 4.5%. 1 μg standard deviation 0.02 recovery 48.9 ± 14.3%.	System I. Whatman No. 1 paper, hexane 70% methanol elution with methanol II. Whatman No. 1 paper, heptane–formamide III. Whatman No. 1 paper, Shell sol. 140: 32% ethanol.	149	1
Procedure up to chromatography behind ¼ in. lead shielding. Counted in pulse-height analyzer.	Recoveries at 10 mμg level. 80% urine. 95% plasma. Sensitivity 1 mμg. Accuracy ±3%.	Column 1 × 7.5 cm Whoelm alumina in benzene; gradient elution 1:7 benzene; 50% ethanol in benzene.	124	2
Counting on Al planchettes with argon amyl alcohol flow counter.	Sensitivity 0.1 μg with TTC 0.01 μg with BT. Errors: 10% for 0.01 μg, 5% for 0.05 μg.	Venous blood from rat adrenal, 2 × paper chromatography. Peripheral blood. Column chromatography on silica gel; paper chromatography.	93	3
	Qualitative only. Constant isotope ratio.	System I. Ethyl acetate/cyclohexane/ethanol 45:45:10. System II. Ethyl acetate/water-saturated *n*-hexane/absolute ethanol 80:15:5. System III. Ethyl acetate/cyclohexane 50:50. System IV. Alumina, Brockman activity II deactivated 10% water. Set up in petrol. Gradient elution petrol/benzene.	106	4

steroids or mixtures. For the determination of individual compounds, the mixtures are separated by paper chromatography and treated with a standard solution of the ^{14}C-tetrazolium salt, either before or after elution of the spots.

This method, although of interest as an isotope derivative procedure developed from a colorimetric estimation, is severely limited by the availability of the reagent. The procedure is also outlined in Table V.

d. Qualitative Determination of Esterogens with Dimethyl Sulfate-^3H. Estrone and estradiol were identified in the plasma of laying hens by O'Grady and Heald (106). These workers used ^{14}C-labeled estrogens and dimethyl sulfate-^3H as the reagent. Their procedure for the separation of the estradiol and estrone from estriol was that developed by Svendsen (131) (see Section III-2-B-c and Table II, no. 4). Thin-layer chromatographic separation of the derivatives was carried out and final purification was effected by column chromatography by the method of Brown (21), using alumina of Brockman activity II further deactivated with 9–10% of water.

This method shows an interesting application of the double isotope derivative method to the qualitative identification of submicrogram amounts of materials by the use of constant isotope ratios and could easily be made quantitative by the use of standardized reagent.

Full details of the procedure are outlined in Table V.

3. Other Compounds

A. INTRODUCTION

The double isotope-derivative method of analysis has been confined almost exclusively to the determination of amino acids, peptides, and steroids; nonetheless it has found considerable application in these fields alone.

Labeled reagents have, however, found small application in other fields, but in a much more limited and simple way. The methods for the determination of sugars and polysaccharides (101) and for the estimation of organic acids (76) involve mainly the simple isotope-derivative procedure, in some cases with reverse dilution. The reason for the simplicity of the procedure arises because the level of estimation is of the order of milligrams, not micrograms, and not from the inapplicability of the more complex double isotope method.

B. REDUCING SUGARS AND REDUCING END GROUPS IN POLYSACCHARIDES

The need for a micromethod for the estimation of sugars and polysaccharides arose from the widespread use of chromatographic procedure in the separation of sugars at the submilligram level. The use of the cyanydrin reaction originally developed as a micromethod by Lippich in 1929 (87) and later further developed (37,53,96) was proposed for this purpose by Moyer and Isbell in 1958 (101). These workers had already devised a procedure for the standardization of ^{14}C-labeled cyanide using the reaction with a reducing sugar (99) and applied the same labeled reagent to the structural analysis of labeled dextrans (100).

The polymeric aldehydes resulting from the periodate oxidation of the dextrans were hydrolyzed and the D-glucose and D-erythrose from the 1,3′ and 1,4′ linkages were estimated. The nitriles produced by reaction of the hydrolysis products with sodium cyanide-^{14}C in the presence of barium carbonate, were converted *in situ* to the aldonic acids which were isolated as their salts after the addition of carriers. These materials were counted and then compared with the values obtained for standard solutions of the two sugars.

From the procedure for the standardization of sodium cyanide, Moyer and Isbell developed a general method for the determination of small quantities of reducing sugars and for the study of polysaccharides (101).

The sugar solution (20 μl containing less than 0.001 mmole), 0.6*M* ammonium chloride solution (10 μl), and an alkaline solution of sodium canide (0.0025 mmole containing 0.005 mmole of sodium in 10 μl of water, previously standardized with D-glucose) are sealed in a tube. After the contents have been mixed, the tubes are kept at 50–55°C for 24 hr. They are then opened, a few drops of 10% aqueous formic acid added, and the contents evaporated at 60°C in a stream of air. Removal of the hydrogen cyanide is completed by a further evaporation of two small portions of water. The residue is then counted and the quantity of sugar determined.

Moyer and Isbell reported a standard deviation of ±1.4% for all their determinations, and a satisfactory reproducibility for samples from 0.04–0.2 mg. However, they only applied their method to the determination of pure sugars and to the estimation of the number of carbonyl groups in polysaccharides. The development of suitable chromatographic techniques would enable the method to be applied to the complete determination of sugars in mixtures. Since moderately high specific activity sodium cyanide-^{14}C and tritiated sugars are available, the same reagent could be used in a double derivative analysis accurate at the microgram level.

De Wulf and his co-workers (39) attempted to use cyanide-^{14}C as a reagent for the determination of glycogen but nonspecific reactions led to the abandoning of this procedure. They were, however, successful with sodium borotritide as a reagent for this material.

The solution of glycogen (0.2 ml containing 2.0 mg) and sodium borotritide (5 μmoles, 10 μc) is adjusted to pH 9.5, covered with liquid paraffin, and incubated at 60°C for 24 hr. The reaction is stopped by the addition of glacial acetic acid and the polysaccharide is isolated and purified by precipitation with alcohol. This material is then hydrolyzed and counted.

De Wulf and his colleagues found that the activity obtained after hydrolysis and scintillation counting was solely in sorbitol.

C. ORGANIC CARBOXYLIC ACIDS

The process known as "tritiating methylation" was first applied by Melander (95) and Verly and his co-workers (140) for the synthesis of labeled methanol. The latter group of workers prepared carboxyl-tritiated β-naphthoic acid by exchange with tritiated water in dioxane, isolated the dry tritiated acid, esterified it with diazomethane in the usual way, and hydrolyzed the ester using a slight excess of powdered potassium hydroxide in the presence of carrier methanol. In the analytical method for carboxylic acids the first part of this procedure is an alternative to the direct esterification with diazomethane-^{14}C, which is a complicated and expensive procedure although it has been carried out (89).

Baumgartner, Lazer, and Dalziel applied the procedure of "tritiating methylation" to the determination of gibberellins in a number of systems at a concentration of 10–50 μg/ml and in treated plant tissues (7). The method was first used for the determination of the ratios of giberellic acid to dihydrogiberellic acid in fermentation liquors.

The crude mixture was treated with tritiated water and diazomethane as described above and the tritiated esters separated by paper chromatography. The proportion of the two esters is then determined by scanning or scintillation counting.

The same group of workers also used this method for determination of the absolute concentration of giberellic acid and its dihydro derivative.

The procedure is exactly as described above but a duplicate sample containing a known amount of the acid or the dihydroacid is carried through the analysis to serve as an internal standard. An alternative to this method is esterification with diazomethane-^{14}C. The samples are also treated as described above and then esterified using a ^{14}C reagent at specific activity 200 μc/mmole and 0.003 mmole/sample. This method gave a quick determination of the acids within the range 5–150 μg with an error of 2–5%.

The latter procedure would be much more accurate and easy to carry out if tritiated diazomethane were stable enough to handle. However, such a method could be applied using tritium-labeled diazoethane or diazopropane.

A double isotope derivative procedure was carried out in the determination of giberellic acid or dihydrogiberellic acid in plant tissues.

The tissue samples are extracted with a solvent containing a known amount of giberellic acid-^{3}H (100–200 mc/g) and the crude extract is purified. This extract is then esterified with diazomethane-^{14}C and a pure crystalline sample of the double labeled ester isolated for counting after the addition of 50–200 μg of the unlabeled ester. These reagents enable the acid or its dihydro derivative to be determined at 5–10% accuracy in the concentration range of 50–200 μg per sample.

The "tritiating methylation" procedure was applied to the quantitative chromatographic analysis of organic acids by Koch and Jurriens (76).

The anhydrous acid (up to 10 mg) in anhydrous diethyl ether, acetone, or dioxan is treated with an ethereal solution of tritiated water (0.2 ml of a solution containing 1 μl of H_2O, 6.67 μc in 2 ml of ether). After 15 min the acids are esterified with diazomethane and the esters separated by thin-layer chromatography. Silica gel plates are sprayed with 1% Hetraphor (BASF) in ethanol, but where silica gel plates impregnated with 20% silver nitrate are used, the reagent is 0.1% 2, in 7-dichrofluorescein in ethanol is used. The spots are then located under ultraviolet light, scraped from the plate, extracted with ether–hexane (7:3 v/v) for 4 hr, and counted.

Recoveries of 95% were reported in the 1–5-mg range, although at levels in the microgram range results were still satisfactory if carrier esters were added before chromatography. When the method was applied to more polar compounds, special isolation and/or counting techniques were necessary. Quantitative extraction of dimethyl tartrate could only be attained by the addition of 50 mg of carrier material before chromatography, and benzoic and phthalic acids were determined by suspension counting of the silica gel.

D. ALKALOIDS

Wiley and Metzger have recently reported the application of the double isotope derivative technique in toxicological analysis (146). They describe a method for the determination of strychnine in blood using methyl iodide-^{3}H as the reagent and strychnine methiodide-^{14}C as the tracer compound.

Strychnine methiodide-^{14}C was prepared from strychnine (430 mg) in absolute ethanol and a twofold excess of methyl iodide-^{14}C. This mixture was heated under reflux for 2 hr and then cooled in Dry Ice. The precipitate which separated was collected. Additional product was obtained by evaporation of the solvent. Human citrated blood (1 ml) was diluted with water (1 ml) and a solution of strychnine methiodide-^{14}C in ethanol (0.2 ml 9600 cpm) added. The solution was concentrated on a rotary evaporator at 60°C and 15 mm following the addition of ethanol (2 ml and then 20 ml). Methyl iodide-^{3}H (0.1 g, 30 μc in ethanol, 1 ml previously standardized as strychnine methiodide) was added and the mixture sealed in a glass tube and heated at 60°C for 4 hr. When the tube was opened the solution was evaporated to 0.1 ml and chromatographed on Whatman No. 1 paper using ethyl acetate–absolute methanol–28% NH_4OH (100:10:5). The methiodide spot was located by spraying a strip of the paper where unlabeled derivative had been applied with potassium iodoplatinate. The area of the paper corresponding to this R_f, where the doubly labeled derivative had been applied was then cut out and counted in a liquid scintillation counter. Correction was made for the counting efficiency by counting a standard sample of strychnine methiodide-^{14}C which had been run on the same paper chromatogram.

This method showed much better results than the spectrophotometric procedure despite the fact that blank values were high.

E. BIOCHEMICAL INTERMEDIATES

Baldessarini and Kopin (4) describe a method for the estimation of tissue levels of *S*-adenosylmethionine using the ^{14}C-labeled compound and *N*-acetylserotonin-acetyl-^{3}H as the labeled reagent. In this method the derivative is prepared by an enzymic reaction giving melatonin-methoxy-^{14}C-acetyl-^{3}H.

$$\text{HO-indole-}CH_2CH_2NHCOC^3H_3 \xrightarrow[\text{Hydroxyindole-}O\text{-methyltransferase}]{^{14}C\text{-}S\text{-adenosylmethionine}} {}^{14}CH_3O\text{-indole-}CH_2CH_2NHCOC^3H_3$$

Tissue is homogenized with trichloracetic acid (10% in 0.05*N* HCl, 5–10 vol), centrifuged at 4°C, and the supernatant extracted with ether (3 × 5 vol) at 0°C. The aqueous phase (0.5–2.0 ml) is mixed with *S*-adenosylmethionine-methyl-^{14}C (1 μg, specific activity 5.5 mc/mmole) and an equal volume of phosphate buffer (pH 6.8) added. This mixture is incubated at 37°C for 1 hr with *N*-acetylserotonin-^3H (2 μg, specific activity 35 mc/mmole) and purified hydroxyindole-*O*-methyltransferase (0.03 unit).

The reaction is stopped by the addition of 1*N* sodium hydroxide (2.0 ml) and the doubly labeled derivative extracted into chloroform (3 vols). Aliquots of this solution are washed with 1*N* sodium hydroxide (3 × 0.25 vols), evaporated to dryness, and counted.

In this method the chromatographic steps are eliminated since it has been shown that the only labeled material in the chloroform extract is the melatonin. Such results show the high specificity of enzymic reaction in the double derivative method.

F. PESTICIDES

Bogner and his co-workers (16) have reported an extensive investigation of the application of double isotope derivative analysis to the determination of pesticides in tissues and foodstuffs. Their work involved a study of the possible methods for the determination of DDT, DDE, Dieldrin, Systox, and Diazinon. DDT and DDE together with the ^{14}C compounds were tetranitrated and the dianilides formed by reaction with aniline-^3H. The derivatives were purified to constant isotope ratio by paper chromatography and then counted. Determinations at levels down to 1 μg were described, but no details were given of the application of the procedure to materials of biological origin.

Dieldrin was determined using ^3H-acetic anhydride as the reagent which, in the presence of 40% HBr led to the formation of the 6-acetoxy-7-bromo derivative. These workers made an extensive investigation of the chromatographic purification of the dieldrin derivative, but were unable to isolate a material of the required radiochemical purity. Multiple recrystallization of the derivative from 70 μg of dieldrin in the presence of 50 mg of carrier gave a product of the required purity.

The same group of workers proposed to determine both the thiol- and thionoisomers of Systox as the 2,4-dinitrobenzoyl derivative. They prepared ^{14}C-thionosystox but were unable to complete the analysis because of the nonavailability of the tritiated reagent.

It was also proposed that Diazonin should be cleaved to the hydroxy-pyrimidine and the derivative prepared using acetic anhydride-^{3}H. This, however, proved impracticable.

DDT

^{14}C-DDT, specific activity 4.15×10^{6} dpm/mg, was purified by partition chromatography on paper disks using dimethylformamide–2-phenoxyethanol–ethyl acetate (10:2:88) as stationary phase and 2,2,4-trimethylpentane as mobile phase. This was followed by a further chromatography on paper strips impregnated with 5% soya oil in ether and developed with acetone–water (3:1). ^{3}H-Aniline, specific activity 50 μc/mg was chromatographed on a 14×1.5 cm cellulose column to check its purity. Development was with *n*-butanol saturated with 1*N* hydrochloric acid. DDT-^{14}C (1–50 μg) in acetone and 1% stearic acid in acetone (1 ml) were taken to dryness and nitrated by the addition of fuming nitric acid–concentrated sulfuric acid (1:1) (2.5 ml) at 0°C. The mixture was heated at 100°C for 1 hr, cooled, diluted with water (10 ml) and extracted with benzene (15 ml). This extract was washed with 5% sodium hydroxide (10-ml portions) until clear and dried over anhydrous sodium sulfate. Aniline-^{3}H (64.7 μg) in benzene (1 ml) was added to the solution of the trinitrated DDT, the solvent was evaporated, and the residue chromatographed according to one of the following. (All systems apply to the same paper strip and all developments over the same distance.)

1. 3 × water; 1 × 20% acetone in water; 2 × heptane followed by elution with chloroform. The whole procedure is repeated three times.

2. heptane–chloroform (4:1) saturated with 20% ammonia solution, and heptane saturated with methanol.

These workers (16) describe many other chromatographic systems, but give no details for the application of the method.

Dieldrin

Dieldrin-^{14}C of specific activity 19.5 μc/mg was used with acetic anhydride-^{3}H, 2.07×10^{10} dpm/ml.

Dieldrin-^{14}C (70 μg) with a 2:1 mixture of the acetic anhydride-^{3}H and 40% HBr (0.1 ml) were heated at 130–140°C for 1 hr, cooled, made alkaline with 5% sodium hydroxide solution and extracted with benzene.

Unlabeled 6-acetoxy-7-bromo dieldrin (50 mg) was added to the reaction mixture and the derivative purified by three precipitations from methanol with water, and two crystallizations from hot methanol.

Again, no details of the application of the method are described and no figures for the accuracy and reproducibility of the results are given.

IV. CONCLUSION

Isotope derivative analysis has now established itself as a recognized technique in biochemical analysis. Much development in method, apparatus, and technique has taken place since the original work of Keston and his collaborators, but the principles applied remain the same. Although the field of amino acid analysis has been further developed from that original work by the use of tritiated and ^{35}S-labeled reagents, in conjunction with ^{14}C-tracers, the methods have not been so widely used as in the steroid field. Here new methods are still appearing in the literature (e.g., 116) since it is in this field that the double isotope derivative method really has all the required characteristics. It is a technique which is concise, reproducible, highly sensitive, and very selective even if the procedures involved are lengthy and rather costly.

It is, however, surprising that the method has not found wider application in other fields. This may be due to the unwillingness of workers to change to a new method of analysis, especially one known to be rather involved, when an existing method is proving to be satisfactory; but the disadvantages of the isotope derivative procedures are far outweighed by the advantages where the special characteristics are required. The recent developments in the analysis of sugars and carboxylic acids by the derivative procedures may be the beginning of a widening in the field of application of this method of biochemical analysis.

References

1. Akabori, S., K. Ohno, and N. Narita, *Bull. Chem. Soc. Japan, 25,* 214 (1952).
2. Andersen, H. A., E. Bojesen, P. K. Jensen, and B. Sorensen, *Acta Endocrinol., 48,* 114 (1965).
3. Avivi, P., S. A. Simpson, J. F. Tait, and J. K. Whitehead, *Proc. Radioisotope Conf., 2nd, 1,* 313 (1954).
4. Baldessarini, R. J., and I. J. Kopin, *Anal. Biochem., 6,* 289 (1963).
5. Bambas, L. L., U.S. Pat. 2,450,406 (October, 1948).
6. Bardin, C. W., and M. B. Lipsett, *Steroids, 9,* 71 (1967).
7. Baumgartner, W. E., L. Lazer, and A. Dalziel in *Advances in Tracer Methodology,* Vol. 1, S. Rothschild, Ed., Plenum Press, New York, 1963, p. 257.
8. Bayly, R. J., and H. Weigl, *Nature, 188,* 384 (1960).
9. Beale, D., and J. K. Whitehead, *Clin. Chim. Acta, 5,* 150 (1960).
10. Beale, D., and J. K. Whitehead, in *Tritium in the Physical and Biological Sciences,* Vol. 1, G. R. M. Hartcup, Ed., International Atomic Energy Agency, Vienna, 1962, p. 179.

11. Benraad, T. J., and P. W. C. Kloppenborg, *Clin. Chim. Acta, 12*, 565 (1965).
12. Berliner, D. L., *Federation Proc., 15*, 219 (1956).
13. Berliner, D. L., *Proc. Soc. Exptl. Biol. Med., 94*, 126 (1957).
14. Berliner, D. L., D. B. Dominguez, and G. Wesenkov, *Anal. Chem., 29*, 1797 (1957).
15. Blaedel, W. J., and M. A. Evenson, *Radiochemical Methods of Analysis*, Vol. 2, International Atomic Energy Agency, Vienna, 1965, p. 371.
16. Bogner, R. L., N. S. Domek, S. Eleftheriou, and J. J. Ross, Jr., Nuclear Science and Engineering Co., U.S.A., Report N.S.E.C. 85, May 1963.
17. Bojesen, E., *Scand. J. Clin. Lab. Invest., 8*, 55 (1956).
18. Bojesen, E., O. Buus, R. Svendsen, and L. Thunberg, in *Qualitative and Quantitative Analysis of Steroid Hormones*, H. Carstensen, Ed., Dekker, New York, to be published.
19. Bojesen, E., and H. Degn, *Acta Endocrinol., 37*, 541 (1961).
20. Bojesen, E., A. S. Keston, and M. Carsiotis, *Abstr. Commun. XIX, Intern. Physiol. Congr., Montreal*, 220 (1953).
21. Brown, J. B., *Biochem. J., 60*, 185 (1953).
22. Brown, W. G., L. Kaplan, A. R. Van Dyken, and K. E. Wilzbach, *Proc. Intern. Conf. Peaceful Uses At. Energy, 15*, 16 (1956).
23. Burger, H. G., J. R. Kent, and A. E. Kellie, *J. Clin. Endocrinol. Metab., 14*, 432 (1964).
24. Bush, I. E., *J. Biol. Chem., 183*, 763 (1950).
25. Bush, I. E., *Federation Proc., 12*, 186 (1953).
26. Bush, I. E., *The Chromatography of Steroids*, Pergamon Press, London, 1961.
27. *Catalogue of Radioactive Products*, No. R C 10, The Radiochemical Centre, Amersham, Buckinghamshire, England.
28. Catch, J. R., *Carbon-14 Compounds*, Butterworths, London, 1961.
29. Cejka, V., and E. M. Venneman, in *Advances in Tracer Methodology*, Vol. 3, S. Rothschild, Ed., Plenum Press, New York, 1966, 173.
30. Cherbuliez, E., B. Bachler, M. C. Lebeau, A. R. Sussmann, and J. Rabinowitz, *Helv. Chim. Acta, 43*, 896 (1960).
31. Cherbuliez, E., B. Bachler, and J. Rabinowitz, *Helv. Chim. Acta, 43*, 1871 (1960).
32. Cherbuliez, E., A. R. Sussmann, and J. Rabinowitz, *Helv. Chim. Acta, 44*, 319 (1961).
33. Christensen, N. H., *Proc. Conf. Use Radioisotopes Phys. Sci. Ind., Copenhagen, 2*, 131 (1961).
34. Christensen, N. H., *Acta Chem. Scand., 15*, 219 (1961).
35. Clark, L., and L. J. Roth, *J. Am. Pharm. Soc., 46*, 646 (1957).
36. Coghlan, J. P., M. Wintour, and B. A. Scoggins, *Australian J. Exptl. Biol. Med. Sci., 44*, 639 (1966).
37. Coombs, R. D., III, A. R. Reid, and C. B. Purves, *Anal. Chem., 25*, 511 (1953).
38. Dahlerup-Petersen, B., K. Linderstrøm-Lang, and M. Ottesen, *Acta Chem. Scand., 6*, 1135 (1952).
39. De Wulf, H., N. Lejeune, and H. G. Hers, *Arch. Intern. Physiol. Biochim., 73*, 362 (1965).

40. Dray, F., *Bull. Soc. Chim. Biol.*, *47*, 2145 (1965).
41. Drayer, N. M., and C. J. P. Giroud, *Steroids*, *5*, 289 (1965).
42. Dulmanis, A., J. P. Coghlan, M. Wintour, and B. Hudson, *Australian J. Exptl. Biol. Med. Sci.*, *42*, 385 (1964).
43. Edman, P., *Acta Chem. Scand.*, *4*, 277 (1950).
44. Edman, P., *Acta Chem. Scand.*, *4*, 283 (1950).
45. Edman, P., and J. Sjoquist, *Acta Chem. Scand.*, *10*, 1507 (1956).
46. Engell, H. C., F. Bro-Rasmussen, and O. Buus, *Danish Med. Bull.*, *5*, 176 (1958).
47. Evans, E. A., *Nature*, *109*, 169 (1966).
48. Evans, E. A., *Tritium and Its Compounds*, Butterworths, London, 1966.
49. Evans, E. A., and F. G. Stanford, *Nature*, *197*, 551 (1963).
50. Evans, E. A., and F. G. Stanford, *Nature*, *199*, 763 (1963).
51. Fraenkel-Conrat, H., and J. Fraenkel-Conrat, *Acta Chem. Scand.*, *5*, 1409 (1951).
52. Fraenkel-Conrat, H., J. I. Harris, and A. L. Levy, in *Methods of Biochemical Analysis*, Vol. 2, D. Glick, Ed., Interscience, New York, 1955, p. 359.
53. Frampton, V. L., L. P. Foley, L. L. Smith, and J. G. Malone, *Anal. Chem.*, *23*, 1244 (1951).
54. Gann, D. S., and R. H. Travis, *Am. J. Physiol.*, *207*, 1095 (1964).
55. Gerber, G. B., and J. Remy-Defraigne, *Anal. Biochem.*, *11*, 396 (1965).
56. Gut, M., and M. Hayano, in *Advances in Tracer Methodology*, Vol. 1, S. Rothschild, Ed., Plenum Press, New York, 1963, p. 60.
57. Gut, M., R. Underwood, S. A. S. Tait, J. F. Tait, A. Riondel, A. Southern, and B. Little, *Intern. Congr. Hormonal Steroids, Milan, 1962, Excerpts Med. Intern. Congr. Ser.*, *51*, 129 (1962), Abstr. No. 126.
58. Henderson, H. H., F. Crowley, and L. E. Gaudette, in *Advances in Tracer Methodology*, Vol. 2, S. Rothschild, Ed., Plenum Press, New York, 1965, p. 83.
59. Hesselbo, T., *Intern. J. Appl. Radiation Isotopes*, *16*, 329 (1965).
60. Hillman, D. A., and C. H. P. Giroud, *J. Clin. Endocrinol. Metab.*, *25*, 243 (1965).
61. Hollander, V. P., and J. Vinecour, *Anal. Chem.*, *30*, 1429 (1957).
62. Horton, R., *J. Clin. Endocrinol. Metab.*, *25*, 1237 (1965).
63. Horton, R., and J. F. Tait, *Proc. Intern. Congr. Endocrinol. 2nd Congr., Excerpta Med. Intern. Congr. Ser.*, *83*, 268 (1964).
64. Hudson, B., J. Coghlan, A. Dulmanis, M. Wintour, and I. Ekkel, *Australian J. Exptl. Biol. Med. Sci.*, *41*, 235 (1963).
65. Kabara, J. J., N. R. Spafford, M. A. McKendry, and N. L. Freeman, in *Advances in Tracer Methodology*, Vol. 1, S. Rothschild, Ed., Plenum Press, New York, 1963, p. 76.
66. Keston, A. S., and J. Lospalluto, *Federation Proc.*, *10*, 207 (1951).
67. Keston, A. S., S. Udenfriend, and R. K. Cannan, *J. Am. Chem. Soc.*, *68*, 1390 (1946).
68. Keston, A. S., S. Udenfriend, and R. K. Cannan, *J. Am. Chem. Soc.*, *71*, 249 (1949).

69. Keston, A. S., S. Udenfriend, and M. Levy, *J. Am. Chem. Soc.*, *69*, 3151 (1947).
70. Keston, A. S., S. Udenfriend, and M. Levy, *J. Am. Chem. Soc.*, *72*, 748 (1950).
71. Kirschner, M. A., M. B. Lipsett, and D. R. Collins, *J. Clin. Invest.*, *44*, 657 (1965).
72. Klein, P. D., in *Advances in Tracer Methodology*, Vol. 2, S. Rothschild, Ed., Plenum Press, New York, 1965, p. 145.
73. Klein, P. D., and J. C. Knight, *J. Am. Chem. Soc.*, *87*, 2657 (1965).
74. Kliman, B., in *Gas Chromatography of Steroids*, M. B. Lippsett, Ed., Plenum Press, New York, 1965, pp. 101–111.
75. Kliman, B., and R. E. Peterson, *J. Biol. Chem.*, *235*, 1639 (1960).
76. Koch, G. K., and G. Jurriens, *Nature*, *208*, 1312 (1965).
77. Kodding, R., H. P. Wolff, J. Karl, and M. Torbica, *Symp. Deut. Ges. Endokrin.*, *8*, 321 (1961).
78. Kowarski, A., J. Finkelstein, B. Loras, and C. J. Migeon, *Steroids*, *3*, 95 (1964).
79. Laragh, J. H., J. E. Sealey, and P. D. Klein, *Radiochemical Methods of Analysis*, Vol. 2, p. 353, International Atomic Energy Agency, Vienna, 1965.
80. Laumas, K. R., and M. Gut, *J. Org. Chem.*, *27*, 314 (1962).
81. Laver, W. G., *Virology*, *14*, 499 (1961).
82. Laver, W. G., *Biochim. Biophys. Acta*, *53*, 469 (1961).
83. Levy, M., in *Methods in Enzymology*, Vol. IV, S. P. Colwick and N. O. Kaplan, Eds., Academic Press, New York, 1957, p. 238.
84. Levy, M., A. S. Keston, and S. Udenfriend, *J. Am. Chem. Soc.*, *70*, 2289 (1948).
85. Levy, M., and E. Slobodian, *J. Biol. Chem.*, *199*, 563 (1952).
86. Lim, N. Y., and R. V. Brooks, *Steroids*, *6*, 561 (1965).
87. Lippich, E., *Z. Chem.*, *9*, 91 (1929).
88. Lukas, D. S., and R. E. Peterson, *J. Clin. Invest.*, *45*, 782 (1966).
89. Mangold, J. K., *J. Am. Oil Chemists' Soc.*, *38*, 708 (1961).
90. Markova, Yu. V., A. M. Pozharskaya, V. I. Maimind, T. F. Zhukova, N. A. Kosolapova, and M. M. Shchukina, *Dokl. Akad. Nauk SSSR*, *91*, 1129 (1953).
91. Marshall, L. M., and R. E. Cook, *J. Am. Chem. Soc.*, *84*, 2647 (1962).
92. Marton, J., T. Gosztonyi, and L. Otvos, *Acta Chem. Hung. Acad. Sci.*, *25*, 115 (1960).
93. Marton, F., T. Gosztonyi, P. Weisz, and F. Meisel, *Abstr. Congr. Endocrinol. 1st Congr., Session XIb*, 522 (1960).
94. Mechanic, G., S. J. Skupp, L. B. Safier, and A. C. Kibrick, *Arch. Biochem. Biophys.*, *86*, 71 (1960).
95. Melander, L., *Arkiv Kemi.*, *3*, 525 (1951).
96. Militzer, W., *Arch. Biochem.*, *9*, 91 (1946).
97. van der Molen, H. J., B. Runnebaum, E. E. Nishizawa, E. Kristensen, T. Kirschbaum, W. G. West, and K. B. Eck-nes, *J. Clin. Endocrinol. Metab.*, *25*, 170 (1965).
98. Moye, C. J., *Australian J. Chem.*, *3*, 436 (1961).

99. Moyer, J. D., and H. S. Isbell, *Anal. Chem.*, *29*, 393 (1957).
100. Moyer, J. D., and H. S. Isbell, *Anal. Chem.*, *29*, 1862 (1957).
101. Moyer, J. D., and H. S. Isbell, *Anal. Chem.*, *30*, 1975 (1958).
102. Murray, A., and D. L. Williams, *Organic Synthesis with Isotopes*, Parts I and II, Interscience, New York, 1958.
103. Murray, A., and D. L. Williams, *Organic Syntheses with Isotopes*, Part II, Interscience, New York, 1958, p. 1991.
104. Niu, C. I., and H. Fraenkel-Conrat, *J. Am. Chem. Soc.*, *77*, 5882 (1955).
105. Ogle, J. R., Inorganic Department, Radiochemical Centre, Amersham, Buckinghamshire, England, personal communication.
106. O'Grady, J. E., and P. J. Heald, *Nature*, *205*, 390 (1965).
107. Okita, G. T., J. J. Kabara, F. Richardson, and G. V. LeRoy, *Nucleonics*, *15*, 111 (1957).
108. Pearlman, W. H., and E. Cereco, *J. Biol. Chem.*, *203*, 127 (1953).
109. Peterson, R. E., in *Advances in Tracer Methodology*, Vol. 1, S. Rothschild, Ed., Plenum Press, New York, 1963, p. 265.
110. Peterson, R. E., *Aldosterone*, E. E. Banlion and P. Robel, Eds., Blackwell, Oxford, 1964.
111. Porter, R. R., *Methods Med. Res.*, *3*, 256 (1951).
112. Redfield, R. R., and C. B. Anfisen, *J. Biol. Chem.*, *221*, 385 (1956).
113. Riondel, A., J. F. Tait, M. Gut, S. A. S. Tait, E. Joachim, and B. Little, *J. Clin. Endocrinol. Metab.*, *23*, 620 (1963).
114. Riondel, A., J. F. Tait, S. A. S. Tait, M. Gut, and B. Little, *J. Clin. Endocrinol. Metab.*, *25*, 229 (1965).
115. Riondel, A., J. F. Tait, S. A. S. Tait, B. Little, and M. Gut, *Proc. Endocrine Soc.*, *44th, Chicago*, 1962, 16.
116. Rivarola, M. A., and C. J. Migeon, *Steroids*, *7*, 103 (1966).
117. Richin, P., *Chem. Rev.*, *65*, 685 (1965).
118. Ryle, A. P., F. Sanger, L. F. Smith, and R. Kitai, *Biochem. J.*, *60*, 541 (1955).
119. Saez, J. M., S. Saez, and C. J. Migeon, *Steriods*, *9*, 1 (1967).
120. Sanger, F., *Biochem. J.*, *39*, 507 (1945).
121. Shedl, H. P., P. S. Chen, Jr., G. Greene, and D. Redd, *J. Clin. Endocrinol. Metab.*, *19*, 1223 (1959).
122. Simpson, S. A., and J. F. Tait, *Mem. Soc. Endocrinol.*, *2*, 9 (1953).
123. Sirchis, J., Ed., *Proc. Conf. Methods Preparing Storing Marked Molecules, 1963*, Brussels, European Atomic Energy Community, 1964, p. 1346.
124. Slaunwhite, W. R., Jr., and L. Neely, *Anal. Biochem.*, *5*, 133 (1963).
125. Slobodian, E., and M. Levy, *J. Biol. Chem.*, *201*, 371 (1953).
126. Sommerville, J. L., Ed., *The Isotope Index*, Scientific Equipment Co., Indianapolis, Indiana.
127. Southern, A. L., E. N. Tractenberg, and M. Gut, in press.
128. Spedding, F. H., J. E. Powell, and H. J. Svec, *J. Am. Chem. Soc.*, *77*, 1393 (1955).
129. Sprinson, B. D., and D. Rittenberg, *J. Biol. Chem.*, *198*, 655 (1952).
130. Stachenko, J., C. Laplante, and C. J. P. Giroud, *Can. J. Biochem.*, *42*, 1275 (1964).

131. Svendsen, R., *Acta Endocrinol.*, *35*, 161 (1960).
132. Svendsen, R., and B. Sorensen, *Acta Endocrinol.*, *47*, 237 (1964).
133. Tait, J. F., B. Little, S. A. S. Tait, A. Riondel, C. Flood, E. Joachim, and M. Gut, in *Advances in Tracer Methodology*, Vol. 2, S. Rothschild, Ed., Plenum Press, New York, 1965, p. 227.
134. Technical Bulletin, 63/13, The Radiochemical Centre, Amersham, Buckinghamshire, England.
135. Thunberg, L., *Intern. J. Appl. Radiation Isotopes*, *16*, 413 (1965).
136. Thunberg, L., in *Advances in Tracer Methodology*, Vol. 3, S. Rothschild, Ed., Plenum Press, New York, 1966, p. 45.
137. Udenfriend, S., *J. Biol. Chem.*, *187*, 65 (1950).
138. Udenfriend, S., and S. F. Velick, *J. Biol. Chem.*, *190*, 733 (1951).
139. Velick, S. F., and S. Udenfriend, *J. Biol. Chem.*, *190*, 721 (1951).
140. Verly, W. G., J. R. Rachele, V. du Vigneand, M. L. Eindinof, and J. E. Knoll, *J. Am. Chem. Soc.*, *74*, 5941 (1952).
141. Werbin, H., and G. V. LeRoy, *J. Am. Chem. Soc.*, *76*, 5260 (1954).
142. Whitehead, J. K., *Biochem. J.*, *68*, 653 (1958).
143. Whitehead, J. K., *Biochem. J.*, *68*, 662 (1958).
144. Whitehead, J. K., and D. Beale, *Clin. Chim. Acta*, *4*, 710 (1959).
145. Wieland, T., H. Merz, A. Rennecke, *Ber.*, *91*, 683 (1958).
146. Wiley, R. A., and J. L. Metzger, *J. Pharm. Sci.*, *56*, 144 (1967).
147. Wilzbach, K. E., *J. Am. Chem. Soc.*, *79*, 1013 (1957).
148. Wolff, H. P., and M. Torbica, *Lancet*, *1963-I*, 1346 (1963).
149. Woolever, C. A., and A. Goldfien, *Intern. J. Appl. Radiation Istopes*, *14*, 163 (1963).
150. Zaffaroni, A., and K. B. Burton, *J. Biol. Chem.*, *193*, 769 (1951).
151. Zaffaroni, A., in *Recent Progress in Hormone Research*, Vol. 8, G. Pincus, Ed., Academic Press, New York, 1953, p. 51.

Bioluminescence Assay: Principles and Practice

BERNARD L. STREHLER, *Department of Biological Sciences, University of Southern California, Los Angeles, California, and Aging Research Laboratory, Veterans Hospital, Baltimore, Maryland*

I. INTRODUCTION

Many biochemical analytical methods make use of the interactions of light and matter in the determination of the concentrations of important intermediates. Of these, spectrophotometric methods are probably the most widely applied for the assay of both the simple mixtures present in extracts of biological material and in intact organelles, cells, and tissues of living things. The dual-wavelength spectrophotometric technique developed by Chance (1) has extended the usefulness of spectrophotometric methods, as has the availability of commercial spectrophotometric devices capable of high resolution and possessing great stability and accuracy. Another facet of light–matter interaction is used in the methods which employ bioluminescence in biochemical analysis.

The first such application of bioluminescence to the study of a biological process was the use of luminous bacteria by the late E. Newton Harvey (2) in 1928. Harvey measured the production of oxygen by plants under anaerobic conditions. He showed (using the luminous bacterium, *Photobacterium fischeri*) that leaves of *Elodea* which had been immersed in a thick suspension of luminous bacteria and kept there until the light disappeared, would immediately produce luminescence if, while immersed in the bacterial suspension, the plant were briefly exposed to light. It was Harvey who laid the foundation for most of the substantial progress made during the last 20 years in the field of bioluminescence. He and his associates also used luminescence in studies of the effects of narcotics, respiratory inhibitors, pressure, and other factors on enzymes and enzyme systems (3).

Following the initial discovery by McElroy in 1947 (4), that adenosine triphosphate will cause extracts of firefly lanterns, which have grown dark, to luminesce, Strehler and Totter in 1952 (5) developed a method which made use of this reaction in the assay of a number of important biological energy-transducing intermediates. This method and various refinements and modifications of it have been widely employed in a variety of different biochemical studies including photosynthesis (6), neural function (7), radiation effects (8), oxidative phosphorylation, muscle biochemistry (9), differentiation (10), cell death (11), and space biology (12). This chapter, which constitutes a summary both of methodological procedures and of their underlying principles, as well as a variety of specific applications which have been made during the intervening years, falls into three general subdivisions. The first deals with the principles and techniques of light measurement most useful in such studies, the second with the biochemical and physical background of bioluminescent and chemiluminescent reactions, and the final portion deals with specific applications of the various biochemical systems employed.

II. LIGHT EMISSION AND MEASUREMENT

1. The Emission of Light

The emission or absorption of electromagnetic radiation occurs as the result of the interaction of a light quantum with electrically charged particles. In the absorption process, the energy contained in the light quantum is converted into a temporarily stored form of energy (as an excited state of the atom or molecule which absorbed it). This excited state results from the displacement of paired electrons into more improbable orbits (possessing higher than "ground state" potential energy) and is called the singlet excited state.

The energy residing in such a system can be redistributed in one of four ways. The first mechanism through which the stored energy is redistributed is called *fluorescence* and consists essentially of the reemission (within 10^{-8} sec or less) of the light quantum plus or minus any thermal energy which has been dissipated to or absorbed from the environment.

The second alternative is the transformation of the excited state into another energetic electronic configuration in which electrons have, in contrast to the singlet state, become unpaired. This state is referred to as a triplet state and because of the improbability of transitions from this state to the singlet state, it is relatively stable compared to the lifetime of the excited singlet state and may reemit stored energy as light within a few thousandths of a second to several thousand seconds after absorption. Such emission is called *phosphorescence.*

The third possibility is that a portion of the energy in the excited state is converted into some other more stable form. In the case of photosynthesis, it is transformed into *stored potential chemical energy* which is then used to drive coupled reactions that result in the storage of carbohydrates.

Finally, the energy may be transformed into thermal movements of the molecules comprising the system, that is, into *heat energy.*

For light energy to be absorbed by an oscillator, such as the electrons in a pigment molecule, there must be a correspondence between the energy in the light quantum and the energy gained in the electronic transition involved. Electronic transitions in isolated molecules or atoms are extremely concise in their energy demands. Therefore, atomic absorptions in the gas phase consist of very narrow spectral lines. These absorption lines also correspond to the emission spectra of the atom in question. But, in polyatomic systems, liquids and solids, the spectral absorption and emission lines are greatly broadened because of the interactions of the atoms in the molecule. The absorption spectra of the polyatomic molecules involved in bioluminescent reactions, therefore, are characteristically broad by comparison with atomic absorption and emission spectra. Moreover, the types of transitions from ground state to excited state and vice versa are subject to specific interactions with other molcules in the environment. For example, the absorption spectrum of the visual pigment, retinine, in solution is considerably altered when it is combined as rhodopsin, the photochemically active protein-pigment complex involved in visual processes. Conversely, the emission spectra of bioluminescent reactions, for example, the light emitted by firefly lantern extracts, is modified by environmental factors such as pH, temperature, etc.

Atoms and molecules may, in principle, be brought into the excited state by any process which can channel energy into the appropriate electronic oscillations. Thus, in addition to the absorption of light mentioned above, transitions to higher electronic states can result from energetic collisions between atoms or molecules. This occurs both in an incandescent filament and in the sun. Electrons can be raised to higher energy states by mechanical forces (triboluminescence), or higher energy states may be populated by leading electrons through a conducting material in such a manner that they lose potential energy suddenly as they move from a position of high potential to one of low potential.

Finally, excited states and consequent light emission may be produced through the production of improbable electronic arrangements (excited states) as a consequence of chemical reactions. This latter group of reactions is called collectively "chemiluminescence." Although it is not certain that some bioluminescences are not produced by transitions of electrons within so-called conduction bands, bioluminescent reactions are generally thought to be specific types of chemiluminescent reactions.

The energy source of bioluminescent reactions is, with few exceptions, the energy released when oxygen combines with a material of moderately high reducing potential. Energy liberated at some particular step in the oxidation sequence is then channeled into a molecule capable of excitation. This complex molecule consequently emits light as it goes from an excited state to its ground state.

2. The Detection of Light

Light emitted as a consequence of any of the processes outlined in the previous section may be detected and measured through the use of one or more of the following devices. Each of the detection methods discussed subsequently depends upon the amplification of the energy absorbed when an electron in an atom, molecule, or complex is raised to a higher energy state. The primary reaction is always the same in principle: the movement of electrons within an absorber to a more improbable position, followed by the coupling of the disturbance by induced light absorption in the system to the release of a much larger amount of energy by the amplifier portion of the detector. Among the devices in use are the human eye, photographic emulsions, photocells of various types, photomultipliers, heat-sensing devices, and image intensifiers (special kinds of photomultipliers).

A. VISUAL MEASUREMENT

Although the human eye is an exceedingly refined instrument and is connected with a very precise and complex amplifier-recorder, the nervous system, it is not well suited to quantitative measurements of great precision because the sensitivity of the eye tends to vary inversely with the amount of light which it receives, a process called adaptation. In consequence, in order to make reasonably precise measurements of light intensity with the human eye, the measurements must be made against some kind of comparison standard. For example, the intensity or color of a standard light source or field may be matched to the intensity of an unknown source. Such a system is not suited to kinetic measurements of very small changes in light intensity versus time, but is certainly useful, even without comparison standards in measurements of a qualitative nature.

The eye, after sufficient dark adaptation, is capable of detecting quite small numbers of photons [probably in the range of 1–10 per resolving time (about $\frac{1}{20}$ sec)] if the source has a relatively small surface area.

B. PHOTOGRAPHIC RECORDING

A photographic image has certain advantages over a visual image as a means of detecting small amounts of light. For one thing, inasmuch as suitable emulsions do not vary greatly from the reciprocity rule, by extending the exposure interval, photographic images may be obtained from light sources which are very dim. The amplifier portion of the photographic process depends on the fact that electrons, which have been moved to improbable positions as the result of light absorption within a photographic grain, act essentially as catalysts for the reduction of millions of additional ions existing within that photographic emulsion grain. The energy for this amplification comes from the reaction between the emulsion components (silver salts) and the reducing materials present in the developer. Although photographic recordings of the time course of bioluminescent reactions have been made, such procedures have largely been superceded by the advent of photoelectric detection devices.

C. ELECTRICAL TRANSDUCERS OF LIGHT ENERGY

a. Thermocouples. Although this device is not, strictly speaking, a photoelectric device, but rather a thermoelectric device, it can be used to detect light of moderately high intensity. Thermocouples are employed in calibrating other light-measuring devices against a standard light source of known spectral distribution and intensity. They are particularly suitable for this application because the thermocouple produces a signal proportional to the energy absorbed by it irrespective of the wavelength of the absorbed light. Light energy is first transformed into heat. Bolometers and thermistors which make use of other effects of heat on a sensing element also require the conversion of light energy into thermal energy before its amount can be measured.

b. Photoelectric Devices. True photoelectric devices possess absorbing substances in which electrons can be moved to improbable positions as a result of the absorption of light. The simplest photoelectric devices, such as the lead sulphide conductivity cell, possess electrons that, under the influence of light, are transferred from a trapped locus into a so-called conducting band within the photosensitive crystal. Although these devices have certain advantages, they also possess certain disadvantages. First is the fact that their conductivity is very temperature sensitive, for thermal energy as well as light energy can dislodge electrons from stable positions; second, under normal operating conditions they are not high internal-resistance devices as are the photoemissive types of detectors which will be discussed shortly. This means that one must discern small per cent changes in resistance in making measurements. In practice, photoconductive devices are usually built into the arms of a Wheatstone bridge circuit. Because of the large and sometimes unpredictable changes in background conductivity, the most sensitive applications usually require that the light be converted into an ac signal by appropriate chopping of the light source, and that an ac amplifier tuned to the frequency of the chopped incident light be used to detect the signal. Photoconductive cells may approach to within about 1% of the sensitivity of photoemissive devices such as photocells or photomultipliers.

Alkaline metals and certain of their alloys have a particularly low affinity for electrons. Photoemissive types of light detectors make use of the fact that absorbed light quanta may eject electrons from surfaces to which they are weakly bound. At the center of the tube is the strongly charged positive electrode, the anode, whereas the peripherally located cathode is coated with an alkaline metal (e.g., cesium). Such a detector may either operate as a vacuum tube or it may be filled with a suitably ionizable gas and be operated as a Geiger-Muller tube. In the latter, the acceleration of the primary ejected photoelectron toward the anode may produce successive and multiplying ionizations which produce an electric pulse of very short duration, an ionic barrage which is discharged at the electrodes.

The most commonly used photodetector is the photomultiplier, which is essentially a vacuum-type photocell containing a cascade of additional amplifying stages, called dynodes. Figure 1 illustrates the structure of a typical photomultiplier. The absorbed light quantum causes the ejection of a photoelectron from the photocathode. This photoelectron is accelerated over a very substantial voltage drop (ca. 100 V) toward the first dynode. When this electron hits the dynode, its accumulated kinetic energy in turn causes the ejection of a number of additional electrons, these move as a cascade toward the second dynode where the process is repeated. At the terminus of the chain (the anode of the photomultiplier tube), the photoelectrons and their progeny are collected and cause a small negative pulse to be generated there. The negative current resulting from successive pulses is of sufficient magnitude to be coupled directly to the grid of an electrometer tube in a dc amplifier or it may be directed into an ac amplifier such as is used in pulse counting.

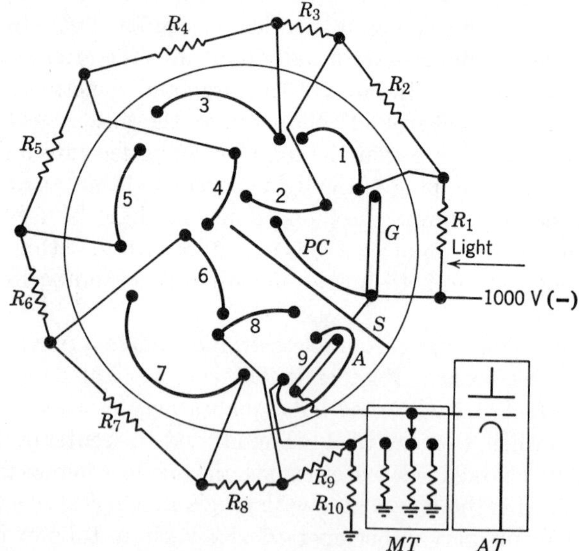

Fig. 1. Schematic diagram of structure of photomultiplier. (*A*) anode; (*G*) grill; (*1–9*) dynodes; (*S*) shield; (R_{1-9}) 0.68 meg; (R_{10}) 0.56 meg; (*PC*) photocathode; (*AT*) amplifier tube; (*MT*) meter multiplier.

A special adaptation of the photomultiplier is the image intensifier. In this unique phototube, a primary photoemitted electron is accelerated toward the first dynode (which in this case consists of a membranous material impregnated with substances capable of secondary emission.) The cascades of electrons proceed from dynode to dynode toward the anode in the same two-dimensional spatial array that the primary pattern of photoemission possessed. At the anode, which is impregnated with a phosphor, the intensified original optical pattern is projected.

The light-detecting devices mentioned above possess certain general qualities of which the user should be aware. The first such property is the ability of the electrons to move from positions of high probability to positions of low probability (i.e., in this case, ejection from the photoelectric surface) through the action of any suitable energy source. Thermal energy is particularly important in this regard, for collisions between atoms or molecules in a photosensitive surface are capable of ejecting electrons (thermal, not photoelectrons). Thus, it is often desirable to use photomultipliers or other photosensitive devices at reduced temperatures. particularly if the tube is to be used to detect very small signals. Use at 0°C rather than room temperature (with due precautions against condensation, moisture, etc.) or even better, at the temperature of Dry Ice or of liquid nitrogen, will substantially reduce the thermal dark current of a tube.

A second feature which photoelectric devices possess is unequal sensitivity to light of different wavelengths. Thus, most photosensitive surfaces employed in commercially available tubes have maximum sensitivity in the blue region of the spectrum and decrease to less than 1% of the maximum sensitivity at about 700 mμ. Special photoemissive surfaces with extended red sensitivity have been developed in recent years and are now available commercially. However, special surfaces which confer sensitivity in the red or infrared region also have greater inherent dark currents, since enhanced red sensitivity results from the loose binding of potential thermoelectrons as well as potential photoelectrons. It is therefore particularly desirable to operate red-sensitive tubes at quite reduced temperatures, particularly when very weak sources are to be measured.

The third feature of photoemissive devices which must be taken into account, especially when working with very low light levels, is the fact that light quanta arrive at the photocathode randomly in time. Therefore, as the number of photoelectrons generated decreases per unit time, random fluctuations in signal will be generated (the so-called Schott noise). In order to obtain statistically significant measurements of low-intensity light sources, it is therefore necessary to use statistical procedures similar to those used in radioactive counting techniques.

D. EQUIPMENT

a. Housing Apparatus. In order to make measurements of bioluminescent reactions it is necessary to purchase or construct (*1*) an apparatus in which to house the photomultiplier, (*2*) a shutter arrangement which serves to protect the photomultiplier from light while the sample is being inserted, and (*3*) a sample holder (constant position) which can also be conveniently and effectively shielded from extraneous light during measurement.

An important further consideration in the design of a luminescence-measuring chamber is the efficiency with which light is transmitted from the emitting material to the detecting surface. Inasmuch as the fraction of light impinging on the photosensitive surfaces depends essentially on the solid angle (of the sphere whose center is the sample and which passes through the photosensitive surface) that the photosensitive surface subtends, it follows that the sample and detector should be as close to each other as is mechanically feasible. Second, the entire sample should be exposed to the photomultiplier. Third, if convenient, a mirror arrangement which can reflect light coming out of the rear of the sample should be incorporated as well as such lens systems as may increase the effective solid angle subtended by the photodetector.

We have found the simplest design to be that illustrated in Figure 2 (13). It consists essentially of two tightly fitting pieces of metal tubing, the inner of which is capable of rotating in the outer. Holes are carefully milled in each of these tubes as indicated in the diagram, and the shutter is opened or closed by rotating the tube. A light-tight cap covers the end of the inner tube. The cuvette may conveniently extend into the recessed cavity in the cap.

When the light emitted is of such low intensity that a cooled photomultiplier is desirable, a means must also be included to protect the sample from the coolant, and to protect the photomultiplier from the warm sample. Figure 3 shows one such device (14). The shutter

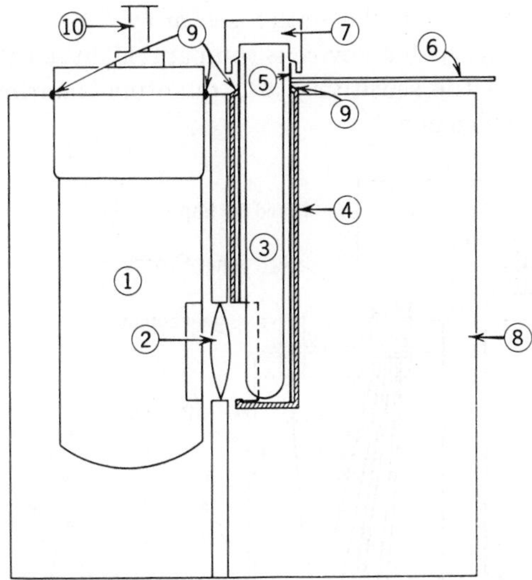

Fig. 2. Design of high aperture photomultiplier housing suitable for routine applications. The lens is optional (13). (*1*) Side-on type photomultiplier; (*2*) lens in holder; (*3*) cuvette; (*4*) outer tube; (*5*) inner tube; (*6*) shutter handle; (*7*) light-tight cap; (*8*) light-tight housing; (*9*) light seals; (*10*) lead to photometer.

consists of a rotatable disk. The photomultiplier is housed in a Dewar flask which can be filled with liquid nitrogen through a light-proof port. Since condensation of moisture within the chamber and on the leads of the photomultiplier may cause electrical difficulties, it is important to guard against such moisture accumulation. When mounting the photomultiplier, it and the Dewar port should be carefully aligned with the shutter. The photomultiplier should be located as close as is feasible to the unsilvered window.

A highly efficient light-collecting device constructed in our laboratory is shown in Figure 4 and consists of an end-window type photomultiplier mounted with its photosensitive surface uppermost, in a light-tight box. The shutter arrangement consists of a photographic plate holder with a round hole appropriately cut in it. The sample is placed in a depression slide centered over this aperture. Room light excluded from the photomultiplier during sample insertion by sliding the plate holder shutter into position and is prevented from reaching the sample and photomultiplier during measurement by closing a light-tight cover over the entire assembly. Because of its high optical efficiency, such a device is comparable, in sensitivity at room temperature, to the sensitivity of a quantum-counting device with lower optical efficiency.

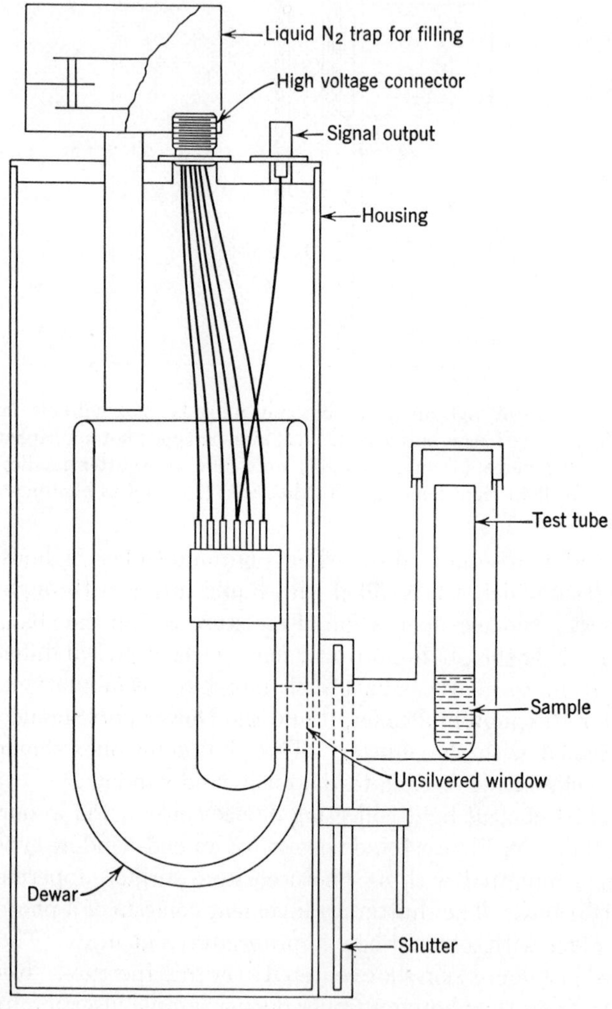

Fig. 3. Construction of quantum-counting photomultiplier housing.

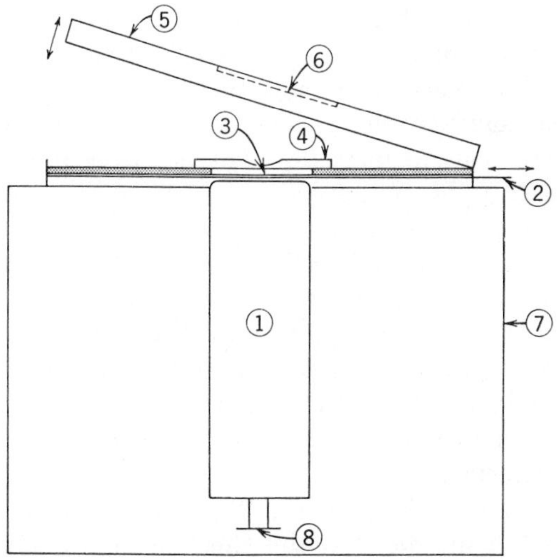

Fig. 4. Photomultiplier housing suitable for measurement of extremely small volumes of sample with high optical efficiency. Note the close position of sample to photomultiplier. (*1*) End-on type photomultiplier; (*2*) shutter; (*3*) glass window; (*4*) depression slide; (*5*) light-tight cover; (*6*) mirror mounted on cover; (*7*) light-tight housing; (*8*) lead to photometer.

b. The Voltage Supply and Load Resistors. Because the amplification of a photomultiplier is an exponential function of the voltage applied to it, it is important that the voltage applied to the photomultiplier divider network be extremely stable. Different tube manufacturers recommend different voltages between the various dynode stages, and the so-called voltage divider network should be constructed according to these recommendations. In principle, a divider network consists of resistors arranged in series and connected to a stable high voltage supply. The photocathode usually is connected directly to the negative terminal of the voltage supply and successive dynodes are connected between resistors in the series. A final resistor is inserted between the last dynode and ground. The positive terminal of the voltage supply is also grounded.

Finally, the anode of the photomultiplier is connected to (*a*) the grid of the input tube of the associated amplifier and (*b*) to ground through a "load resistor." In practice, there is an arrangement for switching one of several high resistances (load resistors) into the circuit so that the sensitivity (voltage developed across this resistor) may be varied conveniently in known increments.

c. Amplifiers and Recorders. The detection of light emitted by luminescent reactions requires (in addition to the primary sensing element, such as a photomultiplier) a suitable device for measuring the current changes induced by the light as well as an appropriate means of displaying or recording the signal developed. A variety of such devices is available commercially or may be constructed.

Amplifiers: The primary qualities which a photocurrent amplifier must have are: (*a*) sufficient sensitivity to respond to the signal developed, (*b*) stability and linearity, and (*c*) a high-input impedance (i.e., the amplifier must not drain off a significant portion of a signal during measurement).

(*1*) *dc Amplifiers.* In practice, the load resistance may be as high as 10^{10}–10^{12} Ω and therefore this effective input impedance must be at least 100 times greater (10^{12}–10^{14} Ω) than this value if the signal is not to be altered by more than 1%. Devices which measure very small currents are called electrometers and the most common ones usually employ special tubes at their first stage having a very low grid current; i.e., the resistance between the grid of the tube and its other electrodes is very high during operation. Since the function of the grid in a vacuum tube is to change the amount of current passing from the cathode to plate, and since the electrons pass through the wire mesh in the grid in moving between cathode and plate, it is clear that some electrons will strike the grid and that such electrons will change the potential of the grid. Electrometer tubes are specially designed to minimize such a transfer of electrons from cathode to grid.

Many commercial devices are available which will function well in such applications. Among these are high-quality vacuum tube voltmeters, the photometer-photomultiplier manufactured by the American Instrument Company, Silver Spring, Maryland, and vibrating reed electrometers such as Model 3810, manufactured by the Applied Physics Corporation, Monrovia, California.

(*2*) *Vibrating Reed Electrometers.* This type of instrument employs a special principle to achieve extremely high ("infinite") dc input impedance. The voltage to be detected is applied to a capacitor. This capacitor is, however, of a very unusual type in that the distance between the plates is made to change at a high frequency; hence the name vibrating reed. The consequent alternating change in capacitance causes current to flow into and out of the condenser in proportion to the signal voltage applied to it. This ac current generated can be amplified and its magnitude measured. Thus, the signal voltage developed is measured without any such current drain as is inherent in the use of even the best electrometer tubes.

(*3*) *Pulse Counting Equipment.* The photocurrent developed by photomultiplier tubes may be measured by quite different means, called pulse counting. Since each photoelectron (or thermal electron) liberated by the photocathode results in a cascade of about 1,000,000 electrons which arrive nearly simultaneously at the anode, it follows that these barrages generate discrete pulses of negativity on the anode. The magnitudes of the voltage changes at the anode are determined by the capacitance of the anode and their duration is determined by the rate at which the charge is lost through the load resistor. In order to detect these pulses and to measure their number and size, a suitable pulse amplifier, discriminator, shaper, and counter or counting ratemeter must be employed. Such circuitry was developed by P. R. Bell of the Oak Ridge National Laboratory for the measurement of pulses generated in scintillation counting. The device was called the Oak Ridge A-1 linear amplifier, and preamplifier, and circuitry analogous to it is employed in commercial scintillation counters. The principles of operation of such devices are as follows. The small photoelectron pulse arriving at the anode of the photomultiplier is transmitted to the first stage of an amplifier through a small condenser, following which its magnitude is enormously amplified. These amplified pulses are then transmitted to a discriminator circuit which allows the passage of only those pulses which are within a preselected size range. Pulses too small or too large to arise from single photoelectrons are rejected and the passed pulses are then used to generate pulses of uniform size. These final shaped pulses then activate a counting circuit such as a scaler, or, if it is desired to record the rate at which pulses arrive, the output of the linear amplifier–discriminator circuits is transmitted to a counting ratemeter which generates a dc voltage proportional to the counting rate.

(*4*) *Recorders.* The voltage developed by any of the above amplifiers may be applied to any recorder whose sensitivity and impedance are suited to the output of the electrometer or count-rate meter employed. We have found the Varian Model G-14 and the Sargent SRL to be quite satisfactory. The recorder should have a maximum sensitivity of about 1 mV or less full scale, a variable speed chart drive, and a rapid response time (full-scale travel in 1 sec or less).

d. Some Commercially Available Equipment. In addition to the devices mentioned above, several kinds of commercially available equipment directly suited to the measurement of bioluminescence are available. These include, in particular, the Farrand photofluorometer, the Turner fluorometer, Model 110, and with slight adaption, the Aminco photometer-photomultiplier. With suitable corrections for wavelength versus sensitivity, the Aminco-Bowman spectrophotometer may be employed in the measurement of bioluminescent or chemiluminescent spectra.

e. Specialized Equipment. (*1*) *Low-Level Emission Spectrometer.* The wavelengths of light emitted by bioluminescent reactions vary from species to species and are influenced by the physical and chemical environments of the enzymes involved. For example, McElroy and Seliger (15) have shown that the pH of the medium influences the color of light emitted by fireflies (see Fig. 5), and Harvey (16) demonstrated that the light emitted by the firefly shifts strikingly toward the red region of the spectrum when the lantern is

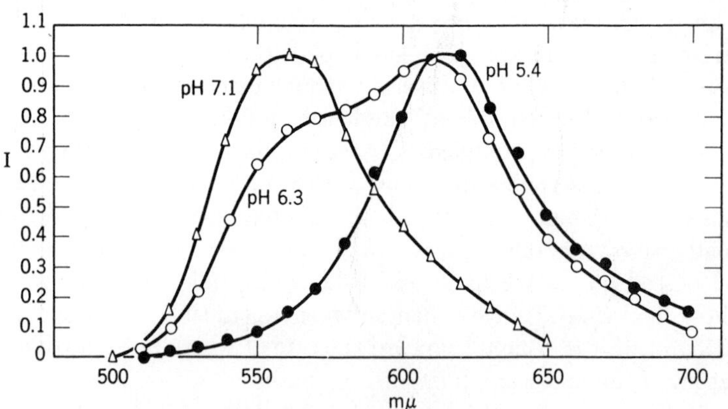

Fig. 5. Effect of pH on the color of light emitted by firefly extract in glycylglycine buffer. Spectra measured with an RCA 7326 phototube sensitive to 8000 Å (15).

heated. Many bioluminescences *in vitro* are quite dim. Because of this, plus the decrease in sensitivity of photomultipliers in the red region of the spectrum as well as the loss in signal caused by a finite slitwidth limited monochromator, it is usually necessary to use detectors with high-gain and low-noise characteristics.

The Leitz quartz monochromator, combined either with a pulse counter or sensitive electrometer, is suitable, as is the Farrand quartz monochromator. Figure 6 shows the emission spectra of luminous bacteria and the bacterial luciferin–luciferase reaction *in vitro* as measured with the latter instrument in combination with a pulse counter.

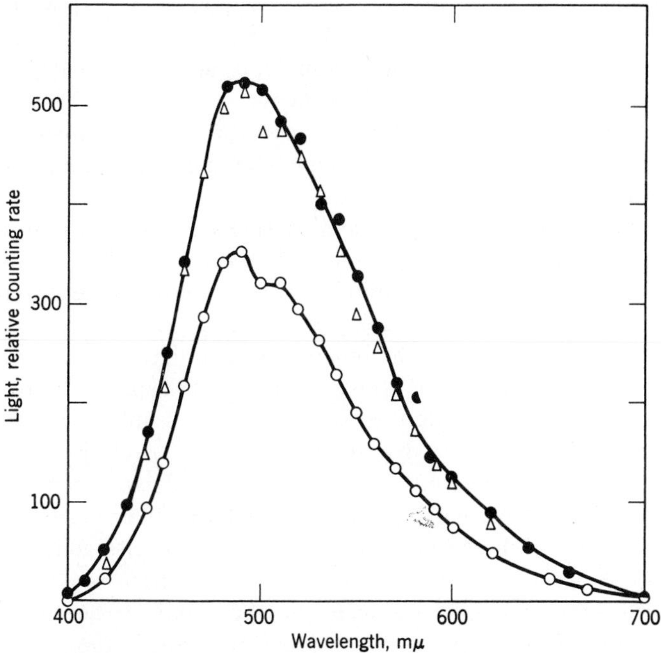

Fig. 6. Comparison of luminescence emission by intact *Photobacterium fisherii* (●) and acetone extracts (○) as measured in a Farrand quartz monochromator (slit width, 0.5 mm). (△) Acetone extract emission normalized to intact bacterial emission at 490 mμ. Luminescence emission of intact *P. fisherii* and 0.2 ml of a 2% extract in 0.3 ml 0.01*M* phosphate buffer determined with liquid nitrogen cooled photomultiplier and quantum counter discussed in Section II-2-D-e-1.

In the measurement of absolute emission spectra, due account must be taken of the facts that (*1*) the detecting system does not have a uniform response to quanta of different wavelengths; (*2*) at constant slit size, the monochromator dispersion curve results in the transmission of different band widths of wavelengths at different wavelength settings; (*3*) light may be absorbed differentially by the sample or associated glassware; and (*4*) the light absorbed within the sample may be reemitted (either as ordinary fluorescence or phosphorescence, or through the process of sensitized fluorescence). This latter process depends on the fact that the probability of transfer of excitation energy from an emitter to an absorber (which may itself subsequently emit) is much greater, when the distances between the absorber and emitter approach the wavelength of the emitted light, than would be calculated from the inverse square law. Since any of the above factors may distort or displace the maxima, it is necessary where feasible to use appropriate controls and standards.

The first two factors are relatively easy to control and correction curves for any given combination of detector and monochromator can be obtained as follows:

At a given slit setting, the signal developed when the sample is replaced with a standard light source is recorded. The signals recorded at each wavelength are divided by the intensity of light emitted by the source at each corresponding wavelength. This quotient gives the spectral sensitivity of the monochromator-detector. Its reciprocal at each wavelength is used as the multiplier (correction factor) when the signals derived from a source of unknown spectral distribution are to be plotted in terms of absolute or relative emission energy versus wavelength. Calibrated standard lamp sources are available from the National Bureau of Standards, Washington, D. C. at modest cost. These sources must be used at precise voltages and currents as specified and must be spatially oriented as prescribed. The correction factors derived as above apply only to the slit widths employed in deriving the factors. Separate correction factors must be determined for each slit setting.

An important source of potential error when such an instrument is used at wavelengths far from the sensitivity maximum of the detector is light scattered within the monochromator. If the percentage of such scatter approaches the value of the reciprocal of the correction factor, serious errors will be introduced. An ingenious method for automatically correcting for the above factors is incorporated in the new Turner absolute spectrofluorimeter Model 210.

(*2*) *Automatic Comparison Emission Spectrometer.* In order to compare small differences in emission spectra or to obtain the ratio of emission of one bioluminescent reaction to the emission of another reaction, a device has been constructed by the author and Dr. R. Monroe. The device employs a Leitz monochromator Model 083, an RCA 8644 photomultiplier as detector, and a sample housing so arranged that light from two different samples is alternately imaged on the entrance slit of the monochromator. The chopping frequency is synchronous with that employed in the Aminco-Chance dual wavelength spectrophotometer and the signal is directed into the latter machine which automatically determines the phase and amplitude of the signals arising from the two light sources. The wavelength drive is activated by a Bodine constant-speed motor and the emission difference spectrum is displayed on a Varian Model G-14 recorder.

(*3*) *Phosphoroscope for Use with Low-Level Light-Induced Photoemissions.* Background phosphorescence emitted by glassware used to hold samples whose phosphorescence is to be measured, can be overcome by use of a variable-speed phosphoroscope which consists entirely of nonphosphorescent materials. This device is shown in Figure 7. Its features include two shutters, rotated by the same drive shaft, a front surface all-metal mirror system for the introduction of actinic light, as well as the subsequent transmission of reemitted radiation, and a sample cuvette milled into a thermostatted aluminum block, appropriately positioned. The decay curve of green plant luminescence obtained with this device is shown in Figure 8.

(*4*) *Apparatus for the Continuous Measurement of Photosynthetic Phosphorylation.* The system illustrated in Figure 9 was designed and constructed to measure ATP synthesis continuously during photosynthetic phosphorylation. This adaptation consists of a flat rectangular sample cuvette (4 mm $\times$ 4 cm $\times$ 4 cm) inserted between two optical filter systems. A mixture of firefly enzyme and chloroplasts, introduced into the cuvette, may be illuminated with red (photosynthesis-inducing) light from which all blue light has been eliminated. The red light transmitted through the suspension is then removed by a second filter system which transmits a portion of the light emitted by the firefly enzyme (in response to ATP synthesis). This light strikes the photosensitive surface of an end-on photomultiplier located next to the second filter system. Applications of the system will be described in a later section.

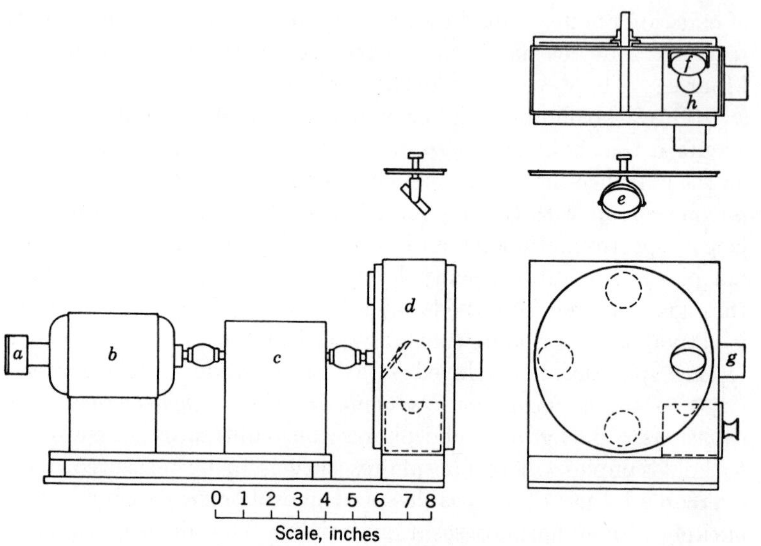

Fig. 7. Design of high aperture variable speed phosphoroscope. (*a*) Tachometer; (*b*) series wound ⅓ hp motor; (*c*) reduction gear box; (*d*) phosphoroscope housing; (*e*) mirror (lid mounted) for illuminating sample; (*f*) mirror for projecting luminescence or phosphorescence out of aperture opposite; (*g*) opening for monitoring fluorescence of sample; (*h*) sample cuvette milled into sliding thermostatted drawer. In operation, light is introduced through shutter *1* at left of *d*, is reflected from mirror *e* onto sample in cuvette *h*; as shutters rotate [each shutter (only one shown at extreme top right) contains two holes 180° apart and the holes in the two shutters are 90° apart]; the illuminating beam is occluded and the second shutter (to the right of *d* in the figure) opens, permitting light reflected by mirror *f* to be projected out of the exit aperture on the detector.

III. PROPERTIES OF BIOLUMINESCENT REACTIONS

1. Energy Sources and Considerations

If a bioluminescent reaction emits light at about 7000 Å, the energy requirement is about 40 kcal, while light emission at about 4800 Å requires nearly 60 kcal per mole of quanta (einstein). It is interesting that there are not many biochemical reactions which are capa-

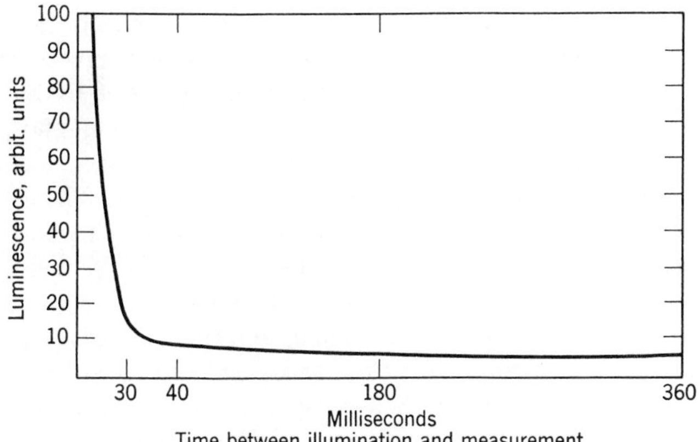

Fig. 8. The time course of green plant luminscence as determined with the phosphoroscope illustrated in Figure 7.

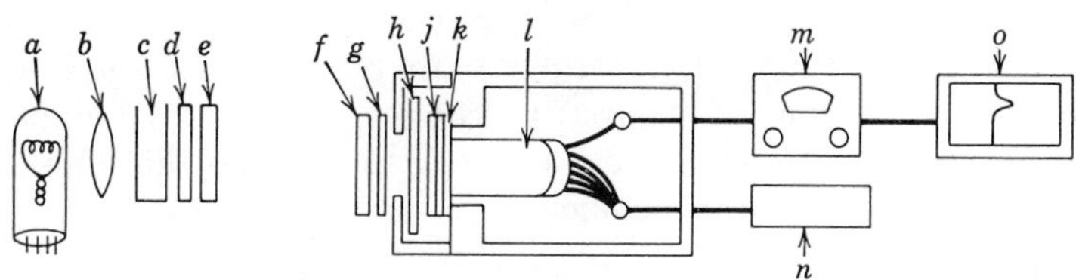

Fig. 9. Design of apparatus suitable for continuous measurement of photosynthetic phosphorylation using the firefly enzyme as a detector in the monitoring system: (*a*) Tungsten filament lamp; (*b*) condensing lens; (*c*) water (heat filter); (*d*) Wratten filter No. 25; (*e*) Wratten filter No. 26; (*f*) Wratten filter No. 29; (*g*) Corning filter No. 2408; (*h*) sample cuvette; (*j*) Corning filter No. 9782; (*k*) Wratten filter No. 65; (*l*) photomultiplier; (*m*) Aminco photometer; (*n*) high-voltage power supply (Baird Atomic No. 312); (*o*) recorder.

ble of yielding this large amount of energy in a single step. For example, the direct transfer to oxygen of two electrons from a substance at the reduction level of carbohydrate would yield only about 50–60 kcal. Moreover, this energy liberation is calculated for the transfer of two electrons over the entire potential span, whereas most biochemical oxidations involve the stepwise transfer of electrons over small potential drops through intermediates such as the pyridine nucleotides, flavins, and cytochromes. It is therefore apparent that either (*1*) some special component which possesses a particularly high reducing potential is involved in light production, or (*2*) that a mechanism is present in bioluminescent systems which permits the transfer of energy conserved in several intermediate step reactions to a final emitting species of molecule. Since luminescence by bacterial luciferase involves the oxidation of reduced riboflavin phosphate, and since the free energy liberated by the oxidation of a mole of reduced riboflavin phosphate by oxygen is only about 35 kcal, this system most probably possesses a mechanism that somehow pools the energy liberated as more than one reduced flavin molecule is oxidized. One possibility is that oxidation of the first flavin molecule yields a peroxide or its bound equivalent, and that the second reduced flavin molecule is oxidized by the "peroxide" produced in the first oxidation. Peroxides are considerably more potent oxidants than molecular oxygen and therefore the use of such an intermediate as the oxidizing portion of the reaction mixture could readily furnish the prerequisite energy for light emission. The dismutation of two peroxide molecules to form water and oxygen, according to the equation $2H_2O_2 \rightarrow 2H_2O + O_2$, produces a free-energy change of about 50 kcal/mole. Since the energy required to produce emission at the short wavelength end of the spectrum of light produced by luminous bacteria is about 60 kcal, it is evident that the peroxide dismutation reaction itself yields nearly sufficient energy to serve as the energy source.

Thermodynamic principles dictate that the free energy liberated as light must be equal to or less than the free energy liberated by the driving chemical reaction. But, there is a further, less obvious fact that may bear on the energetic requirements for bioluminescent emissions: the free-energy content of light varies as a function of its intensity. This relationship is to be expected from elementary physical principles, for even ice emits light quanta at a very low rate, as a result of improbable collisions. However, work cannot be performed (there is no free-energy content) by these quanta unless the receiver is at a lower temperature than the emitter. Thus, the amount of work that can be done per mole of quanta when the light intensity is extremely low is considerably less than the work which can be done when the intensity is high, and, in the present context, the free energy liberated by a reaction driving a bioluminescent emission may indeed be substantially lower than the total energy liberated per einstein provided that the light is of low intensity. On the basis of such considerations, Dr. J. Mayer (17) of the University of Chicago has calculated the minimum energy requirements for bioluminescent reactions at specified emission rates per cubic centimeter of solution, and has concluded that the free energy required as a driving reaction may be as much as 15–20 kcal less than the energy per einstein emitted, provided that the intensity of the light is sufficiently low.

This suggests a mechanism (illustrated in Fig. 10) by which a 50 kcal reaction might yield 60–70 kcal light. The diagram indicates, essentially, that a portion of the activation energy leading to the luminescent reaction and derived from the thermal energy of the medium is available for light emission and is added to the free energy liberated by the total reaction. It can readily be seen that the overall rate of reaction is limited by the height of the activation barrier. However, provided that the activation reaction involves a displacement of electrons on the coordinate characteristic of the excited electronic state that leads to emission, it follows that thermal activation energy may be conserved in the light quantum. An interesting consequence is that the system could act as a refrigerator by converting thermal energy into a part of the total energy emitted as light.

It is not yet clear whether such a mechanism actually operates in biological systems, but the model developed by Mayer extends the range of possible energy-yielding reactions in luminescence excitation

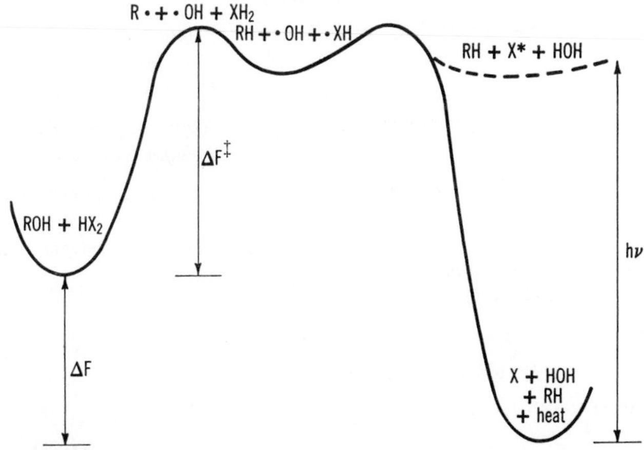

Fig. 10. Schematic diagram illustrating the mechanism whereby free energy of activation may be added to free energy of reaction in distributing total energy liberated by light quantum. ROH could, for example, be HOOH (17).

to systems which may liberate considerably less free energy per mole of product formed than the total energy emitted per einstein.

2. Distribution of Luminescent Reactions

Representatives of nearly every animal phylum and most plants, including all photosynthetic organisms exhibit bioluminescence (18). The phylogenetic distribution of luminescent forms is listed in Table I. Of these biological systems, the most thoroughly studied include the following: (*1*) the firefly, *Photinus pyralis*; (*2*) the luminous bacterium, *Achromobacter fischeri* (photobacterium); (*3*) the sea pansy, *Renilla reniformis*; (*4*) the protochordate, *Balanoglossus*; (*5*) the dynoflagellate, *Gonyaulax*; (*6*) the ctenophore, *Mnemiopsis*; (*7*) the crustacean, *Cypridina hilgendorfii*; and finally, (*8*) green plants.

TABLE I

Phylum	Species	
	Animal Kingdom	
Protozoa	*Noctiluca miliaris* *Peridinium bahamense* *Gonyaulax polyhedra*	
Porifera	Grantia (Calcerous sponges)	
Cnidaria	Various hydrozoans:	*Clytia linearis* *Obelia dichotoma* *Campanularia flexuosa*
	Various scyphozoans:	*Aequorea forskalea* *Pelagia noctiluca*
	Various anthozoans:	*Pennatula phosphoria* *Pteroides griseum* *Renilla reniformis*
Ctenophora	*Mnemiopsis leideyi* *Beroë forskalea*	
Platyhelminthes	None lum.	
Nemertinea	*Emplectonema kandai*	
Aschelminthes	None lum.	
Entoprocta	*Electra piosa*	Note: E. N. Harvey in *Bioluminescence* (18) unable to observe luminescence

(*continued*)

TABLE I (*continued*)

Phylum	Species	
	Animal Kingdom (*continued*)	
Annelida	*Chaetopterus pergamentaceous* *Tomopteris helgolandica* *Odontosyllis enopla* *Henlea ventriculosa* *Eisenia submontana*	
Echiuroidea	None lum.	
Sipunculoidea	None lum.	
Mollusca	*Pholas dactylus* *Rocellaria grandis* *Heteroteuthis dispar* *Nematolampas regalis*	
Brachiopoda	None lum.	
Arthropoda	*Pleuromma abdominale* *Lucicutia flavorcornis* *Heterocarpus alfonsi* *Scoliophanes crassipes* *Lampyris noctiluca* *Photinus pyralis*	
Echinodermata	*Ophiacantha bidenta* *Amphiura squamata*	
Enteropneusta	*Ptychodera bahamensis*	
Chordata	*Pyrosamata fixata* *Salpa pinnata* *Spinar niger* *Porichthys notatus* *Chauliodus sloani*	
	Plant Kingdom	
Thallophyta	Various fungi	*Photobacterium phosphoreum* *Omphalia flavida* *Mycena lux-coeli* *Panus stipticus*
Bryophyta	None lum.	
Pteridophyta	None lum.	
Spermatophyta	None lum.	

3. The Chemistry of Bioluminescence

A. THE FIREFLY (*Photinus pyralis*)

The common early summer firefly of the eastern United States is the most important luminous species in bioluminescence assay. This small beetle, one of the *Lampyridae,* is similar in many respects to other firefly species found in different parts of the world.

In his studies of the classical luciferin–luciferase reaction, Dubois (19) employed the rock-boring clam, *Pholas dactylus*. The light-producing reaction occurred when an extract of a luminous organ, which had been allowed to grow dark (through the exhaustion of its substrate, luciferin), was mixed with hot-water extract (containing luciferin) of the same luminous system. Such a luciferin–luciferase reaction was observed in firefly luminous organ extracts by Harvey (20) 40 years ago; but it was the key observation by McElroy (4) in 1947, that adenosine triphosphate is required for the luminescence of firefly light organs that has made possible the extensive study of the bio-

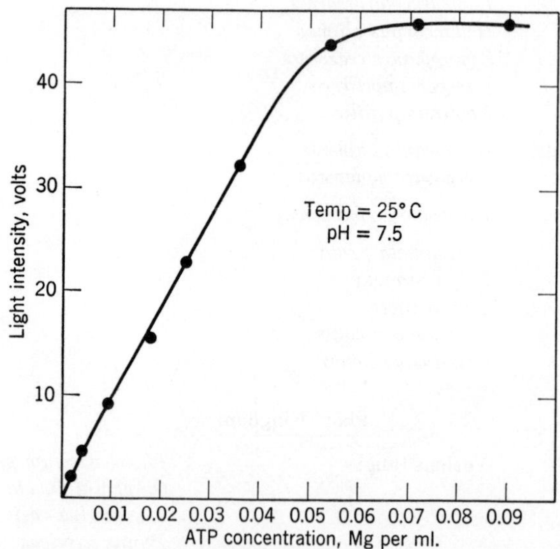

Fig. 11. Effect of ATP concentration on rate of luminescence by firefly enzyme. 1.0 ml of enzyme system (see Section IV-1-C-a) was mixed with $\frac{1}{15}M$ sodium phosphate buffer, and the indicated ATP concentration to give final volume of 10 ml (21).

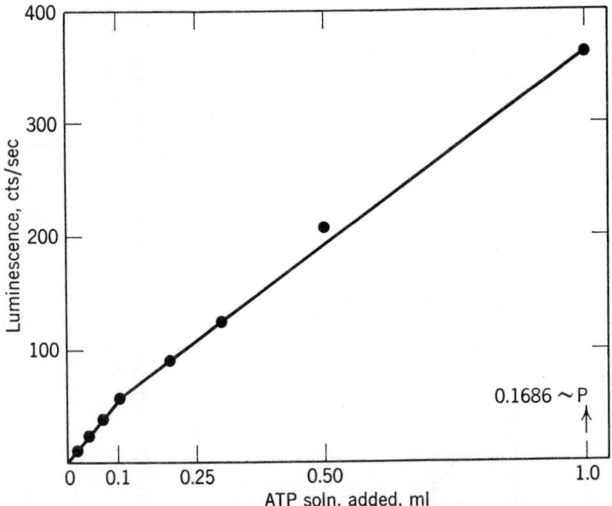

Fig. 12. Relationship between luminescence intensity 5 sec following addition of standardized amount of ATP using a quantum counter described in Figure 3. At zero time a known aliquot of ATP was added to 5 ml of $\frac{1}{15}M$ arsenate buffer pH 7.4 and 1 ml firefly enzyme. (See Section IV-1-C-a.) Final volume 7.0 ml. (See ref. 5.)

chemistry and biophysics of light emission in this system which has taken place since then. Following the discovery of ATP function (see Figs. 11 and 12) in firefly luminescence, McElroy and Strehler (21) defined the general properties of the system and the additional components which are required. They found that the most important of these components is a fluorescent compound which they called firefly luciferin. The structure of luciferin as determined by McCapra and White is illustrated in Figure 13. In addition to luciferin and the enzyme, luciferase, a bivalent metal ion, e.g., magnesium, is required,

```
              H                                   COOH
              |                                    |
       H    C         N      H     H        N — C — H
         C       C                |     |      //     |
         |       ||       C — C = C — C          |
     HO— C       C                           \      |
              C       S                        S — C — H
              |                                    |
              H                                    H
```

Fig. 13. Structure of firefly luciferin as determined by McCapra and White (34).

TABLE II[a]

Effect of Inorganic Ions on Luminescence[b]

Compound added	Concentration, M	Maximum light intensity, V	Compound added	Concentration, M	Light intensity, V
None	—	7.0	$CoSO_4$	1×10^{-3}	80.0
$MgSO_4$	1×10^{-4}	50.0	$CoSO_4$	2×10^{-3}	74.0
$MgSO_4$	1×10^{-3}	90.0	$FeSO_4$	1×10^{-3}	21.0
$MgSO_4$	2×10^{-3}	110.0	$FeSO_4$	2×10^{-3}	32.0
$MnSO_4$	3×10^{-4}	32.0	$NiCl_2$	1×10^{-3}	26.0
$MnSO_4$	1×10^{-3}	90.0	$ZnSO_4$	1×10^{-3}	15.0
$MnSO_4$	2×10^{-3}	92.0			

[a] From ref. 21.

[b] 1 ml of the purified enzyme system was added to an Na_2HPO_4 buffer containing the appropriate concentration of inorganic ion. ATP was added to give a final concentration of 0.08 mg/ml. The final volume of the reaction mixture was 7 ml.

although other ions can substitute for magnesium as is shown in Table II. Oxygen is also needed and this requirement was studied quantitatively by McElroy and Seliger (15). Figure 14 shows the response of the firefly system to the concentration of added luciferin. Magnesium concentration for the reaction is optimal at about 2 μmole/ml (see Fig. 15), and the oxygen saturation curve yields a dissociation constant, according to McElroy and Seliger, of 5×10^{-10} mole/liter (see Fig. 16). The firefly luminescent reaction has a temperature optimum of 25° C at pH 7.4 (21) and an activation energy of 18.5 kcal as is shown in Figure 17. The pH optimum for the reaction is at about pH 7.4, although the system may be used for bio-

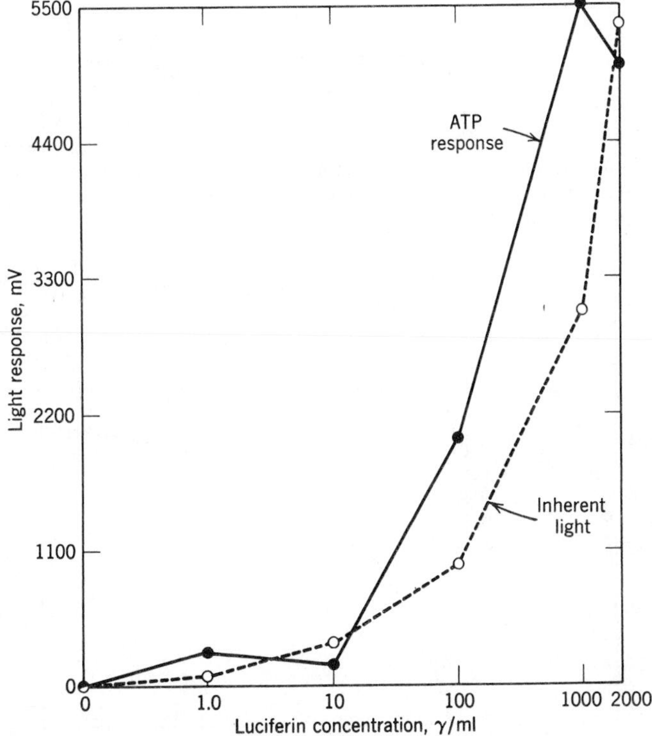

Fig. 14. The effect of luciferin concentration on light emitted by the firefly enzyme. The reaction mixture contained 0.1 ml of 0.01M $MgSO_4$ and 0.001 γ ATP. The inherent light would be produced by bound ATP (12).

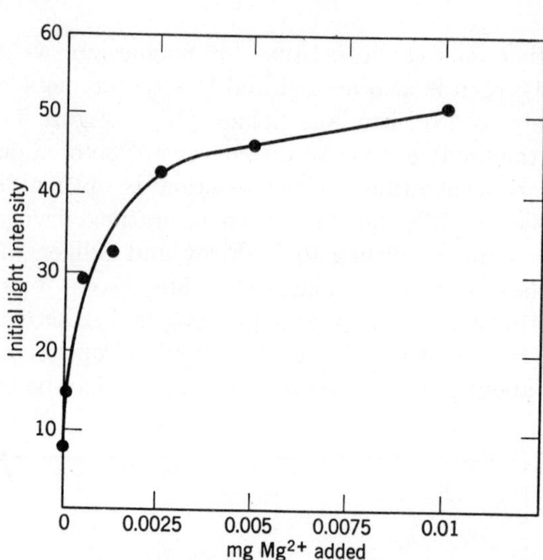

Fig. 15. Effect of added magnesium concentration on initial light intensity emitted by firefly extract with ATP concentration constant. 0.1 ml crude luciferase, 1 ml luciferin, 10 μmoles ATP, 0.05*M* glycine buffer pH 8.0 [$MgSO_4$] as indicated. Total volume 10 ml (61).

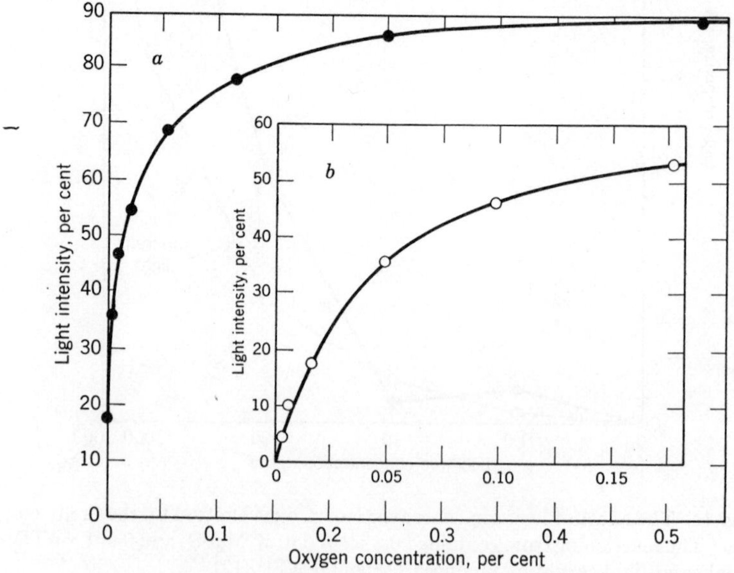

Fig. 16. Effect of oxygen concentration on luminescence intensity (15).

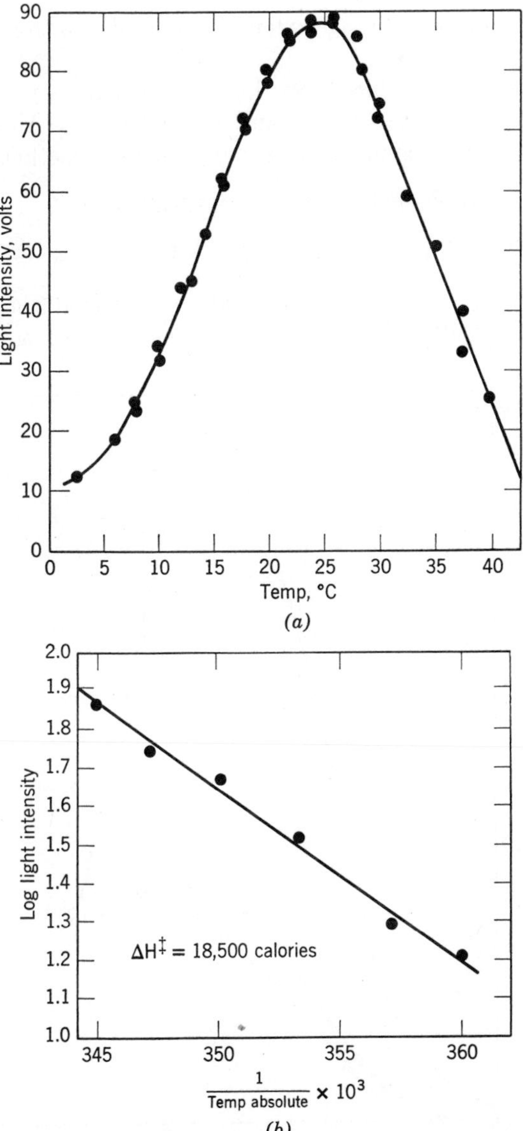

Fig. 17. (*a*) Effect of temperature on relative luminescence intensity (gross data). (*b*) Arhennius plot effect of temperature on relative luminescence emitted by firefly extract (see Section IV-1-C-a); activation energy: 18,500 cal. 1 ml of firefly enzyme was added to $\frac{1}{15}M$ sodium phosphate pH 7.5 containing saturating amounts of ATP to a final volume of 10 ml at the temperature indicated (21).

assay over a substantial range extending from pH 6.5 to 8.0 (see Fig. 18).

The enzyme, firefly luciferase, has been extensively studied by McElroy and associates (22,23) and crystallized by McElroy and Green (24). This enzyme catalyzes at least two reactions, viz,

$$LH_2 \text{ (luciferin)} + ATP \rightleftharpoons AMP \cdot LH_2 + P{-}P \quad (1)$$

$$LH_2 \cdot AMP + \tfrac{1}{2}O_2 \rightarrow L \cdot AMP^* + H_2O \rightarrow L \cdot AMP + h\nu + H_2O \quad (2)$$

The light-producing molecule (as might be predicted from its name) is excited L·AMP*. This molecule is bound to the enzyme very

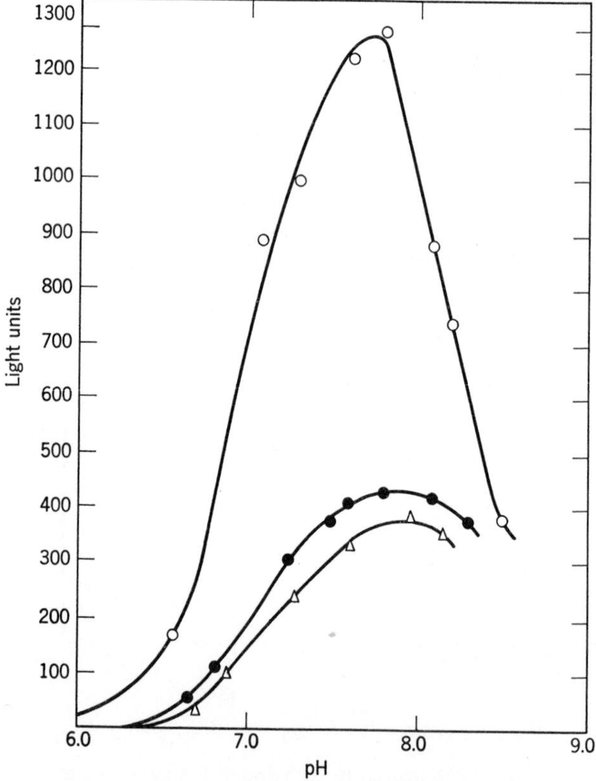

Fig. 18. Effect of pH on intensity of luminescence of firefly extract: (○) Glycine buffer; (●) sodium phosphate buffer; (△) potassium phosphate buffer (15).

tightly and acts as a competitive inhibitor for LH^2 and $LH^2 \cdot AMP$. In kinetic studies employing the system, it is therefore necessary to take into account inhibitory effects of product accumulation. This does not appear to be an important consideration when very low rates of reaction are involved, such as those in which millimicromolar quantities of ATP are assayed. However, those who employ the system for assay purposes should take account of this factor by appropriately designed controls and internal standards.

A component of crude firefly lantern extracts which can influence the kinetics of the reaction is the very active adenylic kinase (myokinase) which is present. In the presence of substantial amounts of adenosine monophosphate or adenosine diphosphate, the reaction catalyzed by myokinase may produce artifactual kinetic measurements. However, since the response of the firefly enzyme is essentially instantaneous (less than a second to peak intensity) to added ATP, the relatively slow myokinase equilibrium does not substantially affect the accuracy of ATP determination if the measurements are made within a few seconds after the sample is added.

A second component which plays a role in the yield of luminescence in response to added adenosine triphosphate is the highly active pyrophosphatase present in crude firefly lantern extracts. This enzyme tends to drive the equilibrium described in reaction (1) to the right in consequence of its destruction of one of the reaction products. In the presence of large amounts of added pyrophosphate, the equilibrium of the reaction will be shifted to the left (lower light output), but as pyrophosphate is hydrolyzed, the equilibrium moves to the right with a consequent increase in luminescent intensity. These effects are shown in Figure 19, taken from unpublished data of Dr. Robert Monroe.

Pyrophosphate exerts another important influence which is due to the fact that, in its presence, luciferase can cleave adenyloxyluciferin, with the regeneration of adenosine triphosphate and oxidized luciferin. ATP thus produced is available for a new cycle of $LH_2 \cdot AMP$ formation. The addition of pyrophosphate to a system containing substantial amounts of LH_2 and $L \cdot AMP$ will cause a flash of luminescence by favoring the synthesis of ATP from adenyloxyluciferin and its reutilization in the formation of adenyl luciferin. This phenomenon is illustrated in Figure 19.

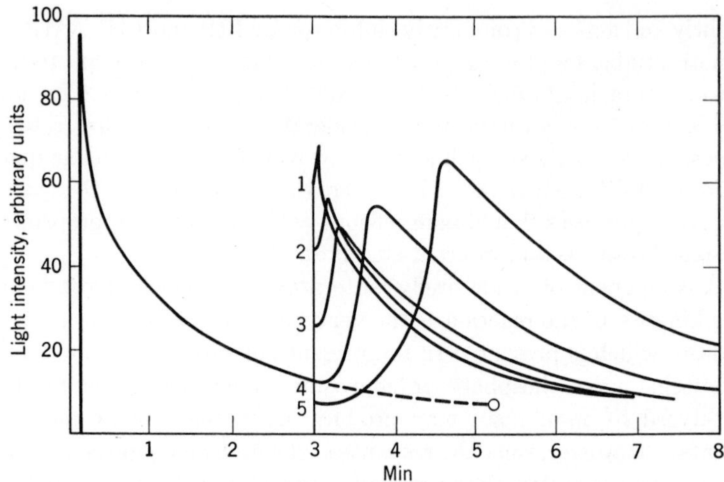

Fig. 19. Effect of addition of different amounts of pyrophosphate to firefly ATP system; concentration added, 1, 2, 4, 8, and 16 $\times 10^{-5}M$ for *1–5*, respectively; ATP, $10^{-6}M$; luciferin, $2 \times 10^{-5}M$; 0.1 ml firefly extract (see Section IV-1-C-c) (unpublished data of R. Monroe).

The firefly enzyme is inhibited by a number of substances including various quenching agents. The most generally present of such substances are monovalent anions, such as chloride, which at high molarity may cause an inhibition of luminescence. The effect of chloride concentration on luminescence intensity is shown in Figure 20.

The above brief resume of some of the main features of the firefly luminescent reaction should be thoroughly familar to those who wish to apply the system to their own special assay problems. Aside from such prerequisite familiarity with the basic reaction mechanism, two indispensable additional tools are the use of an appropriate internal standard and a modicum of intelligence or alertness. Whereas the usual assay procedure includes the standardization of the reaction system against known ATP concentration, it is essential that interference from components within the reaction mixture used for assay be ruled out in determining the absolute concentration of ATP present. For this, the comparison of the level of luminescence produced by the addition of the unknown as compared to the level produced by the addition of unknown plus a known amount of ATP is an *indispensable control* in all pilot studies and is desirable as a routine in many other investigations.

In studying the kinetics of coupled reactions (i.e., ATP formed within a mixture of firefly enzyme and some generator of ATP such as mitochondria), one must also keep in mind differences between the effect of pH on the firefly reaction as well as its effect on the ATP-generating system. Moreover, since the enzyme is unstable at temperatures above 25°C, and since the activation energy of the luminescent reaction itself is about 18.5 kcal, variations in luminescence intensity caused by small changes or differences in the temperature of the reaction mixture must be *stringently avoided*. Finally, it is important that the ATP concentration being assayed lie within the *linear range* of the enzyme's saturation curve. For example, in using an internal standard, it is essential that the sum of the added ATP concentration (known) and the unknown does not exceed half saturation of the firefly enzyme (see Fig. 12). This concentration is reached at about 35 μg of adenosine triphosphate per milliliter.

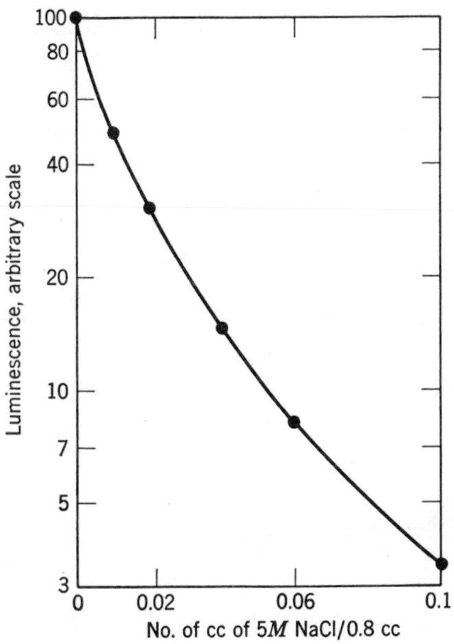

Fig. 20. Effect of chloride concentration on luminescence emitted by firefly extract. 0.2 ml ATP (0.168 μg ~ ppm) $\frac{1}{15}M$ arsenate buffer pH 7.4; 2 mg $MgSO_4$; 0.2 ml firefly extract (see Section IV-1-C-b) and the indicated volume of NaCl in a volume of 0.8 ml (5).

B. LUMINOUS BACTERIA

The luminescence produced by luminous bacteria has been known for over 300 years (first reported by Fabricus ab Aquapendente in 1592). *Photobacterium fisherii* (a species known also as *Achromobacter fischeri*), has furnished the material for most of the recent studies of bacterial luminescence. A strain is available from the American Type Culture Collection (No. 7744).

Early workers, including Harvey (18), were unable to obtain a substantial luciferin–luciferase reaction from luminous bacteria although Gerretsen (25) in 1920 did demonstrate a weak reaction with a species obtained from Java. Unfortunately, his work was not confirmed until 1953 when Strehler and Shoup (26) showed that lyophilized powders of *P. fischeri* would produce light when they were added to dilute buffered saline. They also demonstrated that the addition of peroxide to such a preparation would restore a small amount of luminescence after the light had decayed, and that the active

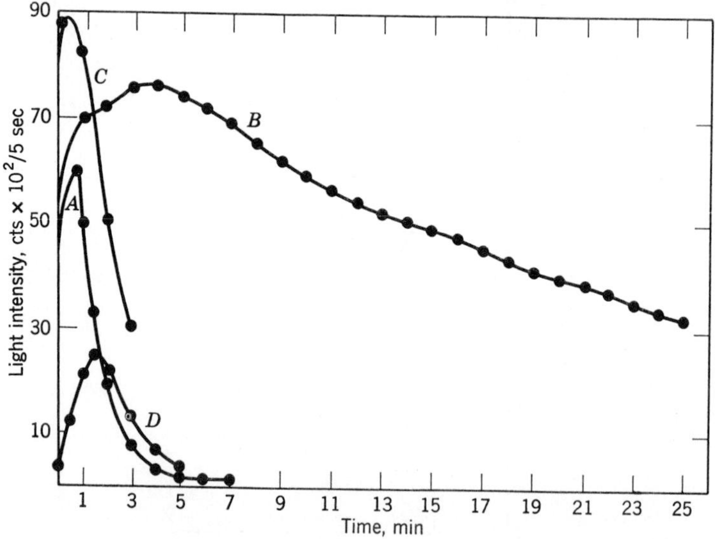

Fig. 21. Initial demonstration of effect of pyridine nucleotide and dihydropyridine nucleotide addition on light production by bacterial luciferase. Luminescence decay curves following addition at zero time of: (*a*) 2 mg/ml acetonized bacterial powder in water; (*b*) 20 mg/ml acetonized bacterial powder in water; (*c*) $DPNH_2$ (20 mg/ml) addition to undialyzed bacterial extract (2 mg/ml); (*d*) DPN (16 mg/ml) addition to undialyzed bacterial extract (2 mg/ml).

component in the bacterial extract supporting this luminescence was a riboflavin compound. Shortly thereafter, Strehler (27) obtained unequivocal evidence of a bacterial luciferin–luciferase reaction in which reduced pyridine nucleotides (DPNH) acted as a substitute for "bacterial luciferin" (see Fig. 21). DPN, in the presence of suitable substrates, is also capable of causing light emission, and evidence for additional components was presented. That riboflavin phosphate (FMN) was one such additional requirement for bacterial luminescence was shown shortly thereafter by McElroy and Hastings (28), but even with the inclusion of these substances in the reaction mixture, the level of luminescence did not approach that of the intact bacteria. Strehler and Cormier (29) had observed (Fig. 22) that the addition of hot-water extracts from a variety of different materials,

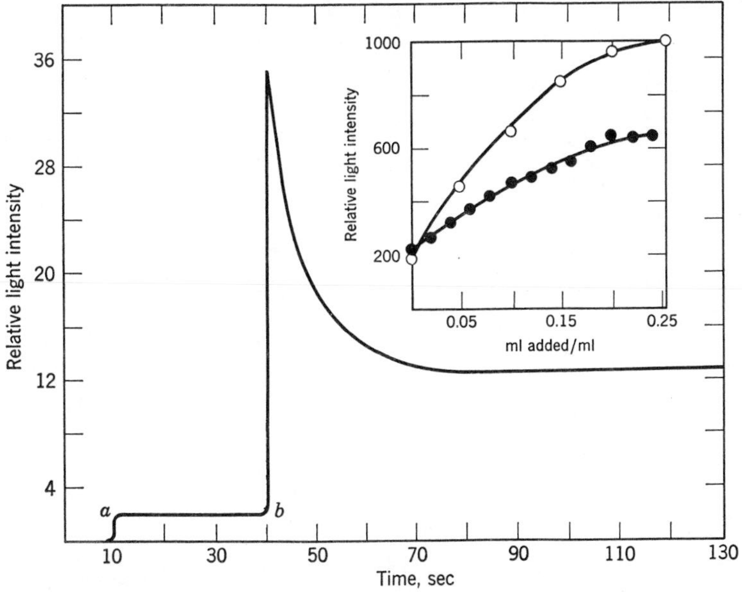

Fig. 22. In the insert is shown the effect on luminescence of the addition of a 10% hot water extract of luminous bacteria (●) and kidney cortex extract (○). The luminescence system substrates were exhausted by incubating overnight with $DPNH_2$ and malate and dialyzing for 24 hr. The system contained 0.5 ml exhausted system, 0.21 μg FMN, 15 mg malate, 500 μg $DPNH_2$, and 0.3 ml 0.1*M* phosphate pH 7.0 in a total volume of 1.25 ml. At (*a*) the above mixture was exposed to the light-detecting system, and at (*b*) 0.2 ml 25% boiled acetonized kidney cortex extract was added.

including luminous bacteria, cause a stimulation of luminescence above and beyond that produced by the addition of flavin compounds and DPNH. A convenient source of such a factor was found to be "defatted" hog kidney cortex powders and the factor, then referred to as kidney cortex factor (KCF), was shortly identified by Strehler and Cormier (30,31) as long-chain fatty aldehyde (see Fig. 23). Various aldehydes from C_6 to C_{18} were shown to be effective, and the active principle in the hog kidney powders was shown to be palmitic aldehyde (C_{14}). Strehler, Harvey, Chang, and Cormier (32) then demonstrated that reducėd riboflavin phosphate, as well as reduced riboflavin would support luminescence and that under these conditions reduced pyridine nucleotides were not required (see Fig. 24). This observation clarified the role of the pyridine nucleotide and identified it as a

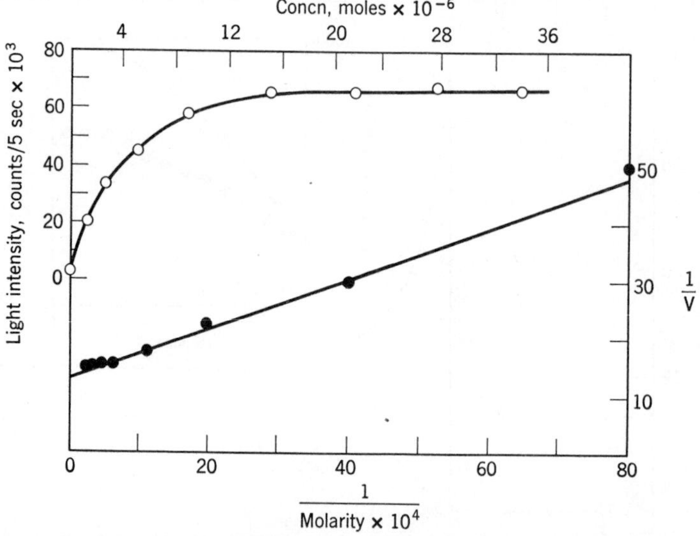

Fig. 23. Effect of concentration of palmitic aldehyde on luminescence of bacterial extract. 0.1 ml of 2% aqueous extract of *P. fisherii* was mixed with 1.5 ml of 0.1*M* phosphate buffer pH 7.0, 1000 μg $DPNH_2$ and 4 μg FMN and the light intensity measured. The indicated amount of palmitic aldehyde was then added and the light intensity measured again. The upper curve shows the relationship between aldehyde concentration and light intensity (upper left-hand scale) and the lower line is a plot of the reciprocal of light intensity versus the reciprocal of the aldehyde concentration (lower and right-hand scale) (31).

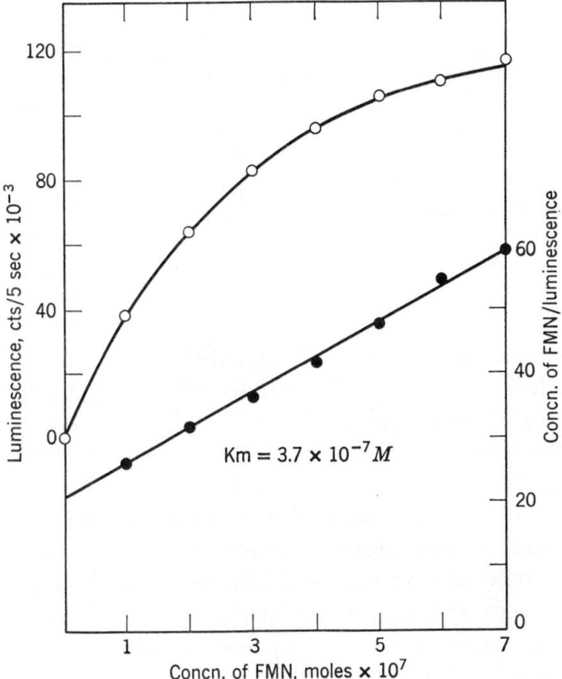

Fig. 24. Effect of FMN concentration on luminescence of cell-free extracts of luminous bacteria. The reaction mixture contained $(NH_4)_2SO_4$ fractionated bacterial luciferase in phosphate buffer pH 6.8. DPNH was added at zero time and FMN at 25 sec (31).

reductant of flavins. A number of studies by Cormier, Strehler, Harvey, Johnson, Hastings, McElroy, and their collaborators rapidly revealed many of the remaining unexplored parameters (33,34). Since (on a molar basis) peroxide adducts of aldehyde appear to be somewhat more effective than the free aldehyde in simulating luminescence (35), Strehler suggested that these compounds were the real intermediates, and that the reaction sequence is as follows:

$$XH_2 + DPN \rightleftharpoons DPNH_2 + X \qquad (3)$$

$$DPNH_2 + FMN \rightleftharpoons DPN + FMNH_2 \qquad (4)$$

$$FMNH_2 + \tfrac{1}{2}\, O_2 \rightleftharpoons FMN + H_2O \qquad \text{(side reaction)} \qquad (4a)$$

$$FMNH_2 + O_2 + R-C\begin{matrix} \diagup\!\!\diagup O \\ \diagdown H \end{matrix} \underset{}{\overset{\text{enzyme}}{\rightleftharpoons}} FMN + \left[\text{enzyme}\cdot R-\overset{\overset{\textstyle OH}{\diagup}}{\underset{\underset{\textstyle H}{\diagdown}}{C}}-OOH\right] \quad (5)$$

$$\left[\text{enzyme}\cdot R-\overset{\overset{\textstyle OH}{\diagup}}{\underset{\underset{\textstyle H}{\diagdown}}{C}}-OOH\right] + FMNH_2 \rightarrow \left[\text{enzyme}\cdot R-C\begin{matrix} \diagup\!\!\diagup O \\ \diagdown H \end{matrix}\right] + FMN + 2H_2O + h\nu$$

Strehler and Johnson (36) showed that the flavin reduction and flavin oxidation steps are differentially affected by high hydrostatic pressures and that the presence of aldehyde is necessary in order to observe the pressure sensitivity of the second reaction. A kinetic analysis of the reaction may be found elsewhere (37).

Like the firefly, the bacterial luminescent reaction can be used to assay a variety of important biological constituents. Included among these are flavin mononucleotide, DPN, and long-chain aldehydes, as well as enzymes and substrates which may be involved in the formation or degradation of these components. The bacterial luciferin–luciferase reaction is particularly sensitive to long-chain aldehydes, as is illustrated by the fact that a luciferase solution containing all necessary components except aldehyde will emit only a very weak luminescence. However, the proximity of a pipet containing a dilute solution of aldehyde will cause a bright luminescence along the miniscus and walls of the test tube due to volatilization of aldehyde into the solution surface.

In addition to the use of the bacterial system for the assay of materials that can be coupled to the generation or destruction of any molecular species required for the reaction, luminous bacteria, as has been noted by Harvey (18), are ultrasensitive indicators of the presence of oxygen and have been used for the measurement of this substance.

C. RENILLA RENIFORMIS

According to studies by Cormier (38) on the colonial coelenterate *Renilla reniformis*, the bluish luminescence of this species is a result of the oxidation of a tryptamine derivative (luciferin) and requires the presence of adenosine 3′,5′-diphosphate. The equation for the *Renilla* luminescence reaction is: luciferin + adenosine 3′,5′-diphosphate + Ca^{2+} + oxygen → light. The reader is referred to Cormier's (39) original publication for further details.

D. BALANOGLOSSUS

A luminous species of the protochordate, *Balanoglossus*, has also been studied by Cormier and co-workers (40). Extracts of the luminous portions of this animal exhibit a typical luciferin–luciferase reaction and the system responds to the addition of hydrogen peroxide with a slight luminous emission. The luciferase in this instance is a "peroxidase" and the luciferin has not been characterized.

E. MNEMIOPSIS LEIDYI

This luminous ctenophore (comb jelly) may be found along the Atlantic coast of the United States from Massachusetts to Georgia. It displays a very bright and localized luminescence under its eight rows of comb plates, the motile organs of these animals. This luminescence and that of several species of Scyphozoa (jelly fish) are the only examples of a light-producing reaction in biological systems which do not require molecular oxygen. Harvey (41) has shown that tissue fragments of this animal will luminesce upon mechanical or electrical stimulation even when submerged for several hours in an anaerobic suspension of luminous bacteria. Since luminous bacteria can be used to detect oxygen in a concentration as low as one part in 40 billion, these experiments make it quite likely that the ctenophore is indeed capable of luminescing under completely anaerobic conditions. Lyophilized powders of *Mnemiopsis* are capable of luminescing when they are added to distilled water, but no luciferin–luciferase reaction has yet been demonstrated (42).

Light production by *Mnemiopsis* homogenates is interesting in other respects (42). Fragments of the animals obtained by pressing them through cheesecloth will luminesce when an electrical current is passed through suspensions of them; luminescence is elicited by the addition of various surface active agents; freezing and thawing of the suspensions causes a burst of luminescence each time that the freeze–thaw luminescence cycle is repeated (up to about seven cycles). *Mnemiopsis* luminescence is unique in that light emission is instantaneously inhibited (both in cell suspensions and in the intact animal) following exposure to visible light of moderate intensities. Ability to luminesce is slowly regained in the succeeding dark period, but only under aerobic conditions.

F. CYPRIDINA HILGENDORFII

This luminous copepod crustacean is found in the waters around Japan. In contrast to the other forms described here its luminescence occurs extracellularly. The posterior of the animal possesses two glands, one of which produces luciferin, the other, luciferase. Under suitable conditions or provocation, the animal secretes these substances, which upon contact with each other in the presence of oxygen, emit a very bright blue luminescence. The properties of the luciferin and luciferase have been extensively studied by Chase (43), Harvey (18), Johnson (44), and others (45). A tentative structure for *Cypridina* luciferin (46) is shown in Figure 25. It is slowly inactivated by air (oxidation) and produces a considerable amount of delayed luminescence (reminiscent of phosphorescence) following its exposure to near ultraviolet light.

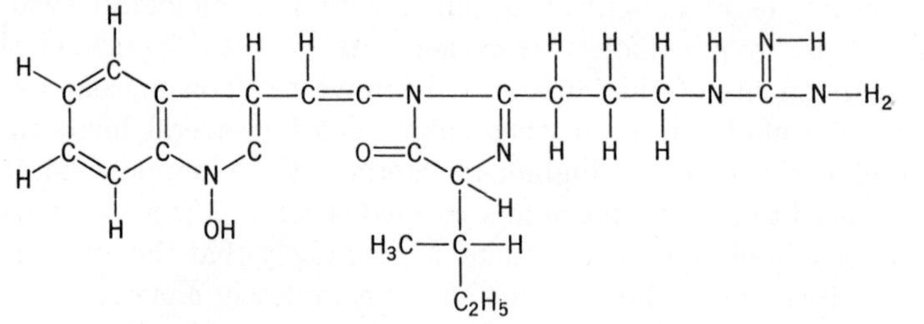

Fig. 25. Proposed structure of *Cypridina* luciferin (46).

G. GONYAULAX

In addition to luminous bacteria, other unicellular organisms produce light. The most studied of these animals is the dynoflagellate *Gonyaulax*. This organism, motile by virtue of the flagellae it possesses, is encased in plate-like armor. These dynoflagellates are photosynthetic. Some species related to the luminous ones grow in great blooms called the "red tide" and such an epidemic may kill large numbers of fish. The poisonous substances synthesized by the dynoflagellate also render the fish toxic.

This species has been investigated by Hastings and Sweeney (47). The individual organisms luminesce when they are stimulated mechanically; for example, they may be caused to glow if the culture medium in which they grow is shaken.

The biochemistry of *Gonyaulax* luminescence is not well understood, but studies by Hastings (48), who has demonstrated a luciferin–luciferase reaction, have clarified some of its properties. An interesting feature of *Gonyaulax* luminescence is its diurnal rhymicity. Samples stimulated at various times of the day or night show systematic differences in emission intensity.

H. GREEN PLANT LUMINESCENCE

In 1950 William Arnold of the Oak Ridge National Laboratory and the author (49) attempted to test the hypothesis that ATP is produced during photosynthesis as a result of a photochemically driven reaction. In order to test the hypothesis, a chloroplast suspension was mixed with firefly enzyme, AMP, Mg^{2+} and then illuminated. The mixture was then quickly placed next to a photomultiplier in the dark. Following such brief illumination, light was emitted by the mixed preparation for several seconds, a result to be expected if ATP were produced photosynthetically. That this luminescence was not primarily due to firefly extract luminescence was demonstrated by omitting the firefly enzyme. In this manner we accidentally discovered that chloroplasts (and green plants generally) emit a luminescence after they are illuminated. Although later studies did establish that the firefly enzyme can be used to measure photosynthetic phosphorylation both in extracts of intact *Chlorella* (50) and in mixture with chloroplasts (51), the luminescent emission measured in the experiments with Arnold represented a newly discovered phenomenon. Certain properties of the reaction strongly indicated its close relationship to the photosynthetic process, and indeed it now appears that the phenomenon represents the reversal of certain early steps in the sequence of reactions following the primary photochemical reaction(s) in photosynthesis. Some of the properties of the luminescent reaction in green plants follow (52).

1. The wavelengths of light leading to luminescence are identical to those required for photosynthesis; i.e., the action spectrum for photosynthesis and green-plant luminescence are identical. This is shown in Figure 26.

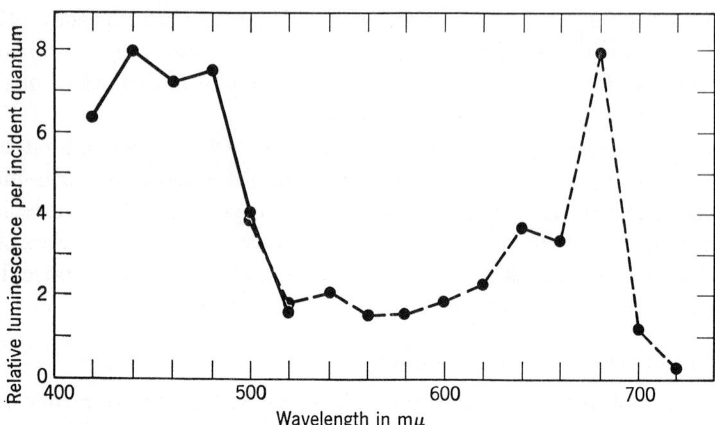

Fig. 26. Action spectrum of *Chlorella* luminescence (50).

2. The rate of luminescence depends on actinic light intensity in a fashion parallel to the light saturation of photosynthesis, but the half-saturation value depends on the portion of the luminescence decay curve measured (see Fig. 27).

3. The decay curve of the luminescence, determined with the variable-speed phosphoroscope described in Figure 7, appears to consist of two main components. The first of these has a half-life of about $\frac{1}{100}$ sec; the latter portion decays much more slowly and has a half-life of $\frac{1}{2}$–2 sec. This feature of the reaction is shown in Figure 8. Arnold and McClay (53) have studied the decay following a single flash and have found it to follow approximately second-order kinetics, as might be expected were the reaction due to a recombination of two different products of the photochemical reaction.

4. The color of the emitted light corresponds closely to the fluorescent emission spectrum of green plants and is therefore due to light emission by chlorophyl *a*.

5. Other features of this reaction have been described in detail elsewhere (52) but it should be pointed out that Arnold and McClay (53), as well as Calvin and Tollin (54), have obtained evidence suggesting that the energy for luminescence may be released as a result of the movement of electrons from a conduction band to a more stable position within the crystal lattice of the chloroplast. Luminescent reactions of porphyrin compounds have been studied in considerable

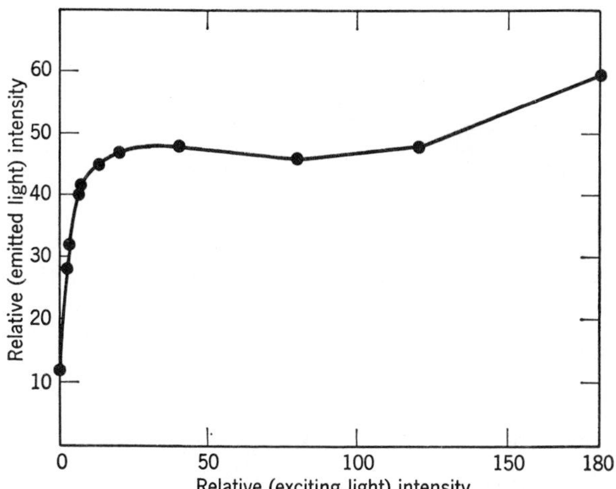

Fig. 27. The effect of exciting light intensity on the intensity of light emitted by suspensions of *Chlorella* (50).

detail by Linschitz and others (55), but it is not yet certain whether the light emitted by green plants involves analogous chemical reactions on the part of chlorophyl or whether chlorophyl acts as a secondary emitter of quanta that are generated as electrons fall into holes in a crystalline matrix.

4. Some Chemiluminescent Reactions

Of the many reactions of organic compounds which can give rise to luminescence, several have been examined in considerable detail. These include the reactions of (luminol) 5-amino-2,3-dihydro-1,4-phthalazinedione and 10,10′-dimethyl-9,9′-biacridinium nitrate, whose chemical structures are shown in Figures 28 and 29. Both of

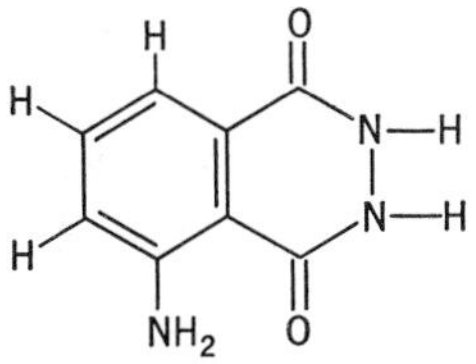

Fig. 28. Structure of luminol. 5-Amino-2,3-dihydro-1,4-phthalidazinedione.

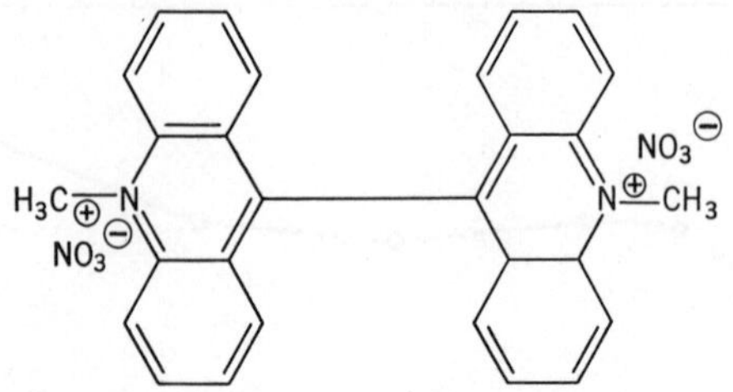

Fig. 29. Structure of 10,10′-dimethyl-9,9′-biacridinium nitrate (lucigenine).

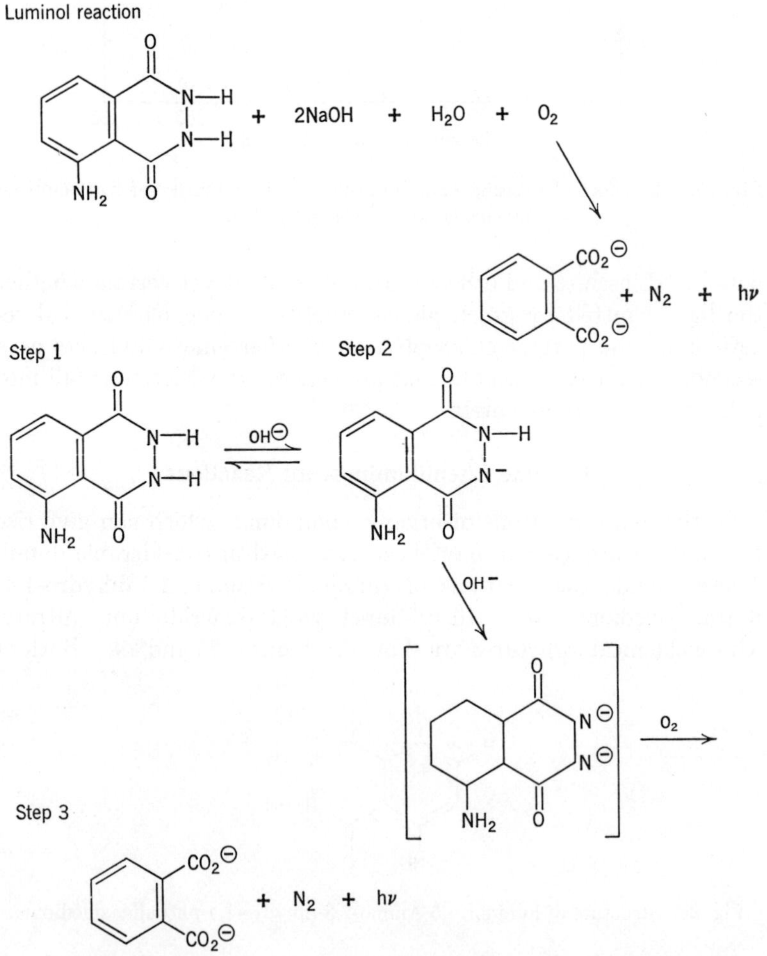

Fig. 30. Proposed sequence of reactions involved in light emission by luminol according to White (56).

these compounds luminesce in the presence of hydrogen peroxide and their luminescence is greatly enhanced by the addition of heme catalysts or ferricyanide.

The luminescence of luminol is thought to occur according to the reaction sequence described by Emil White and co-workers (56) and is illustrated in Figure 30. The crucial step in the reaction appears to be the formation of the excited state of the aminophthalate ion following the substitution of oxygen for the nitrogen atoms in the phthalidazinedione group. In the course of the reaction, nitrogen gas is evolved, amino phthalic acid accumulates, and hydrogen peroxide is consumed. The tentative identification of aminophthalate ion as the emitting molecule is based primarily on the identity of its fluoroscent emission spectrum under alkaline conditions and the emission spectrum of the chemiluminescent reaction.

The luminescence of lucigenine nitrate has been thoroughly studied by Totter and collaborators (57). This compound has the interesting property that it may be reduced by xanthine oxidase and be oxidized on another pathway by molecular oxygen or by peroxide. The latter appears to be a key step in the reaction leading to light emission. Totter has proposed the sequence of reaction in light emission shown in Figure 31. Both of these chemiluminescent reactions are poten-

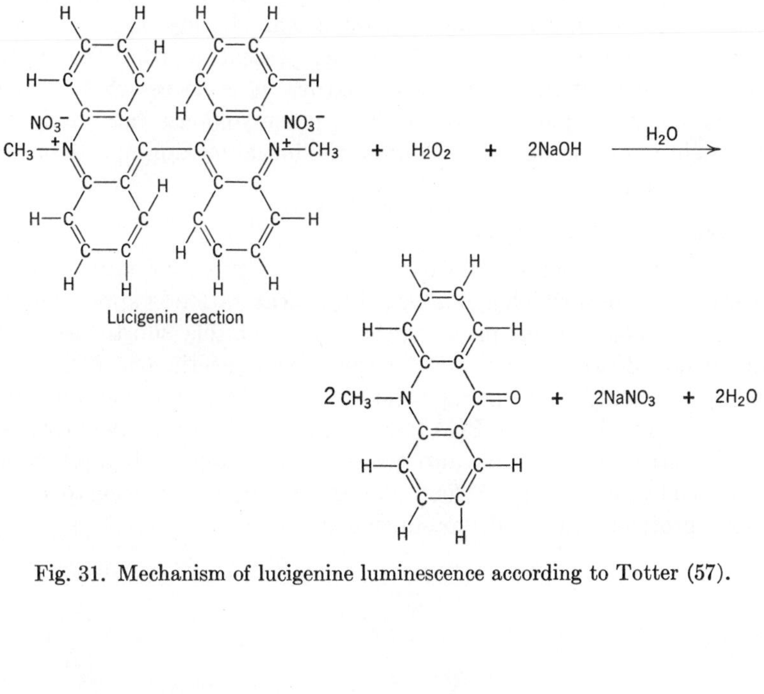

Lucigenin + H_2O_2 + 2NaOH $\xrightarrow{H_2O}$ 2 N-methylacridone + $2NaNO_3$ + $2H_2O$

Lucigenin reaction

Fig. 31. Mechanism of lucigenine luminescence according to Totter (57).

tially applicable to a variety of assays. Both of them may be used as indicators of oxygen concentration since both react with this molecule and produce consequent light even though the rate is low compared to the rate of luminescence in the presence of hydrogen peroxide. Luminol is a particularly effective indicator of small traces of heme compounds.

IV. SPECIFIC APPLICATIONS OF BIOLUMINESCENCE

1. Assays with Firefly Enzyme

A. PRINCIPLES OF ATP INSTANTANEOUS ASSAY

There are, in general, two different procedures of ATP assay involving the firefly system. In the first of these, which is the method of choice for most applications, use is made of the fact that the luminescence immediately following the addition of ATP is proportional to ATP concentration (see Fig. 11). In essence, in such a system, all components of reaction mixture, except the ATP standard or unknown, are added to a sample cuvette. At zero time, the unknown is added and as rapidly as is feasible thereafter (generally within 5 sec), the peak height reached by the system is measured. This method of assay avoids many of the difficulties introduced by myokinase and pyrophosphatase equilibria as well as the inhibition of the reaction by accumulated adenyl oxyluciferin. If very large amounts of adenosine monophosphate (AMP) are present in either the firefly enzyme itself, or in the reaction mixture, it is necessary to make measurements rapidly to avoid the rapid myokinase-catalyzed decrease in luminescence which follows the initial maximum rate of luminescence.

B. PREPARATION OF SAMPLE

Any suitable method which will instantaneously or very rapidly destroy enzymes which can alter ATP concentration is applicable to the firefly assay system provided that any inhibitory substances used are removed prior to assay. For this, trichloracetic acid (TCA) or perchloric acid (PCA) precipitation in the cold is suitable, but one must remove the TCA or PCA following its use. For many purposes, the injection of the sample into two to three volumes of boiling water followed by heating at 100° for 8–12 min, thereby inactivating thermolabile proteins, will yield reproducible results.

Alternatively, a method developed by Carlson (58) for use with muscle and other tissues with very active phosphokinases and transferases may be more suitable for certain applications. Carlson stopped the reaction by quenching the tissue in liquid nitrogen. A suitable fraction of the frozen starting material was homogenized while still frozen with a frozen solution of perchloric acid, in a supercooled mortar. The product was allowed to thaw following homogenization, and was kept at 0° for 10 min. Perchloric acid was neutralized with KOH and the sample centrifuged and assayed with firefly enzyme. Inactivation with boiling ethanol has also been employed with some success but one must remove the ethanol prior to assay. Several of these extraction methods are detailed below.

a. Extraction with Hot Water. The sample is pipetted rapidly into a threefold volume of water at 100° in a boiling water bath. The time of sample pipetting is recorded and 12 min after the beginning of extraction the samples are placed in an ice bath until assay. This method is suitable for extracting ATP from cell suspensions and is the method employed in our discovery of the phenomenon of photosynthetic phosphorylation in *Chlorella* (50). Table III shows the effect of boiling time on ATP level. As a preliminary to any exhaustive survey of ATP levels in a given biological material, the time course of inactivation of enzymes and of ATP extraction should be examined carefully. It should also be mentioned that one need not centrifuge debris from such boiled suspensions prior to assay provided that they are optically homogenous and that the suitable internal standards are employed.

TABLE III[a]

The Extraction of ATP from *Chlorella* Suspension as a Function of Boiling Time

	ATP counts/15 sec	
Boiling time, min	A	B
0	—	—
2	5500	2700
4	3042	2752
10	2872	2760
15	2830	2625
20	2505	2890

[a] From ref. 50.

b. Extraction with Perchloric Acid. One gram of tissue or reaction mixture is rapidly frozen by immersion in liquid nitrogen. The frozen pellet of material to be assayed is mixed with an equal volume of frozen 8% perchloric acid. The mixture is then thoroughly homogenized by grinding with a mortar and pestle while still frozen, and the sample is allowed to thaw at 0°.

After 10–15 min the material is neutralized to pH 7.0 with 1.0*N* potassium hydroxide, centrifuged to remove precipitate, and an aliquot measured with the firefly enzyme as described below.

C. PREPARATION OF FIREFLY ENZYME

Any one of the following methods is suitable for preparing firefly enzyme for ATP assay.

a. Fifty milligrams (10–12 luminous organs) of vacuum-dried firefly lanterns are extracted with grinding at 0° with 5 ml of 0.1*M* sodium arsenate, pH 7.4, for 2–5 min. The suspension is filtered into a test tube kept in an ice bath, after which 50 mg of magnesium sulfate is added. Precipitation of magnesium arsenate may take place but it is not disadvantageous provided that a uniform suspension is maintained.

b. A preparation relatively low in apyrase activity may be prepared by the method of McElroy and Strehler (21) with an ammonium sulfate fractionation procedure. To obtain this active extract, 4 g of vacuum-dried firefly lanterns is ground with sand and extracted twice with a total volume of 100 ml of distilled water. This solution is adjusted to pH 6 with hydrochloric acid. The precipitate is removed after centrifugation and is discarded. Ten grams of ammonium sulfate is added and after 15 min at 0° the solution is centrifuged and the inactive precipitate again discarded. After adjustment of the supernatant solution to pH 7.5, 10 g of ammonium sulfate is added. This solution is cooled to 0° for 15 min, and the inactive precipitate is discarded after centrifuging. The resulting supernatant solution is adjusted to pH 4.5 and 30 g of ammonium sulfate added. The preparation is cooled and filtered and the active precipitate is dissolved in 50 ml of water. This preparation is clear amber in color and may be stored in small convenient vials at −20° for several years. Alternatively, the preparation may be lyophilized and kept in a deep freeze.

c. A firefly-enzyme preparation, suitable for use in studies on phosphorylation by mitochondria is made up as follows: 10 mg of vacuum-dried firefly lanterns is homogenized with 1 ml of 0.05M Tris pH 7.3 at 0 in a Potter homogenizer for 1 min. The homogenate is centrifuged for 15 min at 15,000g to remove cellular debris. The supernatant is removed and stored at 0° until used. Such preparations should be made up on the same day they are used.

D. SAMPLE MIXING

In practice, it is not difficult to mix the sample and reagents outside the light-measuring assembly and to obtain an initial reading within 4 or 5 sec. This suffices for all measurements except those in which an unusually rapid decay of luminescence occurs. Errors potentially arising from the decay of luminescence can be minimized by adding the unknown to the assay system at a precise time just prior to measurement.* An alternative method consists of the injection of the sample into the luciferin–luciferase solution while it is already exposed to the monitoring photomultiplier. This involves different difficulties however, for it is not always possible to mix solutions uniformly, particularly when very minute samples are assayed. The procedures outlined here are suitable for most applications if one applies appropriate internal standards and conducts pilot runs prior to the preparation of critical experimental samples for assay.

E. INTERNAL STANDARDS

The utilization of the firefly method of ATP assay is most effective and efficient if, during the pilot runs, internal standards are carried out. Internal standards may be applied at any stage of the extraction or assay procedures. For example, if one wishes to determine the extent to which a given parameter (e.g., boiling) affects ATP concentration, one may add a known aliquot of ATP immediately after putting the sample into the boiling water and compare the ATP

* If a strip chart recorder is used to indicate the time course of luminescence following mixing, it is most convenient to add the last limiting component of the assay system (e.g., firefly enzyme) precisely as the pen crosses a time line on the strip chart. The intensity of light emission at a specified and reproducible interval after mixing can then readily be determined from the chart directly. This procedure tends to minimize variations in measurement introduced by fluctuations in the schedule of operations prior to the reading of light intensity.

remaining after treatment to that present when the same amount of ATP is added at the end of the boiling period. When internal standards are employed, the response in the presence of added ATP should not be more than 50% greater than the response without added standardizing ATP; moreover, the total ATP concentration present must be kept well below the half saturation concentration for the luciferase system. It may be convenient to divide the sample to be assayed into two aliquots. To one may be added a known amount of ATP and to the other, water or buffer in equivalent volume. If the initial luminescence is measured following the addition of a constant aliquot of firefly enzyme, the amount of ATP in the original sample vessels may be computed by a direct proportionality.

F. ADENOSINE TRIPHOSPHATE EQUILIBRIUM ASSAY

a. Principles. The aliquoting methods described above are suitable for the routine determination of the ATP concentration in biological samples. Measurement of the total high-energy phosphate compounds in equilibrium with ATP may be achieved by allowing the reaction to proceed to a steady level and then proceeding to compare this level with the steady level of luminescence produced by the addition of a known increment in ATP to the system.

Equilibrium studies are particularly useful in systems in which a continuously changing level of ATP in equilibrium with other components is to be studied. Examples of this include studies on the formation of ATP photosynthetically by isolated chloroplasts, and measurements of oxidative phosphorylation by suspensions of mitochondria. Clearly, in such cases the instantaneous concentration of ATP is affected both by the rate of ATP synthesis and by the pool of unphosphorylated acceptors present in the mixture. Thus, when large amounts of AMP plus adequate myokinase are present in a reaction mixture, a large portion of the accumulated high-energy phosphate will exist in the form of ADP and therefore not be assayable by the firefly enzyme (which responds only to ATP). Assuming that the myokinase equilibrium constant is close to 1, the total phosphorylation taking place may be calculated provided that the total adenine nucleotides and the total ATP concentrations are known.

The effect of myokinase addition on the kinetics of ATP accumulation on phosphorylating chloroplast preparations may be seen in

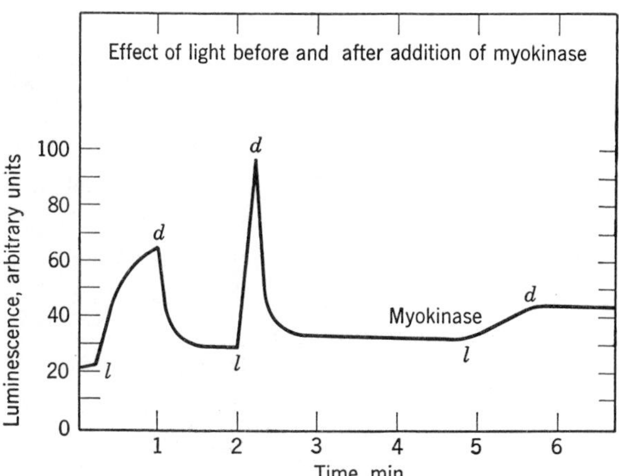

Fig. 32. Use of firefly enzyme to measure photosynthetic phosphorylation. l = illumination, d = dark, myokinase added as indicated. Note absence of spike and overshoot when myokinase equilibrium is permitted to take place. Reaction mixture contained in 2 ml: 1.6 μmoles K_2HPO_4; 17 μmoles NaCl; 3.38 μmoles $MgCl_2$; 19.0 μmoles Tris-HCl pH 8.0; 2 μmoles AMP, 0.1 μmole phenazine methosulfate; 0.2 ml firefly enzyme; 0.2 ml, chloroplast suspension (~700 mg chlorophyl/ml) and 0.1 ml myokinase as indicated.

Figure 32. On the left of the figure, the level of ATP is seen to increase markedly during illumination, and then to decrease rapidly to a new, somewhat higher than original, level in the subsequent dark period. Repeated illuminations cause repeated overshot step increases in the steady level. However, if myokinase is added to the preparation, although the rate of rise of ATP concentration during the illumination period is much reduced, there is no decrease in ATP during the subsequent dark period. This indicates that the myokinase equilibrium is able to keep pace with the rate at which ATP is generated by photosynthetic phosphorylative reactions.

b. Precautions. Although nothing can substitute for common sense in the use of this or any other assay method, there are certain areas where errors may be introduced. It appears provident to be aware of such potential pitfalls before beginning an experiment. Table IV shows a number of such sources of error and the control procedures which we have found to be appropriate.

TABLE IV

Source of error	Control procedure
Precipitation of buffer or activator	Use of arsenate buffer
Oxygen disappearance	Aeration of sample (continuous or intermittent)
Depression of luminescence by negative ions	Internal standard
Effects of turbidity	Internal standard
Absorption of a portion of the emitted light by included pigment	Internal standard
ATP binding	High buffer concentration[a]
Other sources of luminescence	Destroy luminescence of test material or use filters if the color of the second luminescence is different from that of the firefly
Apyrase activity of test material	Inhibit apyrase or destroy by heating

[a] See Figure 61 (page 178) which shows the effect of phosphate and arsenate concentration on the decay of luminescence.

G. ASSAY OF OTHER SUBSTRATES

a. Adenosine Diphosphate

$$\text{ADP} + \text{ADP} \rightleftarrows \text{AMP} + \text{ATP} \tag{6}$$

$$\text{ATP} + \text{LH}_2 \rightarrow \text{luminescence} \tag{7}$$

The concentration of adenosine diphosphate may also be determined with the firefly enzyme. The time course of the response of this system to added ADP is shown in Figure 33. Since the firefly enzyme responds only to adenosine triphosphate, methods for the evaluation of ADP and AMP in a mixture must be based on an indirect evaluation of the ADP concentration. Two methods can be used in principle. The simplest of these is first to assay an aliquot for ATP as outlined above and then to convert the ATP in another aliquot to ADP by treatment with crayfish ATPase. The initial rate of increase of luminescence when such a pretreated system is assayed with firefly enzyme gives a measure of ATP plus ADP in the original mixture.

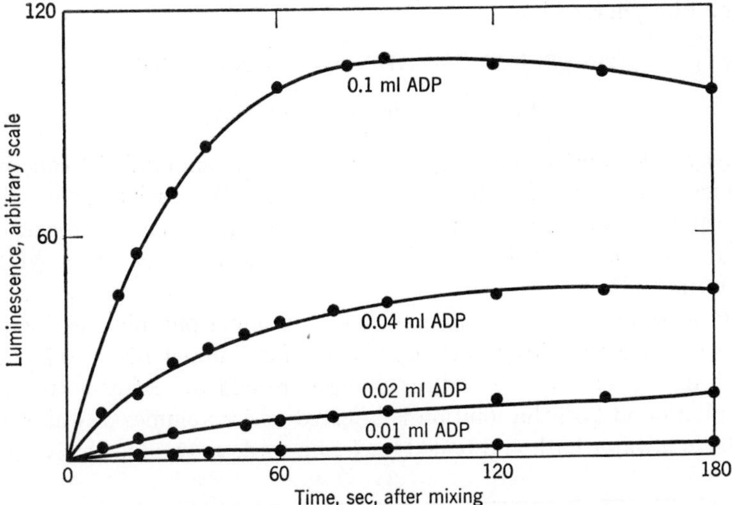

Fig. 33. Effect of added ADP ([ADP] = 202 μg~P/ml) on the time course of firefly enzyme extract luminescence. ADP concentrations indicated were added to 0.2 ml of $\frac{1}{15}M$ arsenate buffer pH 7.4 and 0.2 ml firefly enzyme (see Section IV-1-C-a) in a volume of 0.8 ml. The light produced was measured with a Farrand photofluorometer.

Alternatively, the ADP and AMP concentration can be calculated from three other measurable variables: (*1*) initial ATP concentration, (*2*) ATP concentration after myokinase equilibrium, and (*3*) total adenine nucleotides (measured by converting ADP and AMP to ATP (treatment with phosphocreatine and creative kinase). If small amounts of ADP are to be measured in the presence of very large quantities of ATP or AMP, it is desirable to perform a preliminary separation of the mixture using the ion-exchange method of Cohn (59).*

* This method employs a strong quaternary ammonium anion exchange resin (Dowex 1) and the procedure is as follows: The exchanger is suspended in distilled water and poured into a column 1 cm in cross section to a height of 2 cm. The resin is converted into the chloride form by passing through 10 ml 3N HCl followed by H_2O to neutrality. The sample is adjusted to pH 8–9 and applied to the column. Anion concentration of the sample should be less than 0.01N. Following a wash with 25 ml H_2O the nucleotides are eluted as follows: P_i, adenosine and adenine, 100 ml 0.024M NH_4Cl; AMP, 100 ml 0.003M HCl; ADP, 100 ml 0.01N HCl–0.02N NaCl; ATP, 100 ml 0.01N HCl–0.2N NaCl.

b. Phosphocreatine

$$2Cr{\sim}P + AMP \text{ (or ADP)} \xrightleftharpoons{\text{Cr-Kinase}} 2Cr + ATP \tag{8}$$

$$ATP + \text{luciferin luciferase} \rightarrow \text{light} \tag{9}$$

Any substance capable of producing or consuming ATP can, in principle, be assayed with the firefly enzyme. One such substance is phosphocreatine, which in the presence of the enzyme, creatine kinase, is capable of donating its phosphate group to AMP or ADP to form ATP.

Procedure: 1 cc of a preparation containing phosphocreatine is diluted 2:1 with boiling water and heated for 5 min at 100°. 0.05 ml each of lobster ATPase and myokinase are added. After a 10-min incubation at 25°, the solution is injected into 2 volumes of water at 100° in a water bath and boiled for 5 min to destroy added enzymes.

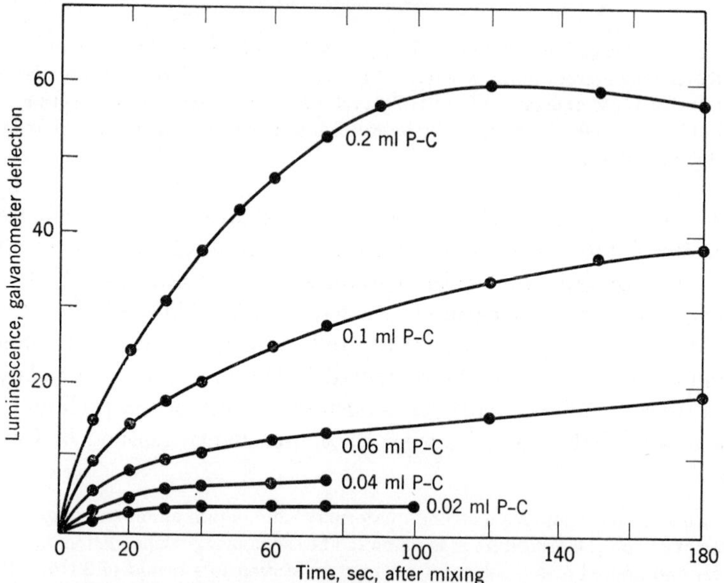

Fig. 34. Effect of concentration of phosphocreatine added on the time course of luminescence of the firefly extract 30 sec following mixing. The reaction mixture contains 0.2 ml firefly enzyme (see Section IV-1-C-a), 0.2 ml 0.1*M* arsenate buffer pH 7.4, 1 mg AMP, 0.02 ml creatine kinase and phosphocreatine (103 μg ${\sim}$P/ml) in a total volume of 1 ml.

To 1.5 cc of this solution add 0.2 μM AMP and 0.1 cc firefly enzyme (see Section IV-1-C-c) and at zero time 0.1 cc of the creatine kinase enzyme. The light level after 10 sec is proportional to Cr$\sim$P concentration. A typical response curve to added Cr$\sim$P is shown in Figures 34 and 35.

c. Total Creatine. This may be assayed through an adaptation of the above reactions:

$$\text{Cr} + \text{ATP} \xrightleftharpoons{\text{Cr-kinase}} \text{Cr}\sim\text{P} + \text{ADP} \tag{10}$$

heat, then:

$$\text{ADP} + \text{myokinase} + \text{ATPase} \rightarrow \text{AMP} \tag{11}$$

heat, then:

$$\text{Cr}\sim\text{P} + \text{AMP} \xrightleftharpoons{\text{Cr-kinase}} \text{ATP} \tag{12}$$

Procedure: To 1 cc of a solution (pH 7.4 2mM $MgSO_4$) containing less than 50 γ creatine (+Cr$\sim$P), add 0.05 ml creatine kinase plus 500 γ ATP. After incubation for 10 min, inject into two volumes of distilled boiling water and heat for 5 min. Treat with 0.2 ml myokinase plus lobster ATPase (0.2 ml) for ½ hr. Heat to 100° for 15 min, add 0.1 ml firefly enzyme (Section IV-1-C-c), and at zero time

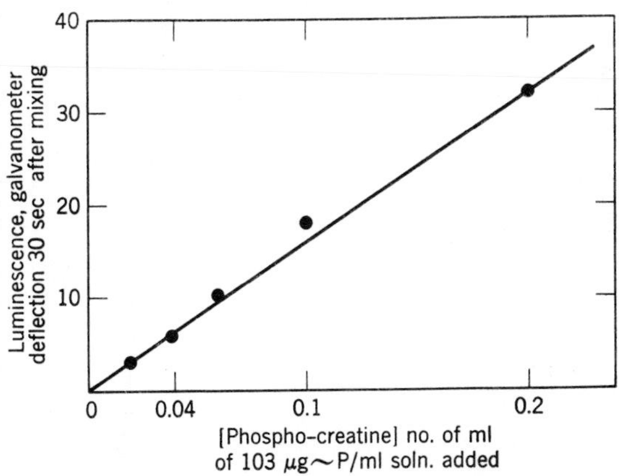

Fig. 35. Luminescence 30 sec following addition of different amounts of phosphocreatine to firefly assay system with added creatine-kinase. The reaction mixture contains 0.2 ml firefly enzyme (see Section IV-1-C-a) 0.2 ml 0.1M arsenate buffer pH 7.4, 1 mg AMP, 0.02 ml creatine kinase and phosphocreatine (103 μg $\sim$P/ml) in a total volume of 1 ml.

0.1 ml creatine kinase. Measure luminescence after 10 sec and compare to standard curve.

d. Glucose

$$\text{Glucose} + \text{ATP} \xrightleftharpoons[\text{myokinase}]{\text{hexokinase or glucokinase}} \text{G6P} + \text{AMP} \quad (13)$$

The assay of this or any other carbohydrate capable of being phosphorylated by ATP in the presence of an appropriate enzyme, is based on the utilization of ATP during the reaction, and the assay of the ATP remaining untitrated with the firefly enzyme.

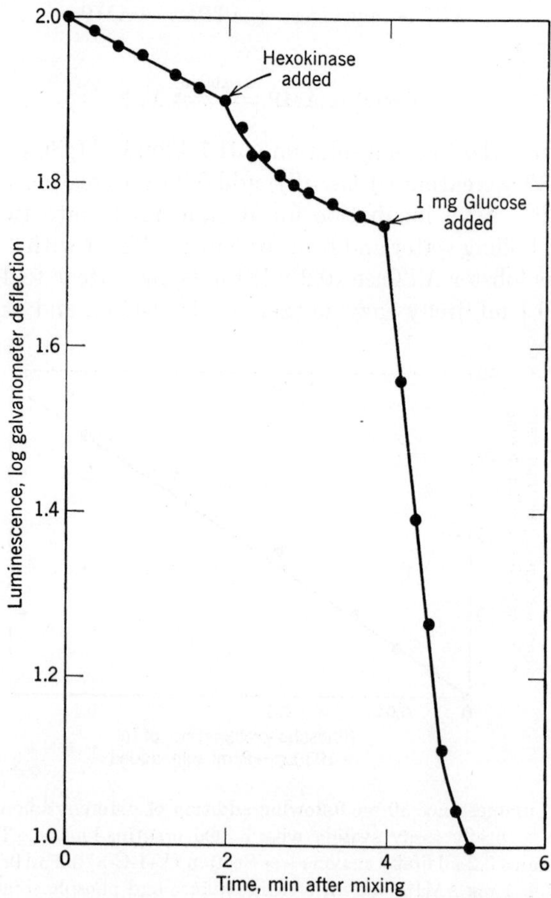

Fig. 36. Effect of hexokinase and of glucose addition on the luminescence of 1 ml firefly enzyme (see Section IV-1-C-a) supplemented with 0.75 mg ATP/ml.

TABLE V[a]

Assay of Glucose with Firefly Enzyme–Hexokinase Preparation

Glucose added, μg	Galvanometer reading[b]	Difference
0	78.0	
2	66.5	11.5
4	54.0	24
6	43.5	34.5
8	31	47
10	19	59

[a] From ref. 5.

[b] One minute after mixing 0.1 ml of water or glucose with 0.7 ml of mixture containing 0.1 ml of firefly enzyme and 0.2 ml of crude hexokinase solution plus 60 μg, ATP. Galvanometer previously set at 100.

A typical result is shown in Figure 36. Amounts of glucose in the range of 2–100 μg may readily be assayed by this method (Table V).

e. Nucleotide Triphosphates. In the presence of sufficient AMP ($10^{-4}M$) crude firefly lantern extracts (Section IV-1-C-c) which contain the appropriate kinases will respond to guanosine triphosphate (GTP), cytidine triphosphate (CTP), or uridine triphosphate (UTP). The rates of these reactions versus concentration of the various nucleotide triphosphates are shown in Figures 44 and 45.

f. Phospho(enol)pyruvic Acid. Firefly lantern extracts do not respond to added phospho(enol)pyruvic (PEP) acid in the presence of added AMP. However, if pyruvic kinase is added, the ATP formed can be assayed readily by the firefly enzyme system. This is shown in Figure 48 in which the response to different added concentrations of PEP is depicted.

H. OTHER REACTIONS

Among the substrates which, in principle, should be assayable with firefly system are pyrophosphate, DPN, and "charged" transfer RNA.

I. ASSAY OF ENZYMES

The firefly enzyme may be employed in the assay of any component which can affect the concentration of ATP. In the preceding we have dealt with substrates which, through the agency of appropriate enzymes, increase or decrease ATP concentration. In general, the determination of a given component requires that it be the limiting reactant, i.e., that other reactants either are present in effective excess, or remain constant during reaction, and that some reproducible relationship exists between the concentration of the unknown and the light emitted by the firefly enzyme. As with substrate determination, the reaction under study may, in the case of enzyme measurement, be assayed either by continuously monitoring the light produced by the firefly enzyme in response to changes in ATP concentration, or, alternatively, the change in ATP level in an incubation mixture may be measured by successively assaying aliquots for their ATP concentration. In either case, it is generally desirable that the enzymes being assayed be saturated with their substrates during the reaction.

The mixed type (luciferase plus assayed enzyme) of reaction is, of course, not appropriate when the component measured is also present in the firefly-enzyme extract in comparable or higher amounts than is present in the unknown. Since the crude firefly enzyme, most conveniently prepared, contains substantial amounts of ATPase, myokinase, pyrophosphatase, and other enzymes, appropriate controls or adaptations of procedures are necessary in special instances.

Examples of enzymes which have been successfully assayed are the following.

a. Apyrase

$$\text{Reaction: ATP} + 2H_2O \xrightleftharpoons{\text{apyrase}} \text{AMP} + 2H_3PO_4 \qquad (14)$$

Provided that the apyrase activity in the unknown is substantially greater than the ATPase of the firefly enzyme, the reaction may be carried out in a mixed system for potato apyrase.

b. ATPase

$$\text{Reaction: ATP} + H_2O \xrightleftharpoons{\text{ATPase}} \text{ADP} + H_3PO_4 \qquad (15)$$

ATPase such as that present in the muscle of the lobster *Homarus americanus*, may most conveniently be assayed by an aliquoting method.

c. Myokinase (Adenylic Kinase)

$$\text{Reaction: ATP} + \text{AMP} \xrightleftharpoons{\text{myokinase}} 2\text{ ADP} \qquad (16)$$

This enzyme may be assayed by measuring either the forward or reverse reaction. The latter procedure is usually preferable since it

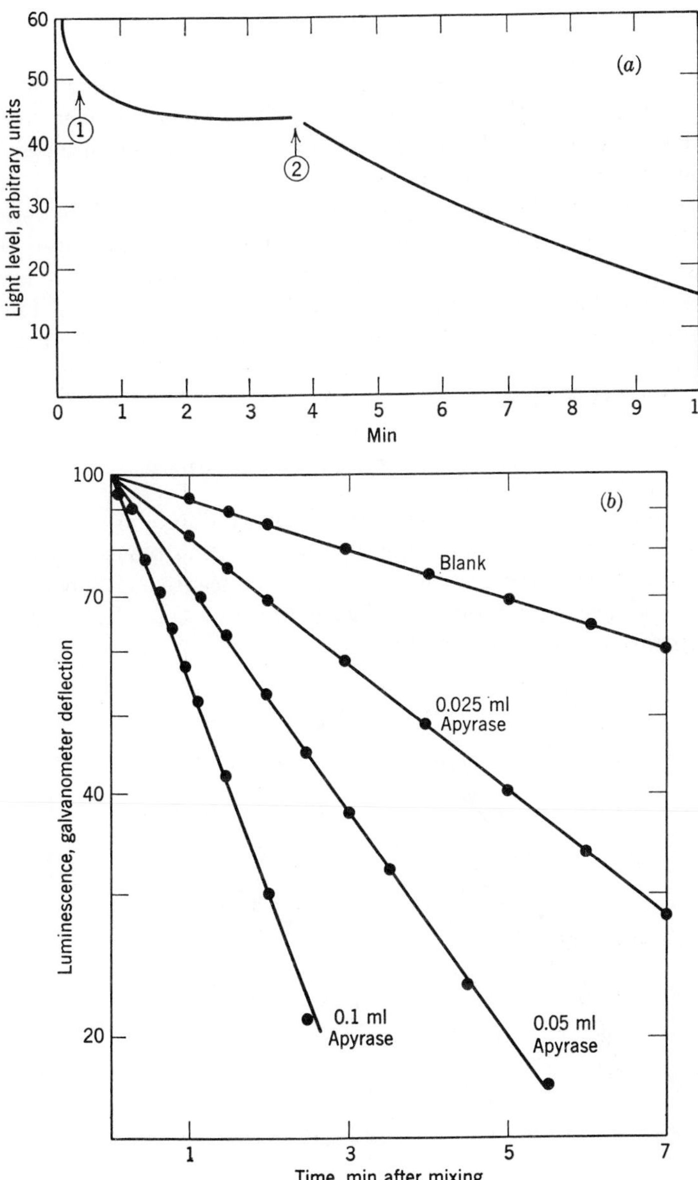

Fig. 37. (*a*) Time course of ATP hydrolysis by apyrase measured in an apyrase-firefly enzyme coupled system. At (*1*) 0.01 μmole ATP added to firefly enzyme; at (*2*) 0.01 ml potato apyrase (5 mg/ml) added to system. (*b*) Effect of various amounts of potato apyrase added to the system described in Figure 37*a*.

permits the use of a high substrate concentration and a high initial rate of reaction. Figure 38 illustrates the response of the firefly enzyme to ADP as well as the time course of the reaction when increments of myokinase are added.

When very low levels of myokinase are to be assayed, it is preferable to use an aliquoting method to measure reaction kinetics. Alternatively, the firefly enzyme may be purified so as to free it of myokinase, but this involves expensive inefficiencies in the use of firefly lanterns, since the recovery of luciferase activity is low during purification.

d. Creatine Kinase.

$$\text{Reaction: Cr}{\sim}\text{P} + \text{ADP} \overset{\text{(AMP)}}{\rightleftharpoons} \text{ATP} + \text{creatine} \qquad (17)$$

This is an enzyme which is very easily assayed in mixture with the firefly enzyme (which does not contain creatine kinase activity). Figure 39 shows the rate of increase of luminescence as a function of creatine-kinase concentration. Figure 40 shows the effect of AMP concentration on the reaction velocity.

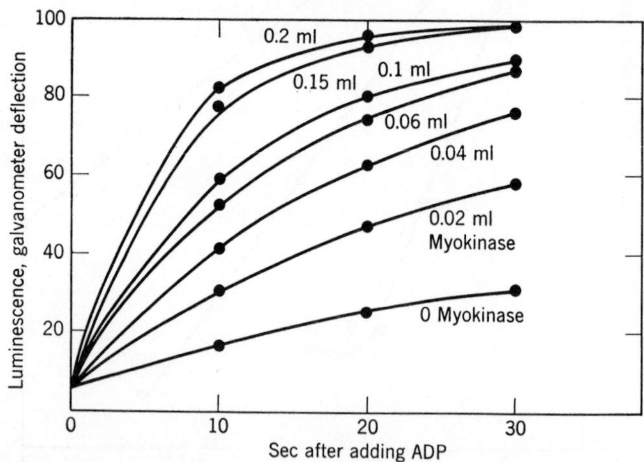

Fig. 38. The effect of concentration of added myokinase on the rate of increase of luminescence in the presence of 2.5 μg ADP/ml added to 5 ml $\frac{1}{15}M$ arsenate buffer pH 7.4 and 1 ml firefly enzyme (see Section IV-1-C-a).

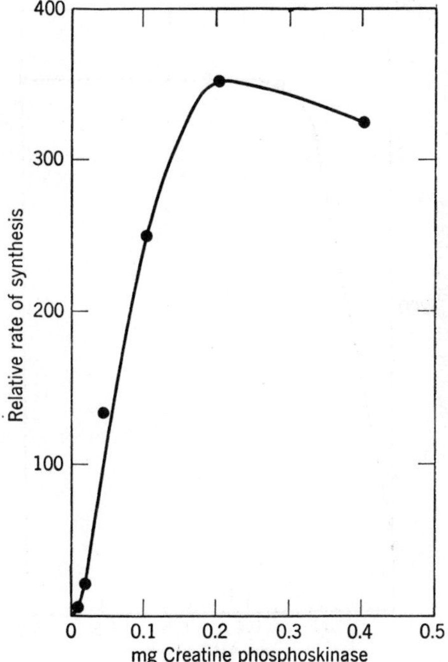

Fig. 39. Effect of creatine kinase on rate of increase of luminescence. Different amounts of creatine kinase added to a medium containing 0.40 ml 0.05*M* arsenate buffer pH 7.3; 0.02 ml 0.1*M* $MgSO_4$; 0.02 ml FF; 0.8 mg AMP, 20 μg phosphocreatine in a final volume of 1 ml. Light output measured with system illustrated in Figure 2.

It should be noted that the reaction appears to proceed with either AMP or ADP as acceptor, but that the former is a preferable substrate because the myokinase activity present in the firefly extract will produce ATP from added ADP. The reaction increases in velocity with time. This increase in rate is probably due to the accumulation of ADP as an additional, more effective acceptor. The level of creatine kinase in developing chick muscle has been determined using this assay and the results of such an application are shown in Figure 42.

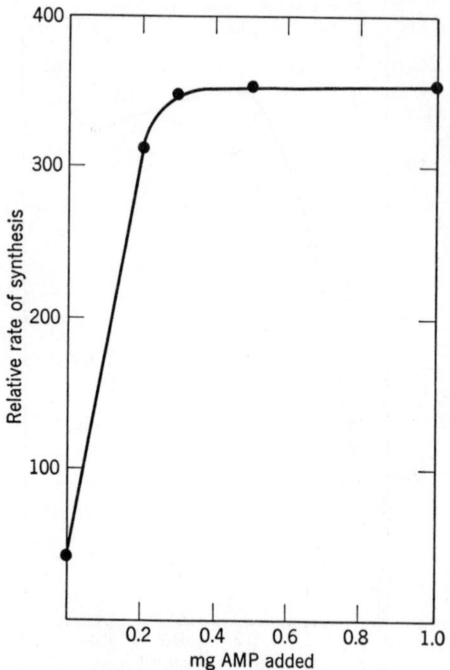

Fig. 40. Effect of AMP concentration on the rate of increase of luminescence by firefly enzyme in presence of phosphocreatine and creatine kinase. Experimental conditions as those listed in legend of Figure 39.

e. Nucleotide Phosphokinases.

$$\text{Reaction: } \begin{matrix} \text{GTP} \\ \text{or CTP} \\ \text{or UTP} \end{matrix} + \text{AMP} \underset{\text{Phosphokinase}}{\overset{\text{nucleotide}}{\rightleftharpoons}} \text{ATP} + \text{NMP} \qquad (18)$$

This important group of reactions is catalyzed by the enzyme(s), nucleotide phosphokinase, present in various biological materials. The evaluation of the concentration of the enzyme may be achieved by coupling the utilization of UTP, GTP, and CTP (which are ineffective in the firefly luminescent reaction) to the formation of ATP from AMP. A typical time course of luminescence assay of this enzymatic activity is shown in Figure 43. Interestingly, there is sufficient nucleotide phosphokinase in lantern extract to permit the undiluted crude system to respond to the trinucleotides in the presence of AMP. Figures 44 and 45 show effect of concentration of CTP and GTP on the rate of increase of luminescence.

f. Hexokinase.

$$ATP + C_6H_{12}O_6 \rightleftharpoons ADP + C_6H_{11}O_6PO_3H_2 \quad (19)$$

This reaction depends on the utilization of ATP (as described above) during phosphorylation of a hexose. The kinetics of ATP utilization following glucose addition was shown in Figure 36, at various concentrations of hexokinase. The relative slopes of the curves versus enzyme concentration are shown in Figure 46. Note the break in the curve at high hexokinase concentration. This discontinuity probably indicates that the rate of the myokinase reaction limits the overall reaction rate when the ATP level is low.

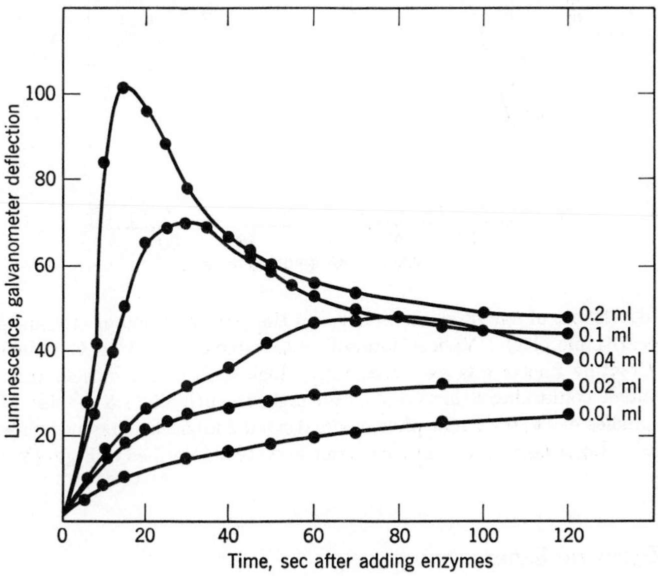

Fig. 41. Kinetics of ATP synthesis when various concentrations of creatine kinase are added to the coupled system containing firefly enzyme, phosphocreatine, and AMP in $\frac{1}{15}M$ arsenate buffer.

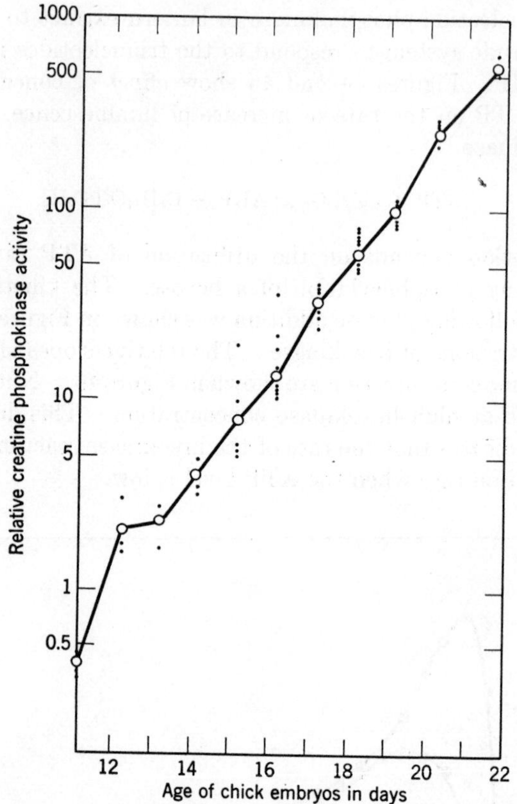

Fig. 42. Effect of age of chick embryo on the concentration of creatine kinase (semilogarithmic plot). Values normalized to value obtained on the nineteenth day. Creatine kinase was extracted from chick embryo leg muscle and added to a system containing 8 μmoles sodium arsenate pH 8.5, 1.5 μmoles $MgSO_4$, 0.0045 μmoles AMP, 0.1 μmole phosphocreatine, 0.2 ml firefly enzyme in a volume of 0.8 ml. Light output was measured in the system described in Figure 2. (From ref. 10.)

g. Pyruvate Kinase.

$$\text{Reaction: PEP} + \text{AMP (ADP)} \rightleftharpoons \text{ATP} + \text{pyruvate} \quad (20)$$

The effect of pyruvate kinase on the time course of luminescence is shown in Figure 47. The effect of substrate and AMP concentrations added are shown in Figures 48 and 49.

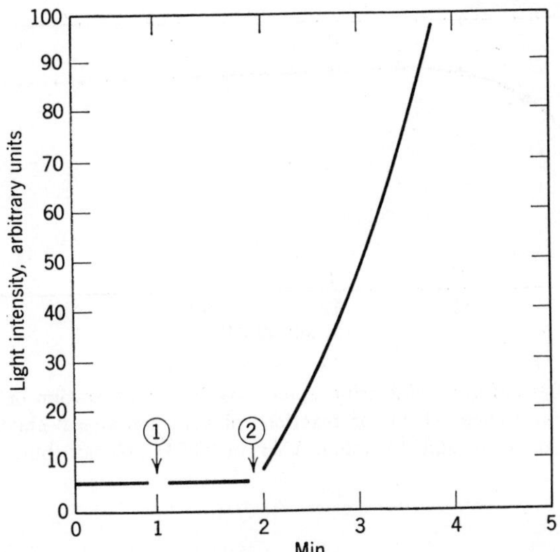

Fig. 43. Effect of addition of crude nucleotide phosphokinase on light emission by 0.1 ml firefly enzyme in presence of 0.5 μmole cytidine triphosphate (CTP) (see Section IV-1-C-c), 0.1 μmole AMP, and 2 μmoles $MgSO_4$ in 0.05*M* arsenate buffer pH 7.4 (total volume of 1 ml). At (*1*) CTP was added and at (*2*) the enzyme was added (0.1 ml).

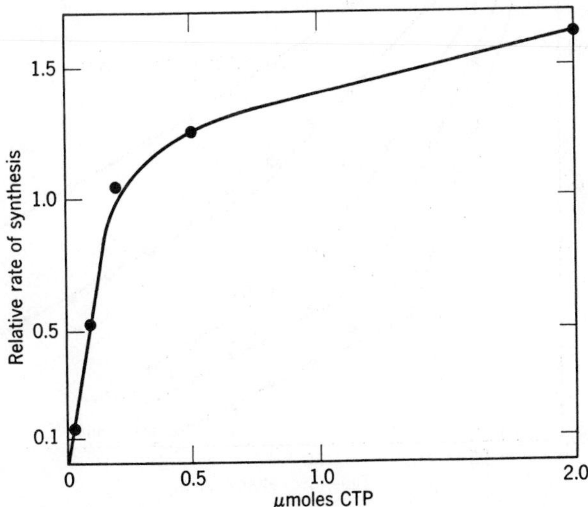

Fig. 44. Effect of cytidine triphosphate concentration on rate of increase in luminescence of firefly enzyme. 1 ml of reaction mixture contains 2 μmoles $MgSO_4$, 0.1 ml firefly enzyme, and 0.1 μmole AMP on 0.05*M* arsenate buffer pH 7.4.

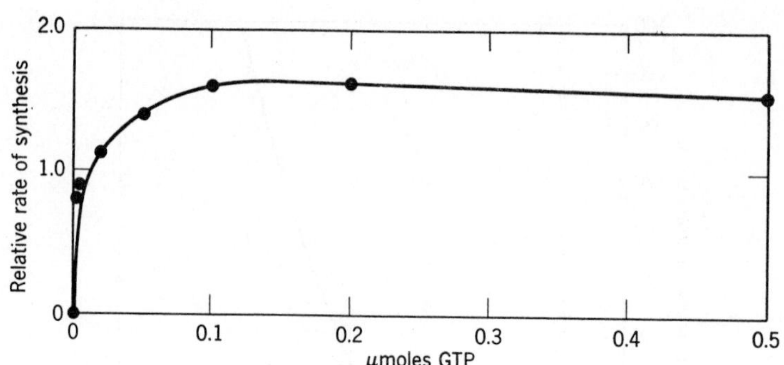

Fig. 45. Effect of guanosine triphosphate (GTP) concentration on rate of increase of luminescence. 1 ml of reaction mixture contains 2 μmoles $MgSO_4$, 0.1 ml firefly enzyme, and 0.1 μmole AMP in 0.05*M* arsenate buffer pH 7.4.

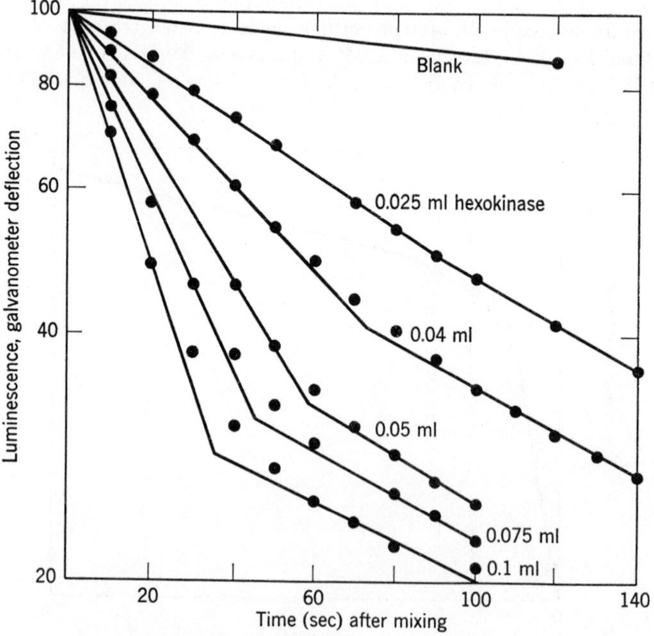

Fig. 46. Effect of concentration on hexokinase at rate of disappearance of luminescence in the presence of 140 μg ATP, 1 mg glucose and 0.2 ml firefly enzyme in $\frac{1}{15}M$ arsenate buffer pH 7.4.

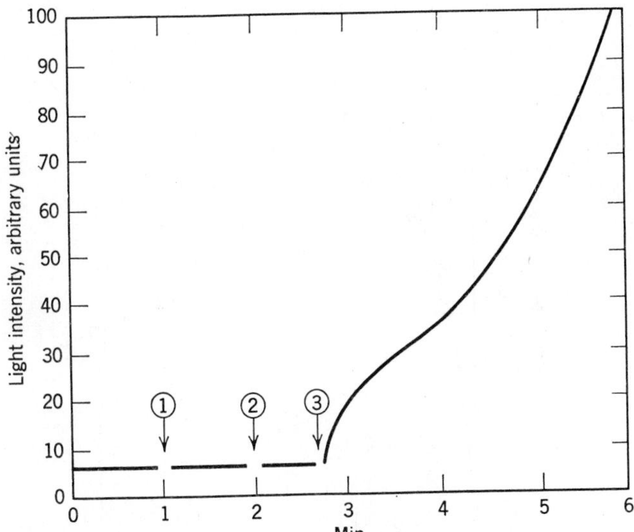

Fig. 47. Effect of 2 μl pyruvate kinase (*3*) addition on time course of luminescence by 0.02 ml of firefly enzyme in presence of 2 μmoles AMP (*2*) and 1 μmole phospho(enol)pyruvate (*1*).

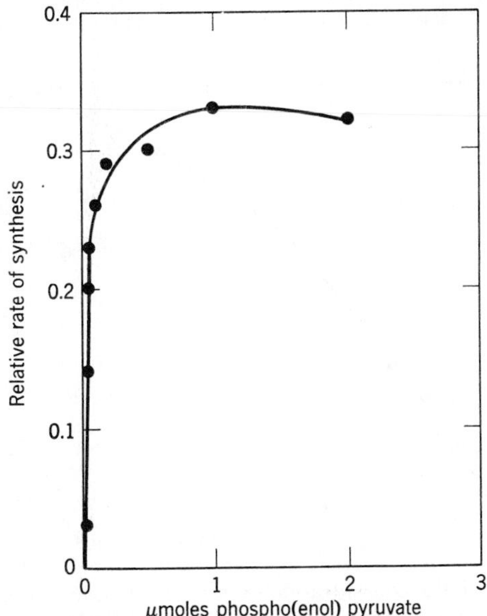

Fig. 48. Effect of added concentration of phospho(enol)pyruvate on rate of increase in luminescence of firefly enzyme (0.02 ml) in the presence of a constant amount of AMP (1 μmole) and pyruvic kinase (2 μl).

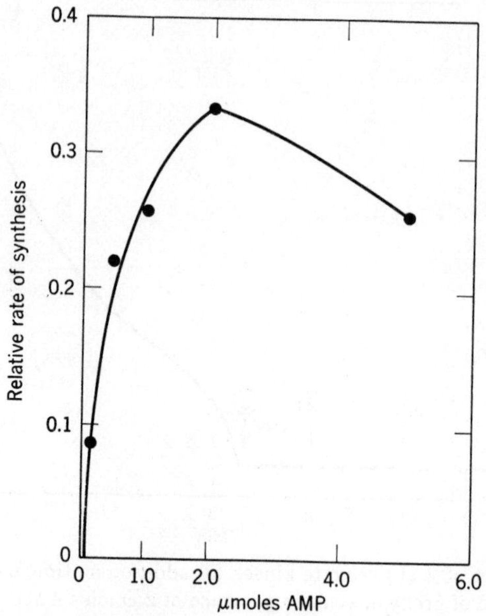

Fig. 49. Effect of AMP concentration on the rate of increase of luminescence in the presence of 2 μl pyruvate kinase, and 0.1 μmole phospho(enol) pyruvate.

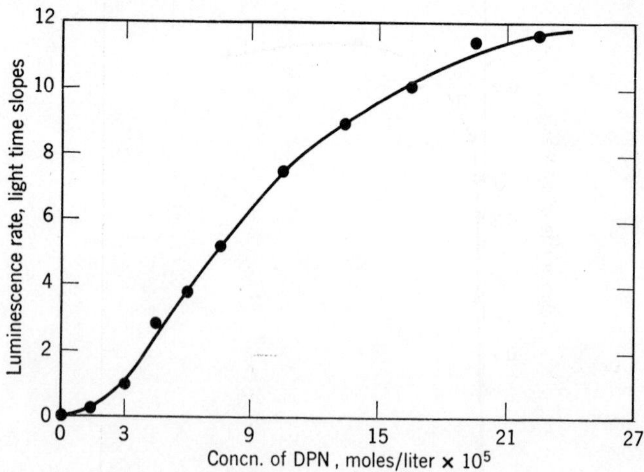

Fig. 50. Effect of DPN concentration on rate of luminescence on luminous bacterial extract. Refer to legend of Figure 22.

2. Bioassay with the Bacterial Luciferin–Luciferase System

Various components necessary for the luminescence of extracts of luminous bacteria may be assayed through appropriate coupling reactions. Of these reactions, the evaluation of DPN, DPNH, FMN, aldehydes, and oxygen are most easily achieved, although any reaction which yields reduced flavin or produces or consumes long-chain aldehydes or oxygen may also be measured under suitable conditions.

A. DPN, DPNH, FMN

The responses of extracts of luminous bacteria to added DPN, DPNH, and FMN are shown in Figures 50, 51, and 24, respectively.

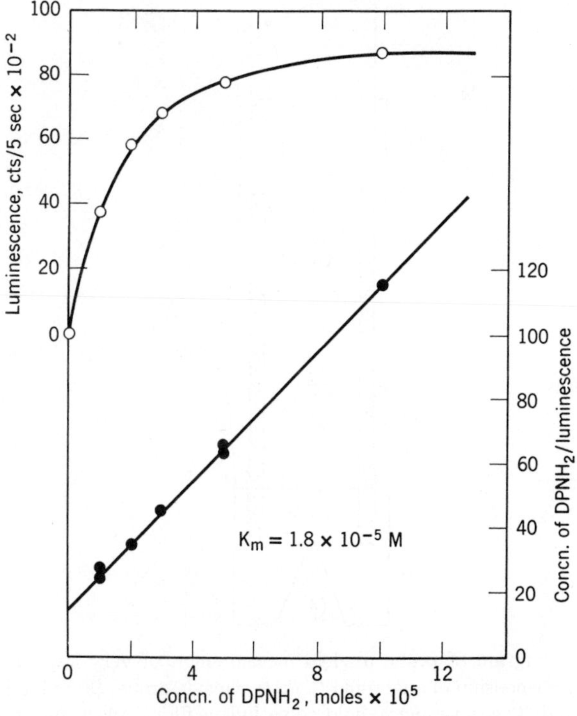

Fig. 51. Effect of DPNH concentration on relative rate of luminescence in system described in legend of Figure 22.

B. LONG-CHAIN ALDEHYDES

Figure 23 illustrates the response of the luminescent reaction to added plasmal (palmitic aldehyde). Half saturation for the system is at $3.5 \times 10^{-6}M$. Various aliphatic aldehydes of chain length greater than C_6 produce the response.

C. OXYGEN DETERMINATION

Intact luminous bacteria may be employed in the measurement of O_2 consumption by respiring enzyme systems. The O_2 probe illus-

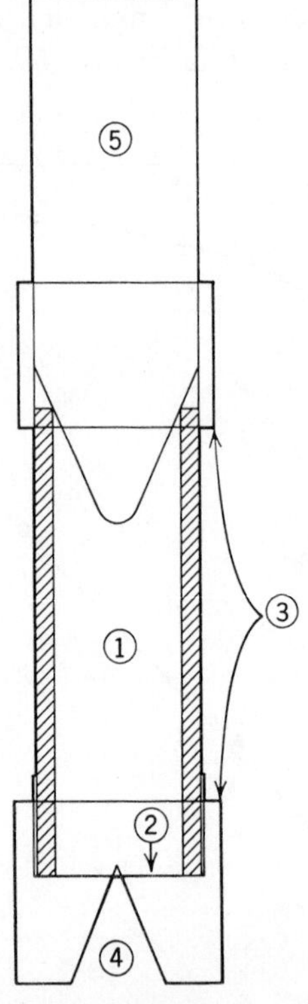

Fig. 52. Diagram of probe used in measurement of very low concentrations of O_2. (*1*) Suspension of anaerobic *P. fischeri* inside probe; (*2*) polyethylene film; (*3*) 5–mm o.d. Tygon tubing to hold polyethylene film in place and probe to glass rod handle (*5*). A notch is cut (*4*) so as to allow trapped air bubbles to escape when the probe is immersed. O_2 diffusing through film causing the bacteria to luminesce. The light emitted is measured in the device described in Figure 2.

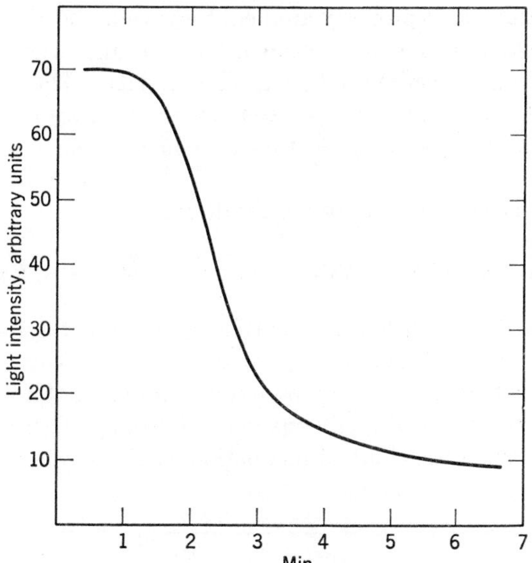

Fig. 53. Mitochondrial respiration as measured with the O_2 probe shown in Figure 52. Very dilute suspension of rat liver mitochondria in a 0.25*M* sucrose medium containing 0.2 μmole AMP and 1 μmole α-keto glutarate. The medium was purged with N_2 to lower the O_2 concentration prior to introducing the probe. (In absence of mitochondria no decrease in luminescence occurs under these conditions.)

trated in Figure 52, was designed for this purpose and the utilization of oxygen by a mitochondrial suspension is illustrated in Figure 53.

3. Complex Systems

A. ASSAY OF ENZYMES AND SUBSTRATES WITHIN HISTOLOGICAL SECTIONS

Specific reactions which produced colored products may be utilized to localize enzymes and reactive substances within histological sections. Many such reactions have been developed in the last decades, but a majority of them utilize substrates which are very different from the naturally occurring substrates. It would, presumably, be advantageous if such reactions could be detected through reactions involving the natural substrates of the enzymes involved. The sensitivity and specificity of bioluminescent and chemiluminescent reactions suggests that properly applied, they would be of considerable aid in the detection and localization of catalysts and reaction products within histological sections of biological materials. The development of such methods is limited at present by the absence of methods of localizing dim luminescences at the subcellular level.

B. MITOCHONDRIAL OXIDATIVE PHOSPHORYLATION

$$\text{Reaction: } XH_2 + \tfrac{1}{2}O_2 + AMP + 2P_i \xrightleftharpoons{\text{mitochondria}} ATP + H_2O + X \quad (21)$$

As shown by the author and Totter (5) in 1951, the formation of ATP during the oxidation of respiratory intermediates by mitochondria can be detected in a reaction mixture consisting of mitochondria, acceptor, substrate, and firefly enzyme. Recently, Batra and Strehler (12) examined the effect of incubation conditions on this reaction

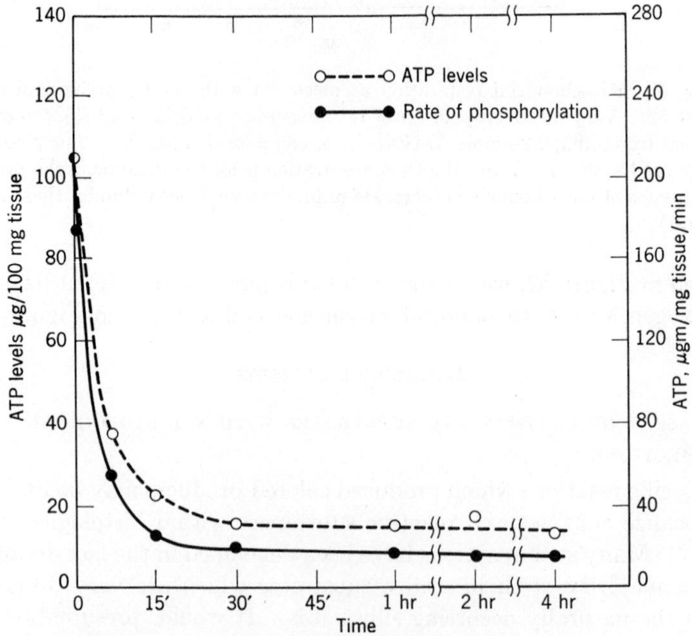

Fig. 54. The decline of ATP levels and rate of synthesis in mousc heart anaerobically incubated as measured with the firefly method (12).

and applied this information to the study of oxidative phosphorylation by mitochondria obtained from heart muscle damaged by ischemia for various periods of time. The effect of anaerobic incubation on both ATP level and capacity to carry out phosphorylation is shown in Figure 54. In the course of this study a striking reversible effect of the concentration of mitochondria on coupled phosphorylation was discovered. This dilution effect is immediately and completely reversed upon addition of serum albumin (1 mg/ml).

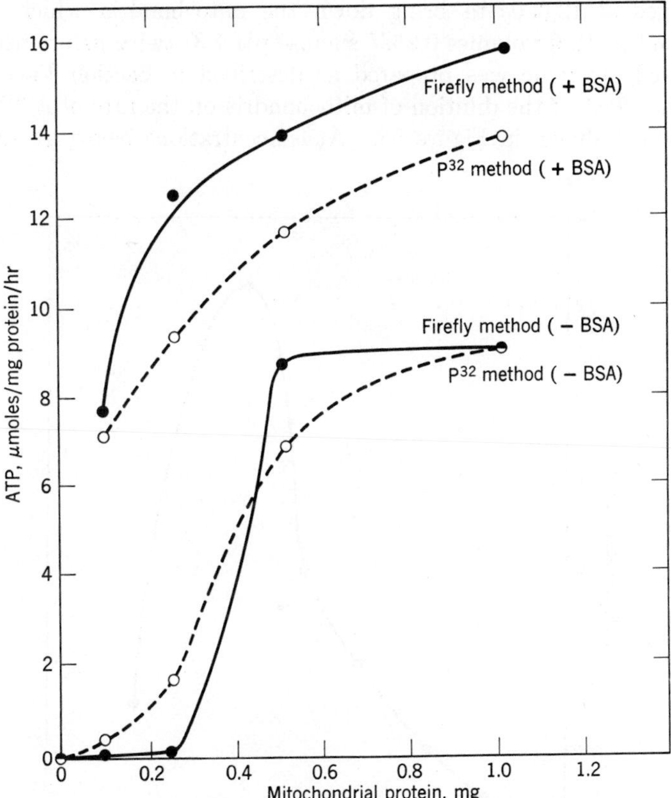

Fig. 55. The mitochondrial dilution phenomena as measured by the firefly method and confirmed with tracer techniques. In a 0.25M sucrose medium pH 7.3, containing 2 μmoles AMP and α-ketoglutarate, 4 μmoles Mg^{2+} and 0.1 ml firefly enzyme, mitochondria were added (final volume of 1.4 ml).

The mitochondria used in these studies, conducted jointly with Dr. Batra and Mr. Nordgren were prepared as follows:

Mature, healthy mice were sacrificed, the hearts quickly removed, anaerobically incubated for the desired time, and then placed in ice cold 0.25*M* sucrose pH 7.3. The tissue was washed, weighed after moval of extraneous tissue and was finely minced with scissors. It was then homogenized in a Potter homogenizer with nine volumes of 0.25*M* sucrose pH 7.3. The homogenate was centrifuged at 3000*g* for 5 min and the supernatant decanted. This supernatant was centrifuged at 12,000*g* to bring down the mitochondria which were washed (with 9 volumes 0.25*M* sucrose pH 7.3) twice before use.

Firefly enzyme was prepared as described in Section IV-1-C-c.

The effect of the dilution of mitochondria on the rate of ATP formation is shown in Figure 55. At concentrations below about 0.1

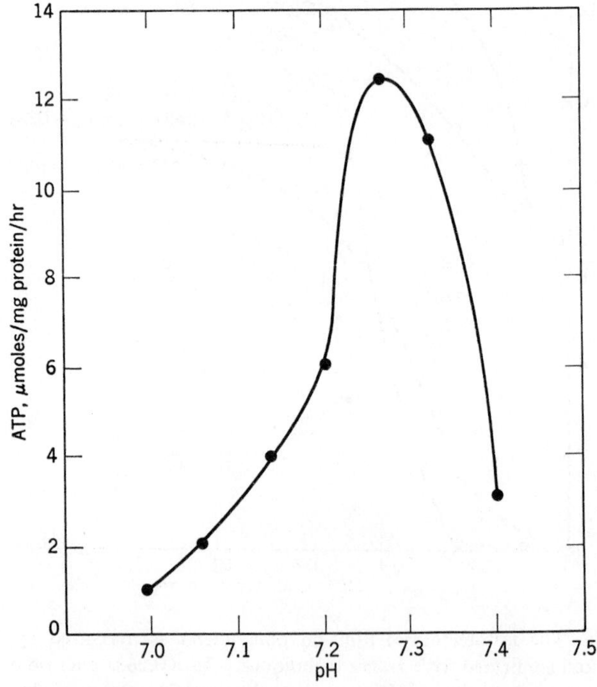

Fig. 56. Effect of pH on the assay described in legend of Figure 55.

mg mitochondrial protein per milliliter the reaction is instantaneously but reversibly inhibited (see legend for details of procedure). The effect of a variety of environmental variables has been examined; the effects of pH, firefly enzyme concentration, and substrate concentration are illustrated in Figures 56–58.

When serum albumin is added to a dilute mitochondrial preparation, there is a rapid initial synthesis of ATP, followed by a decline in synthetic rate to a somewhat lower steady-state level. This is shown in Figure 59.

C. PHOTOSYNTHETIC PHOSPHORYLATION

The firefly enzyme was employed in the assay of ATP synthesis by green plants in 1952 (50). Extracts of suspension of *Chlorella pyre-*

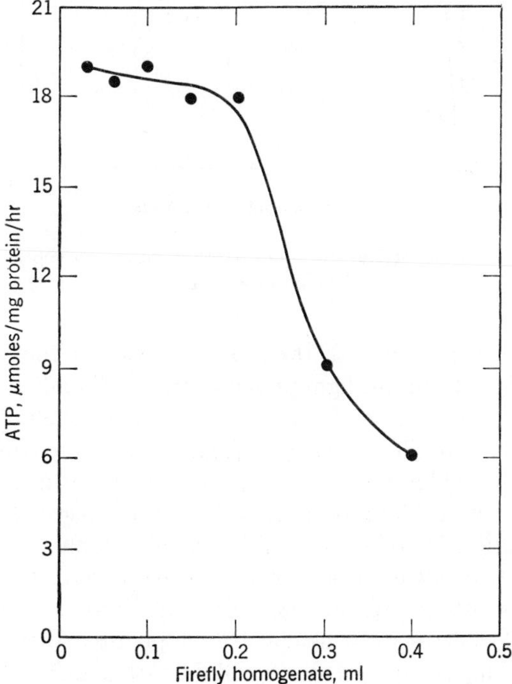

Fig. 57. Effect of firefly enzyme concentration on the assay described in legend of Figure 55.

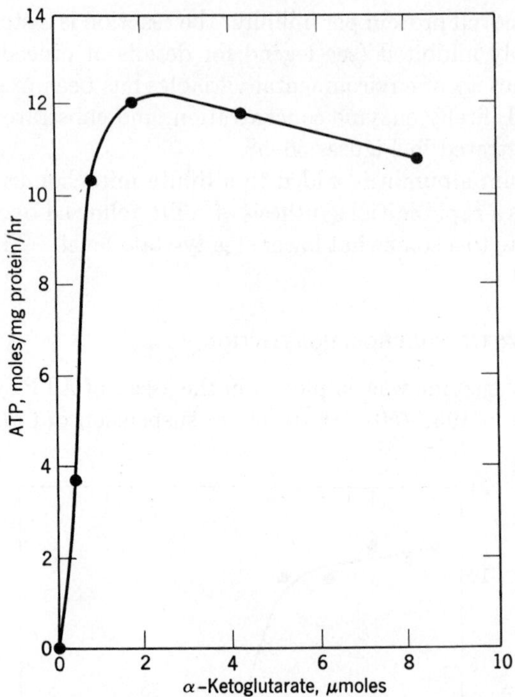

Fig. 58. Effect of α-ketoglutarate concentration added on the assay described in legend of Figure 55.

noidosa were measured with the quantum counter following exposures of the algae to actinic light under a variety of conditions. Figure 60 shows the time course of synthesis of ATP by such algal suspensions in response to illumination. The assay of photosynthetic phosphorylation by chloroplasts was not demonstrable in a mixture of firefly enzyme and chloroplasts (from Pokeweed) because the purified enzyme (Section IV-1-C-b) contained small amounts of $(NH_4)_2SO_4$, a substance which was subsequently shown by Jagendorf (60) to inhibit photosynthetic phosphorylation. In 1961, the author and Hendley (51) successfully demonstrated photosynthetic phosphorylation in a mixed system during illumination using the specially designed apparatus described in Figure 9.

The time course of ATP concentration during and following illumination was illustrated in Figure 32. Addition of myokinase to the

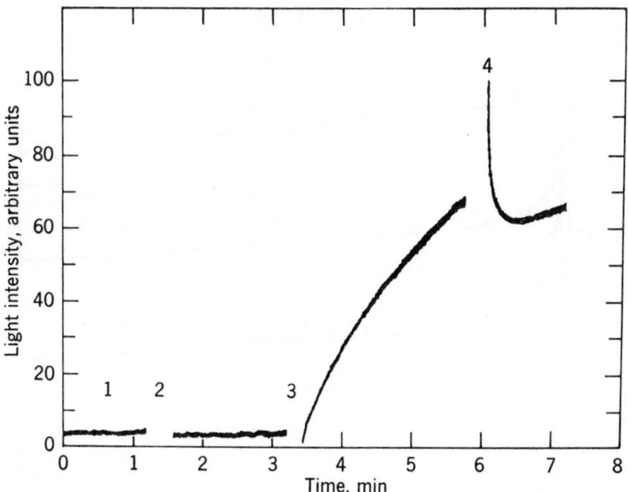

Fig. 59. Typical time course of mitochondrial ATP synthesis assay and the dramatic effect of BSA. (*1*) Basic assay medium containing mitochondria described in legend of Figure 55; (*2*) substrate added; (*3*) 2 mk BSA added; (*4*) ATP standard added (sensitivity of recorder cut 66%).

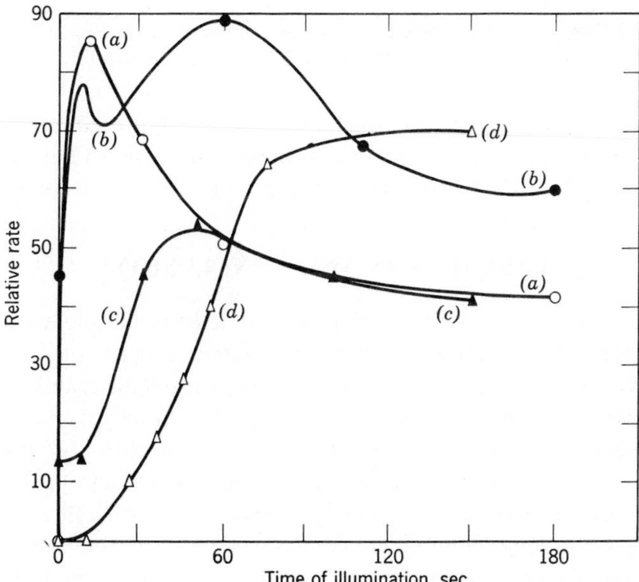

Fig. 60. ATP synthesis by *Chlorella* as measured with the firefly assay method (*c*); fluorescence (*a*); plant chemiluminescence (*b*); and $C^{14}O_2$ fixation (*d*).

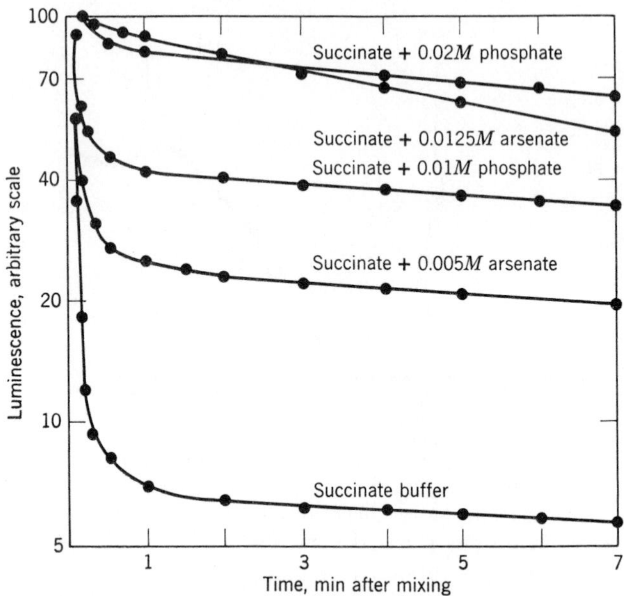

Fig. 61. Effect of different buffers and buffer combinations on response of firefly enzyme [1 ml (see Section IV-1-C-a)] to added ATP.

reaction mixture eliminated the sharp rise and fall characteristic of the reaction, and resulted, rather, in a smooth rise in ATP concentration at a reduced rate without the subsequent rapid decline in the dark.

V. SUMMARY AND CONCLUSIONS

From the foregoing, it is evident that a considerable variety of substances and reactions may be measured directly or indirectly through the application of various bioluminescent reactions. Of these, the firefly enzyme is the most versatile, primarily because of its specificity for ATP and because of the participation of this substance in a great variety of the energy transduction reactions that take place in biological systems. The bacterial system, with its specificity for FMN and long-chain aldehydes, as well as its extreme sensitivity to oxygen, *in vivo* and *in vitro*, is also of considerable potential value in biochemical studies.

Acknowledgments

The author wishes to thank various collaborators and colleagues whose helpfulness over the years has made possible much of what is presented here. In particular the following have contributed in a major fashion: Dr. W. A. Arnold, Dr. John Totter, Dr. W. D. McElroy, Dr. E. N. Harvey, Dr. Frank Johnson, Dr. James Frauch, Dr. Milton Cormier, Dr. Monocher Reporter, Dr. Prem Batra, Dr. R. Monroe, and Dr. Daniel Hendley. Outstanding in his energy, dedication, and contributions is my technical assistant, friend, and colleague, Mr. Richard Nordgren, who carried out many of the new applications recorded here. Thanks are also due my secretary, Miss Gloria Varacalle, whose patience, devotion, persistence, and charm are prodigious.

References

1. B. Chance, *Science, 120,* 767 (1954).
2. E. N. Harvey, *Plant Physiol., 3,* 85 (1928).
3. F. H. Johnson, Ed., *The Luminescence of Biological Systems,* The American Association for the Advancement of Science, Washington, D. C.
4. W. D. McElroy, *Proc. Natl. Acad. Sci. U.S., 33,* 342 (1947).
5. B. L. Strehler and J. R. Totter, *Arch. Biochem. Biophys., 40,* 28 (1952).
6. B. L. Strehler, in *Phosphorus Symposium,* W. McElroy and B. Glass, Eds., The Johns Hopkins Press, Baltimore, 1952.
7. R. Grenell, in *Tranquilizing Drugs,* H. E. Himwich, Ed., The American Association for the Advancement of Science, Publ. No. 46, Washington, D. C., 1957.
8. D. Billen et al., *Arch. Biochem. Biophys., 43,* 1 (1953).
9. L. B. Nanninga, in *Proc. Natl. Acad. Sci. U.S., 46,* 1155 (1960).
10. M. C. Reporter, I. R. Konigsberg, and B. L. Strehler, *Exptl. Cell Res., 30,* 410 (1963).
11. P. P. Batra and B. L. Strehler, *Gerontologia, 13,* 30 (1967).
12. NASA Contractors Report CR-411, Washington, D. C., 1966.
13. B. L. Strehler, in *Methoden der enzymatischen Analyse,* H. U. Bergmeyer, Ed., Verlag Chemie Weinheim Bergstr., 1963.
14. B. L. Strehler, *Arch. Biochem. Biophys., 70,* 507 (1957).
15. W. D. McElroy and H. H. Seliger, in *A Symposium on Light and Life,* W. D. McElroy and B. Glass, Eds., The Johns Hopkins Press, Baltimore, 1961, p. 229.
16. E. N. Harvey, *Living Light,* Princeton University Press, 1940.
17. J. Mayer, in *The Luminescence of Biological Systems,* F. H. Johnson, Ed., American Association for the Advancement of Science, Washington, D. C., 1955, p. 209.
18. E. N. Harvey, *Bioluminescence,* Academic Press, New York, 1952.
19. R. Dubois, *Compt. Rend., 105,* 690 (1887).
20. E. N. Harvey, *Science, 40,* 33 (1914).

21. W. D. McElroy and B. L. Strehler, *Arch. Biochem.*, *22*, 420 (1949).
22. W. D. McElroy and J. Coulombre, *J. Cellular Comp. Physiol.*, *39*, 475 (1951).
23. J. W. Hastings and W. D. McElroy, in *The Luminescence of Biological Systems*, F. H. Johnson, Ed., American Association for the Advancement of Science, Washington, D. C. 1955.
24. W. D. McElroy and A. Green, in *Enzymes, Units of Biological Structure and Function*, O. H. Gaebler, Ed., Academic Press, New York, 1956.
25. F. C. Gerretsen, *Uber die Ursachen des Leuchtens der Leuchtbakterien*, Centr. Bakt. Parasitenk., Abt. II, *52*, 353 (1920).
26. B. L. Strehler and C. S. Shoup, *Arch. Biochem. Biophys.*, *47*, 8 (1953).
27. B. L. Strehler, *J. Am. Chem. Soc.*, *75*, 1264 (1953).
28. W. D. McElroy, J. W. Hastings, V. Sonnenfeld, and J. Coulombre, *Science*, *118*, 385, (1933).
29. B. L. Strehler and M. J. Cormier, *Arch. Biochem. Biophys.*, *47*, 16 (1953).
30. B. L. Strehler and M. J. Cormier, *J. Am. Chem. Soc.*, *75*, 4864 (1953).
31. B. L. Strehler and M. J. Cormier, *J. Biol. Chem.*, *211*, 213 (1954).
32. B. L. Strehler, E. N. Harvey, J. J. Chang, and M. J. Cormier, *Proc. Natl. Acad. Sci. U.S.*, *40*, 10 (1954).
33. S. Kuwababa, M. J. Cormier, L. S. Dure, P. Kreiss, and P. Pfuderer, *Proc. Natl. Acad. Sci. U. S.*, *53*, 822 (1965).
34. W. D. McElroy and B. Glass, Eds., *A Symposium on Light and Life*, McCollum-Pratt Institute of the Johns Hopkins University, The Johns Hopkins Press, Baltimore, 1961.
35. B. L. Strehler, in *Luminescence of Biological Systems*, F. Johnson, Ed., American Association for the Advancement of Science, Washington, D. C., 1955.
36. B. L. Strehler and F. H. Johnson, *Proc. Natl. Acad. Sci. U. S.*, *40*, 606 (1954).
37. B. L. Strehler and M. J. Cormier, *Arch. Biochem. Biophys.*, *53*, 138 (1954).
38. M. J. Cormier, *J. Biol. Chem.*, *231*, 2032 (1962).
39. K. Hori and M. J. Cormier, *Biochim. Biophys. Acta*, *130*, 420 (1966).
40. L. S. Dure and M. J. Cormier, *J. Biol. Chem.*, *239*, 2351 (1964).
41. E. N. Harvey and I. M. Korr, *J. Cellular Comp. Physiol.*, *12*, 319 (1938).
42. B. L. Strehler, *Biol. Bull.*, *107*, 326 (1954).
43. A. M. Chase, F. S. Hurst, and H. J. Zeft, *J. Cellular Comp. Physiol.*, *54*, 115 (1959).
44. F. H. Johnson, E. H. C. Sie, and Y. Haneda, in *A Symposium on Light and Life*, W. D. McElroy and B. Glass, Eds., Johns Hopkins Press, Baltimore, 1961.
45. F. I. Tsuji and R. Sowinski, *J. Cellular Comp. Physiol.*, *58*, 125 (1961).
46. Y. Hirata, O. Shimomura, and S. Eguchi, *Tetrahedron Letters*, *1959* [5], 4.
47. J. W. Hastings and B. M. Sweeny, *J. Cellular Comp. Physiol.*, *49*, 209 (1957).
48. J. W. Hastings and V. C. Bode, in *Light and Life*, W. D. McElroy and B. Glass, Eds., Johns Hopkins Press, Baltimore, 1961.
49. B. L. Strehler and W. A. Arnold, *J. Gen. Physiol.*, *34*, 809 (1951).
50. B. L. Strehler, *Arch. Biochem. Biophys.*, *43*, 67 (1953).
51. B. L. Strehler and D. D. Hendley, in *A Symposium on Light and Life*, W. D. McElroy and H. B. Glass, Eds., Johns Hopkins Press, Baltimore, 1961.

52. B. L. Strehler, *Proc. Intern. Congr. Biochem., 5th, 6,* 82 (1963).
53. W. A. Arnold and J. McClay, in *The Photochemical Apparatus, Its Structure and Function* (Brookhaven Symposium No. 11), C. Fuller et al., Eds., Brookhaven National Laboratories, Upton, New York, 1959.
54. B. Tollen and M. Calvin, *Proc. Natl. Acad. Sci. U.S., 43,* 895 (1957).
55. H. Linschitz and J. Rennert, *Nature, 169,* 193 (1952).
56. E. White, in *A Symposium on Light and Life,* W. D. McElroy and B. Glass, Eds., Johns Hopkins Press, Baltimore, 1961.
57. J. R. Totter, in *Symposium of Chemiluminescence,* Advanced Research Projects Agency, Office of Naval Research, U. S. Army Research Office, Durham, N. C., 1965.
58. F. Carlson and A. Siger, *J. Gen. Physiol., 44,* 33 (1960).
59. W. Cohn, in *Methods of Enzymology,* Vol. 3, S. P. Colowick and N. O. Kaplan, Eds., Academic Press, New York, 1956.
60. A. Jagendorf, in *The Photochemical Apparatus, Its Structure and Function,* (Brookhaven Symposium No. 11), C. Fuller et al., Eds., Brookhaven National Laboratories, Upton, New York, 1959.
61. W. D. McElroy and B. L. Strehler, *Bacteriol. Rev., 18* [3], 177 (1954).

METHODS OF BIOCHEMICAL ANALYSIS VOLUME 16

Utilization of Automation for Studies of Enzyme Kinetics*

MORTON K. SCHWARTZ AND OSCAR BODANSKY, *Sloan-Kettering Institute for Cancer Research and Memorial Hospital for Cancer and Allied Diseases, New York, New York*

*Supported in part by the National Cancer Institute Grant CA 08748, National Institutes of Health, and the American Cancer Society Grant P-164.

I. INTRODUCTION

During the past decade the development of techniques for the automated determination of enzyme activity has proceeded rapidly. In 1963 Schwartz and Bodansky (1) reviewed methods for the assay of enzymes by automated methods. Since then there have been several additional reviews on this subject, as well as the automated determination of biochemical constituents with enzymes as analytical reagents (2–4). It has recently been appreciated by a number of investigators that automated methods may be useful in the study of enzyme kinetics by allowing the planning of a project without undue concern for the time that would be needed to perform a large number of enzyme assays by manual methods. The purpose of the present review is to present automated methods for the study of the effects of changes in enzyme reaction parameters such as time, temperature, substrate concentration, pH, and presence or absence of inhibitors. In addition, methods will be presented for the use of computers in the evaluation of enzyme kinetic data.

II. TIME COURSE OF THE REACTION

1. General

The time course of enzyme reactions can be continuously recorded by the use of first-stage automation equipment such as recording spectrophotometers, the multiple sample absorbance recorder, and recording titrimeters (1). These procedures have been applied chiefly to the measurement of reaction mixture pH changes or of ultraviolet absorbance changes during reduction or oxidation of nucleotides (1).

2. Chromogenic Substrates

Several procedures utilizing chromogenic substrates have been described for the determination of the time course of hydrolytic enzymes. Upon hydrolysis these substrates yield colored products at the pH of the reaction mixture, and the intensity of the color is proportional to the amount of reaction product formed. For example, *O*-nitrophenylbutyrate had been used in this manner for serum cholinesterase (5), phenolphthalein monophosphate or *p*-nitrophenylphosphate for serum alkaline phosphatase (6–8), and L-leucyl-*p*-nitroanilide for serum leucine aminopeptidase (9). Bowers and McComb (8) measured serum alkaline phosphatase activity with *p*-nitrophenylphosphate as substrate at pH 9.8. The liberated *p*-nitrophenol was yellow at this pH. A Perkin-Elmer Model 202 spectrophotometer with a time-drive accessory and an automatic multiple cell holder was used. This instrument allowed simultaneous and continuous recording of the time-course changes in absorbance of five reaction mixtures. The five cuvettes containing reaction mixture were placed in holders on a revolving sample table maintained in a thermostatic compartment. The cells were automatically positioned in turn in the light beam at a time sequence that could be varied from 3 to 120 sec. A mechanism on the recorder lifted the pen from the paper between samples and made one dot on the chart each time a cell was positioned in the light beam.

The disadvantages of the methods described above are that reaction mixtures must be prepared manually, and the enzyme activity values must be calculated by analysis of a curve. However, the advantages, as pointed out by Hausamen et al. (7), are: (*1*) a measurement of the reaction velocity immediately after the start of the reaction with little likelihood of temperature denaturation of the enzyme or inhibition by reaction products, (*2*) measurement of the enzyme activity without adding additional reagents to stop the reaction and perhaps dilute the product beyond the sensitivity of the analytical method, and (*3*) elimination of blank determinations.

3. Two-Point Assays

Blaedel and Hicks (10) have devised an automated system for the assay of lactic dehydrogenase activity in serum in which the absorbance at two time points during the zero-order portion of the reaction was recorded. A recording differential photometer registered the difference in absorbance as the measure of enzyme activity. Two single-channel peristaltic pumps (models PA-56 and PA-6

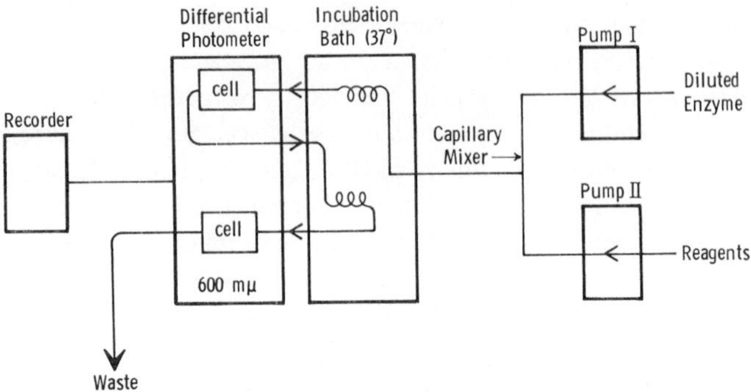

Fig. 1. Schematic outline of apparatus for two time-point assay of serum lactic dehydrogenase (10). See text for details.

obtained from New Brunswick Scientific Co., New Brunswick, N. J.) produced a continuous flow of reagents at a rate of 4.6 ml/min and of diluted enzyme at a rate of 0.9 ml/min (Fig. 1). The two flowing solutions were joined and mixed by passage through a capillary. The reaction mixture then flowed through the 37°C incubation bath to the photometer and then back to the bath for the second cycle. The reagent mixture contained phosphate buffer pH 7.5, sodium lactate, NAD^+, diaphorase, and the dye 2,5-dichlorophenolindophenol. No provision was made for the automated preparation of the enzyme sample, which was diluted manually prior to the assay with a solution of buffer and dye and then placed in position for sampling. The enzymatic reduction of NAD^+ was coupled by the diaphorase to convert the oxidized dye, which has a blue color, to a reduced colorless form. The differential change in absorbance was recorded at 600 mμ. The delays in measurement after beginning of enzyme action were about 50 sec before entry into the first photometer cuvette and about 25 sec before entry into the second. The instrument was calibrated under the same conditions with a reconstituted serum of known lactic dehydrogenase activity, and precise measurements of the assay time were therefore not needed. As Blaedel and Hicks (10) pointed out, the limitation of the method was the need for a standard stable enzyme solution for calibration purposes. About 20 specimens per hour can be analyzed with this method.

Pitot and his associates (11) used the AutoAnalyzer in devising a procedure in which two time points in the reaction course were recorded. In this system, illustrated by the assay for ornithine δ-transaminase, the onflowing enzyme sample was split into two streams. One stream was mixed with substrate and coenzyme (L-ornithine, β-ketoglutarate, and pyridoxal phosphate) and was incubated at 37°C for 5–9 min, whereas the second stream was mixed with the substrate–coenzyme mixture and was passed through a second coil in the 37°C bath and incubated for a longer period, 15–25 min. After incubation each stream was passed through a dialyzer, and the product at each of these two stages of the enzymatic reaction was passed into a color reagent stream. Before passage into a dual differential colorimeter, the short incubation reaction mixture was passed through a time-delay coil of sufficient length so that both reaction mixtures arrived in the flow cells of the colorimeters at the same time. The recorder tracing reflected the enzyme activity between the time periods. Thus in the assay of ornithine δ-transaminase proposed by Pitot et al. (11), the incubation times were 9 and 27 min, giving a recorder tracing representing an 18-min assay. Roodyn (12) has also used an AutoAnalyzer system with a split reaction mixture to obtain two time points in studies of ferricyanide reduction by yeast cells.

4. Complete Enzyme Reaction Rates

Pitot and Pries (13) have combined the Guilford Multiple Sample Absorbance Recorder and cuvette positioner and the Technicon Sampler and constant flow pump to obtain automated simultaneous time-course recordings of four enzyme reaction mixtures. A series of clocks adjusted the sampling time and the operation of the pump to maintain the reaction mixture in the spectrophotometer cuvette. The pump operated only during the sampling period. A schematic drawing of the apparatus is shown in Figure 2. A four-place crook on the sample arm lowered dip tubes into 4 cups of enzyme solution on the sampler. The enzyme solutions were aspirated through tubes *A*, *B*, *C*, and *D* and each mixed with a solution of reagents aspirated from a reservoir through tubes *E*, *F*, *G*, and *H*. Each complete reaction mixture was pumped through a mixing coil and into four flow cuvettes in the spectrophotometer. At this point the clock turned the pump off and the reaction mixture remained in the cuvette for a predetermined time; recordings of the changing ab-

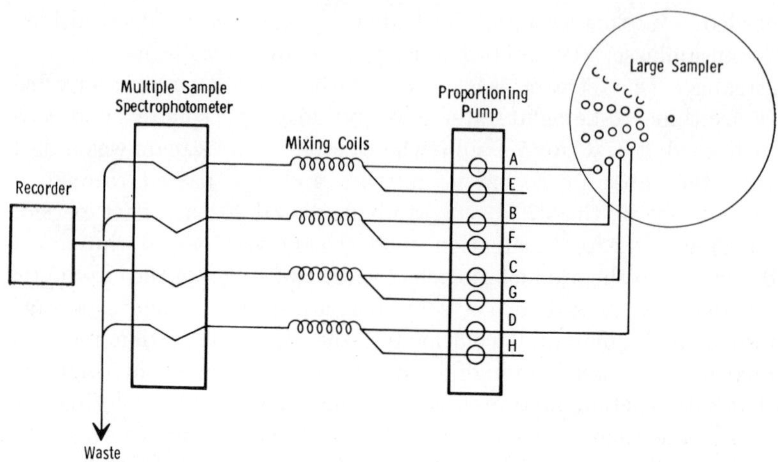

Fig. 2. Schematic outline of apparatus for simultaneous determination of the time courses of four enzyme reactions (13). See text for details.

sorbance were made during this period. At the end of this time, the turntable moved forward to the next set of four enzyme samples; the pump began and washed the cuvettes with substrate buffer solution as the next set of four samples was aspirated and prepared for assay. Using this technique, Pitot and Pries (13) have described methods for the assay of histidase, threonine dehydrase, gluco- or hexokinase, tryptophane pyrrolase, ornithine δ-transaminase, and arginase. In the method for the assay of histidase activity of rat liver, obtained by centrifugation at 105,000g for 90 min, the reagent reservoir contained 70mM histidine pH 9.2, and the enzyme samples were diluted in 20mM sodium pyrophosphate pH 9.2. Reagents and enzyme sample were allowed to flow for 8 min, after which the pump turned off and maintained the reaction mixture in the cuvette for 30 min. The flow rate was 0.8 ml/min, and about 2.4 ml of reaction mixture was maintained in the cuvette. The rate of urocanate formation was monitored at 277 mμ. The 8-min flow time allowed for effective washing of the cuvettes between samples. Maximal rates of enzyme activity were calculated from a line drawn manually between the interval readings on the recorder tracing. The voltage output of the spectrophotometer may be introduced into a computer and rates calculated by digital computation, registering digital readout to a card punch (11). The authors have estimated that as much as 1000 samples per day can be analyzed with this system, and emphasize the importance of continuous time-course assays of enzymes which exhibit lag periods.

The AutoAnalyzer has been primarily designed for the recording of a single time point in a reaction. However, several investigators have utilized the basic AutoAnalyzer system or the Technicon amino acid analyzer for recording the complete time course of enzyme reactions (14–16). Lenard and his associates (15) recorded the time course of the hydrolysis of *N*-acetyl-L-tryptophanamide by α-chymotrypsin using portions of the Technicon amino acid analyzer. An appropriate volume of 20m*M* *N*-acetyl-L-tryptophanamide dissolved in 100m*M* $NaHCO_3$–NaCl buffer, pH 8.83, was placed in a vessel maintained at a thermostatically controlled temperature of 25°C. The enzyme solution (0.0382 mg/ml) was added to the vessel containing the substrate to start the reaction. The reaction mixture was pumped at a rate of 0.42 ml/min, and 10 sec after leaving the reaction vessel began to be segmentally mixed with ninhydrin reagent pumped from a reservoir at a rate of about 1.06 ml/min. The addition of ninhydrin stopped the reaction. The flowing mixture was then interspersed with nitrogen gas. The resulting mixture was passed through an 80-ft coil immersed in a boiling-water bath to develop the color. The hot solution was then cooled by passage through a coil maintained at 25°C and introduced into a tubular flow cell in a colorimeter equipped with 570-mμ interference filters. A continuous recording of the enzyme reaction was obtained. Under the conditions used, a linear reaction was obtained for about 18 min.

Schwartz and Bodansky (16) have utilized the continuous flow system of the AutoAnalyzer to study the time course of the alkaline phosphatase reaction, with β-glycerophosphatase as substrate and colorimetric measurement of the liberated inorganic phosphate. The manifold used in this study is shown in Figure 3. A large volume of reagent mixture containing buffered substrate (0.05m*M* β-glycerophosphate in 80m*M* Tris buffer pH 9.5 and 10m*M* Mg^{2+}) was placed in an Erlenmeyer flask suspended in the AutoAnalyzer dialysis bath at 37°C. This mixture, along with the reagents for phosphate color development, were aspirated for about 5 min to permit calibration of the colorimeter and recorder. At zero time, a human bone phosphatase preparation with an activity of 3.55 IU per milligram of protein was added with rapid mixing to the Erlenmeyer flask containing the buffered substrate. Thirty seconds after this reaction mixture began to be aspirated, an acid molybdate (10 g/liter of ammonium molybdate in 2.3N sulfuric acid) solution joined the flow segmentally, stopping the enzyme reaction and reacting with the liberated inorganic phosphate for the subsequent development of color. The flowing solution was heated to 95°C and mixed with aminonaphtholsulfonic acid. After passage through a delay coil, the intensity of the colored solution was recorded by passage through a colorimeter equipped with 660-mμ filters and a 6-mm constant-flow cell. Inorganic phosphorus standards were treated in a similar fashion after aspiration through the reaction-mixture line.

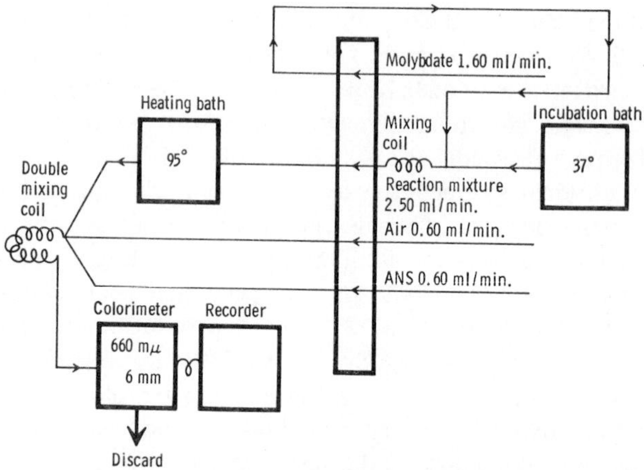

Fig. 3. Schematic outline for automated time-course assay of alkaline phosphatase activity (16). Each vertical grid represents 3 min of reaction time.

The recorder tracing for this experiment is shown in Figure 4. The dashed line drawn on the recorder paper represents a zero-order reaction. This phosphatase preparation obeyed zero-order kinetics through the hydrolysis of about 14% of the β-glycerophosphate. These data are in agreement with results of previously reported manual experiments (17).

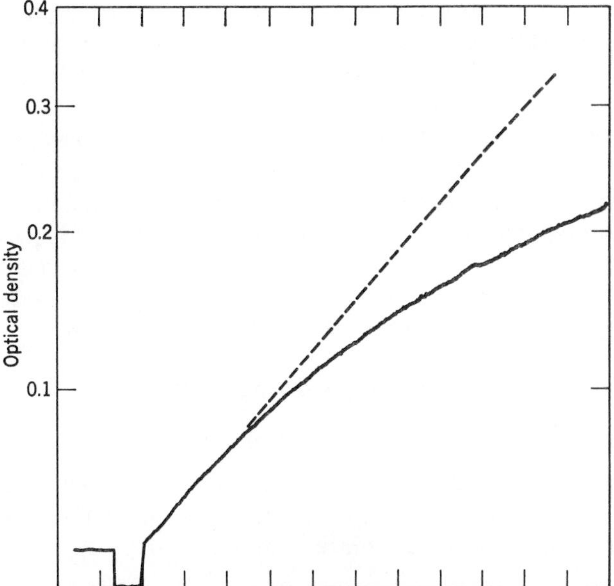

Fig. 4. Recorder tracing of continuous time-course assay of human bone alkaline phosphatase activity (16).

5. Continuous *in vivo* Enzyme Assays

An automated procedure has been devised to monitor whole blood cholinesterase levels *in vivo* in monkeys during investigations of acute parathion toxicity (18). An AutoAnalyzer manifold was prepared with a polyethylene catheter placed in a femoral vein of the animal. Under negative pressure, blood was aspirated continuously at a rate of 0.303 ml/min. The catheter was kept patent, and blood was prevented from clotting by infusing into the catheter, at its exit from the vein, a solution containing 1000 units of heparin per milliliter of saline at the rate of 0.32 ml/min. The blood volume loss was continuously replaced by saline pumped into the antecubital vein at the same rate as the blood was aspirated. The diluted blood, immediately after coming through the catheter, was mixed by continuous flow with a solution of barbital phosphate buffer pH 8.10 and acetylcholine (20 g of acetylcholine iodide per liter) and then introduced into a 37°C heating bath. After incubation, the reaction mixture was passed through a dialyzer and the liberated acetic acid dialyzed into a stream of saline containing phenol red. The decolorization was recorded after passage through a 15-mm tubular flow cell in a colorimeter equipped with 550-mμ filters. There was a delay of 12 min from the time the blood was aspirated until the recording was seen. Control-anesthetized Rhesus monkeys were attached to the continuous flow system for up to 4 hr with no observable change in cholinesterase activity. The oral administration of parathion (20 mg/kg) resulted in a drop in whole-blood cholinesterase activity to 50% of the premedication value. The infusion of pralidoxime chloride could either prevent or reverse the parathion depression of whole blood cholinesterase, but did not prevent the death of the animals.

III. TEMPERATURE

1. General

Automated methods have been used to study the effect of temperature on the catalytic activity of enzymes as well as the differential denaturation effects of temperature on isozymes.

2. Energy of Activation

Tappel and Beck (19) applied the AutoAnalyzer to the determination of the energy of activation of rat liver lysosomal β-glucuronidase. These authors found that the regular 37°C AutoAnalyzer incubation bath, first cooled to a low temperature (about 20°C) and then turned on, heated up a constant rate of about 1°C/min. An AutoAnalyzer manifold devised for the determination of β-glucuronidase activity with *p*-nitrophenyl-β-D-glucuronide as substrate was used and the reaction mixture passed through the incubating bath as it warmed up to 37°C. A thermocouple in the bath recorded the temperature. The thermocouple activated the second pen of a dual-pen recorder. Therefore, the bath temperature and the absorbance reflecting the enzyme activity were recorded on the same chart. The data, replotted as the log of the reaction velocity against the reciprocal of the absolute temperature according to the method of Arrhenius, yielded a linear plot from which an energy of activation of 13.0 kcal was calculated. A similar technique has been used by Boy (14) in his study of the peroxidase activity of haptoglobulins.

Brehmer and his associates (20) have built a constant-flow apparatus which will be described in detail in Section IV. These authors used this instrument to study the kinetics of lactic dehydrogenase. Although they do not present data, they state that effects of temperature change on the enzyme activity can be evaluated by keeping the concentration of the flowing reaction mixture constant and establishing a continuous temperature gradient in the incubation bath. A simultaneous recording of the bath temperature was obtained with a thermoelement.

3. Heat Inactivation

Heat inactivation of enzymes has been useful in distinguishing catalytically similar but structurally different enzymes. Automated methods can be useful in such studies. The rate of heat inactivation of serum alkaline phosphatase has been studied by Posen and his associates (21) with an automated alkaline phosphatase procedure which utilizes phenylphosphate as substrate (22). After calibration of the AutoAnalyzer colorimeter and recorder with the reagents, aspiration of diluted enzyme sample was started at a temperature at which the enzyme was stable. This permitted determination of the activity before denaturation. The enzyme solution was then placed in a bath at 56°C, and samples were withdrawn continuously. The enzyme activity was about 17 King Armstrong units before being placed in the 56°C bath, was reduced to 50% of the original activity in about 8 min, and to negligible levels in 35 min.

Strandjord and Clayson (23) have utilized an automated method to determine the percentage of heat-stable component (LDH 1) of serum lactic dehydrogenase. An AutoAnalyzer manifold (Fig. 5) was prepared for the assay of lactic dehydrogenase activity by recording the absorbency at 340 mμ as NADH was oxidized to NAD^+ during the conversion of pyruvate to lactate. One portion of a manually diluted serum specimen was placed in a water bath at 65 ± 0.1°C for 30 min. At the end of this period, two aliquots of the heated enzyme (cups a', a', Fig. 5) and unheated enzyme (cups a, a, Fig. 5) were placed in the outer row of holes of a specially constructed sampler. The sample plate contained an outer row of holes for the cups containing enzyme and an inner row for cups of substrate. A sample arm with two crooks dipped into an outer and an inner cup simultaneously and withdrew at an equal rate both enzyme sample and substrate, which were then mixed with buffer, passed through a

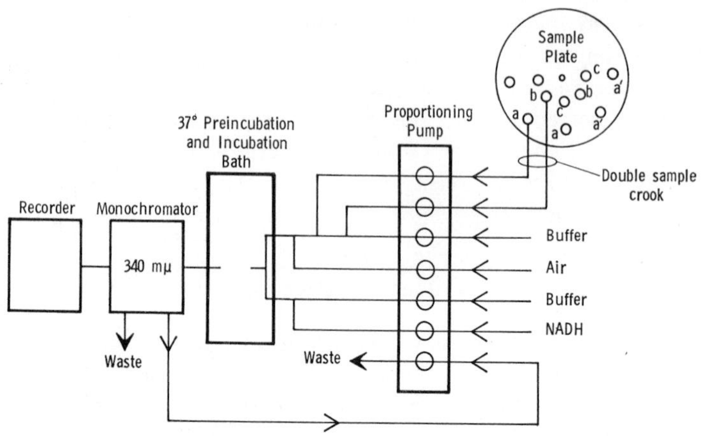

Fig. 5. Schematic outline for the automated determination of lactic dehydrogenase isozymes (23). See text for details.

mixing coil and into an incubation bath. The reaction mixture then passed through a microflow cuvette (Beckman No. 97290) in a monochromator at 340 mμ (Beckman DU with Guilford Recorder). The substrate or inner row of the sample plate had pyruvate (cups *b*, Fig. 5) and oxamate (cups *c*, Fig. 5) solutions in alternate holes. Each sample was mixed first to form a reaction mixture with pyruvate and then with oxamate. Oxamate is a potent inhibitor of lactic dehydrogenase (24) and in the concentration used (0.045m*M*) completely inhibited enzyme activity and reflected the serum blank. With this technique the recorder tracing exhibited alternate peaks of enzyme and blank activity. Thus, if in four outer cups were placed two untreated and two heat-treated specimens, the four recorder peaks reflected the total and the heat-stable lactic dehydrogenase.

IV. EFFECT OF pH

1. General

The optimum pH of an enzyme's activity is necessary information for its characterization. Manual studies of effects of hydrogen ions on enzyme activity require the preparation of reaction mixtures over a wide spectrum of pH values and the assay of large numbers of specimens. A number of automated procedures for such studies have been formulated.

2. Electrode-Containing Cuvette

Taylor and Arora (25) constructed a cell for the study of enzyme kinetics in the Beckman DU–2 spectrophotometer. The apparatus included a stirrer, a salt bridge, and a calomel electrode built into a cuvette as well as a series of pistons to introduce and remove solutions from the cuvette and to create a pH or other kind of gradient. The entire apparatus could be fitted on the slide carriage of the Beckman DU and was maintained in a compartment equipped with specially constructed thermospacers to accommodate the somewhat larger than usual cuvette. Although the authors do not give specific data for any particular enzyme, they state that an enzyme activity–pH curve may be obtained in steps. The pH and the spectrophotometric reading of the reacting mixture were recorded simultaneously. The rate of entrance of the reagents into the cell was obtained by setting a turn screw which controlled the pistons. The cell volume was 3.6 ml. After determining the velocity at any given pH, the cell was washed three or four times with a volume of reaction mixture at another pH before the reaction velocity was determined.

3. Continuous-Flow Systems

A general gradient apparatus has been used with a continuous-flow system to study the effect of pH on rat liver lysosomal β-glucuronidase activity (19). The gradient device consisted of two open-ended glass cylinders bonded on a glass plate and connected to each other by a groove in the plate. A flow outlet was attached to the outer cylinder. The entire apparatus can be placed on a magnetic stirrer and stirring bars placed in each cylinder. The two inner cylinders were surrounded by a large glass cylinder in which ice water was kept. A linear pH gradient was obtained by mixing buffers of the same molarity but different pH, placed in each of the two cylinders. In the described experiment, 20mM acetate buffer was used with a gradient between pH 3.92 and 7.00. The buffer was continuously mixed and was aspirated from the outlet into an AutoAnalyzer manifold at a rate of 0.60 ml/min. Substrate (6.35mM *p*-nitrophenyl-β-D-glucuronide) and enzyme were each introduced at the same rate (0.60 ml/min), producing a total reaction mixture of 1.80 ml. Mixing was accomplished by flow of the reaction mixture through a glass capillary, the diameter of which was widened every 1.5 cm. The reaction mixture was incubated for about 8 min at 37°C and the enzyme reaction was then stopped by introducing ammonium hydroxide, pH 10.7. The absorbance of the yellow color of the liberated *p*-nitrophenol in alkaline medium was recorded during passage through a 15-mm tubular flow cell in a colorimeter equipped with 420-mμ filters. A cam and cycle timer and a dual dipper permitted alternate aspiration of wash water through the lines for 3 min. This procedure was adopted to conserve a limited supply of enzyme and a relatively expensive substrate. It also allowed continuous observation of the reagent base line. A small flow cell containing pH electrodes was introduced into the manifold to monitor the pH values which were recorded on a Radiometer TTT-1 pH meter. In order to use the pH flow cell, it was necessary to alter the flow to produce a stream unsegmented by air. Presumably the enzyme activity course was obtained in one run and the pH values in a second. Optimal β-glucuronidase activity was seen as a maximum between pH 5.15 and 5.30. The authors state that the recorded pH values were precise to better than 0.1 pH unit, but the response of the pH meter decreased in time due to mixing in the flow cell at the low flow rates used in the assay.

Brehmer et al. (20) have built a continuous-flow apparatus for the measurement of enzyme kinetics, particularly that of lactic dehydrogenase, in which one variable such as substrate or hydrogen ion can be continuously changed (Fig. 6). A precision pump, operated by a synchronized motor, drove pistons in three separate cylinders forward against fluid and pushed the three streams toward a micromixer. Two of the solutions, usually the enzyme and coenzyme, entered the mixer directly. In determining the effect of pH, the buffered substrate was allowed to enter the mixer, either directly or by adjustment of a valve, through a gradient device which contained substrate dissolved in a buffer of different pH. As the diluting material entered the gradient chamber, a built-in stirrer insured rapid mixing. The continuously diluted material entered the mixing chamber and joined the enzyme and coenzyme entering from the other lines. The complete reaction mixture was interspersed with

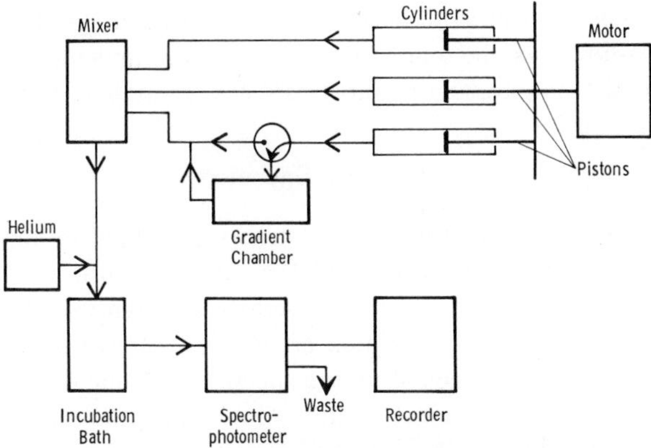

Fig. 6. Schematic outline of apparatus for measurement of enzyme activity during creation of pH or substrate gradients (20). See text for details.

bubbles of helium as it left the mixing chamber and entered a glass coil maintained at a constant temperature. The mixture then flowed through a cuvette in an Eppendorf photometer. The cuvette contained microelectrodes so that the pH and absorbance of the solution could be recorded simultaneously.

The concentration of the diluted material changes as an exponential function and the concentration of hydrogen ion or substrate at any time, C_t, has been expressed by the authors (20) in the following equation:

$$C_t = C_0 \cdot \exp-(V_t/V_g) \tag{1}$$

where C_0 is the concentration before dilution; V_t the flow rate per minute of the diluting material, and V_g the volume of gradient container.

V. EFFECT OF SUBSTRATE

1. General

Automated methods may be used for the evaluation of the relationship between substrate concentration and reaction velocity, and also for studies of substrate specificity. These studies can be conducted by manually preparing solutions of different substrates or different concentrations of the same substrate and introducing them into a continuous-flow system one after the other or devising a gradient system for the preparation of substrate of varying concentrations.

2. Substrate Specificity

The placement of a group of substrates in adjacent cups on an AutoAnalyzer and the aspiration of the substrate into a continuous flow of enzyme and buffer has been used by Roodyn (12,26) in what he refers to as "multiple enzyme analysis." However, in several instances he may have been studying specificity of one or more of the enzymes to several substrates as well as multiple enzymes. In studies of phosphate-liberating activity in homogenates of yeast cells, solutions of adenosine-5′-phosphate, adenosine diphosphate, adenosine triphosphate, glucose-1-phosphate, 3-phosphoglycerate, and fructose-1,6-diphosphate were placed in adjacent cups of an AutoAnalyzer sampler. Each substrate was dissolved in water alone or with the addition of either Mg^{2+} or Ca^{2+}. Exact details of the automated method or of the concentration of substrates and cofactors were not reported, but the phosphatase activity was measured by the determination of the liberated inorganic phosphate. The phosphatase activity pattern with six substrates, each assayed in the presence and absence of two metal ions, was obtained in one run. Presumably all substrates were used at the same pH and concentration. No reference is made to possible difference in activity due to variation in substrate concentration or pH. Greatest activity was seen in the system containing adenosine triphosphate and Mg^{2+}.

Substrate specificity of yeast alcohol dehydrogenase was studied in a similar fashion (26). Methanol, ethanol, *n*-propanol, and *n*-butanol were used as substrates in an automated system recording NAD^+ reduction at 340 mμ. Greatest activity was seen with ethanol as the substrate, and decreasing activity as the chain length of the alcohol increased. Other dehydrogenase activities of the yeast cell homogenate were also studied (12). Aqueous solutions of the following substances at a concentration of 150m*M* were placed in adjacent cups on the sample plate; malate, aspartate, glutamate, alanine, lactate, glycerol, methanol, ethanol, propanol, butanol, fructose, glucose, and sucrose. These solutions were aspirated and mixed with 100m*M* glycine buffer, pH 9.9, and a solution containing 1.1 mg/ml of NAD^+. The highest activities, as determined by the reduction of NAD^+, at 25°C, was seen with malate or ethanol as substrate. Again no reference was made to possible differences in pH or substrate concentration optima.

Roodyn and Wilkie (27) have used this technique in a study of respiratory enzymes in cytoplasmic "petite" strains of *Saccharomyces cerivisiae*. These yeasts are mutations which exhibit a genetic respiratory deficiency and are capable of growing on fermentable substrates but not on nonfermentable ones. The assays measured the enzymatic reduction of ferricyanide and were carried out by an automated two time-point assay at 6.4 and 11.6 min (12). Appropriate dilutions of the following substrates were placed in sample cups to yield final concentrations in the reaction mixture of 0.18m*M* NADH, 1.8m*M* sodium L(+)lactate; 1.8m*M* sodium D(−)lactate or 1.8m*M* potassium succinate. Sample cups containing the substrates were separated from each other by a sample cup of distilled water, and the contents were aspirated at a rate of 40/hr. The final concentration of the other reactants were: 25m*M* potassium phosphate, pH 7.4; 0.05% Triton X-100; 0.4m*M* potassium ferricyanide; 100 m*M* mannitol and a portion of disrupted yeast cells equivalent to 0–50 μg of protein/ml. The data showed that the "petite" strain was able to reduce NADH at 50% and L(+)lactate or succinate at 5% of that of the control cells. A 50% increase in activity was seen with D(−)lactate as substrate.

3. Michaelis Constants

A. NONGRADIENT CONTINUOUS FLOW

Roodyn (12) used the AutoAnalyzer without a gradient-producing device to study the substrate–activity relationship in enzymes of the yeast *Saccharomyces cerevisiae*. The yeast cells were disrupted, and the relation of malate concentration to the activity of malic dehydrogenase in a 500*g* supernatant was determined in the following manner. Solutions of sodium malate in 100m*M* glycine buffer, pH 9.9, were prepared so that the final reaction mixture contained 0.4, 2.2, 4.4, or 9.8m*M* sodium malate. Aliquots of these solutions were placed in cups on the AutoAnalyzer sample plate and aspirated at a rate of 0.42 ml/min. The flowing substrate was mixed with buffer (1.20 ml/min), NAD^+ (0.32 ml/min) and enzyme preparation

(0.10 ml/min). The complete reaction mixture passed through a time-delay coil maintained at 25°C, and then through a continuous-flow cell in a phototube colorimeter. The changes in absorbance at 340 mμ appeared as discrete deflections with each change in substrate concentration. When the reciprocals of the absorbance values were plotted against the reciprocal of malate concentration, a straight line was obtained indicating the suitability of this procedure for the determination of Michaelis constants. The values obtained in this experiment were not given.

B. GRADIENT CONTINUOUS FLOW

Several workers have devised completely automated procedures for the evaluation of the effects of substrate concentration on enzyme reaction velocity. Brehmer and his associates (20) used the continuous-flow apparatus they devised for creating pH gradients to study substrate effects (see Fig. 6 and Section IV). Buffered substrate was placed in the gradient chamber and diluted by a continuous flow of the same buffer without substrate into the chamber. With this technique, the effect of various concentrations of pyruvic acid on the lactic acid dehydrogenase activity in extracts of human liver and heart and rabbit muscle was studied. The Michaelis constants were as follows: rabbit muscle, 0.174m*M*; human liver, 0.145m*M*; and human heart, 0.0425m*M*. The values for the human heart and liver were of the same order of magnitude as those reported for these tissues by Nisselbaum and Bodansky (28).

Tappel and Beck (19) have studied substrate concentration effects on rat liver lysosomal β-glucuronidase, using the gradient device and the AutoAnalyzer manifold devised by them for the production of a pH gradient (Section IV). Exact details were not given, but the authors state that the substrate was introduced into the reaction mixture as a linear gradient and the data replotted from the recorder tracing to yield both the Michaelis constant and the maximum velocity. The K_m value for rat liver lysosomal β-glucuronidase with *p*-nitrophenyl-β-D-glucuronide at pH 5.0 as substrate was 0.094m*M*.

Schwartz and Bodansky (16) have devised a gradient-producing system which utilizes the usual AutoAnalyzer components and manifolds. This system was based on equalizing rates of outward and inward flow from a reagent vessel and was designated as the "equalized continuous flow rate technique."

In Figure 7 is shown a manifold for studying the effect of varying substrate concentrations on alkaline phosphatase activity by the equalized continuous flow rate technique. This manifold had a line that delivered a solution with all reagents except substrate at exactly the same rate as the outward-flowing substrate line. For example, about 30 ml of undiluted substrate was placed in a graduated cylinder. Aspiration was started, and when the volume reached a previously decided level, usually 25 ml, the diluent tube was allowed to discharge into the cylinder. Diluting solution was therefore being added to the cylinder at the same rate as the substrate was aspirated. Mixing, accomplished by the magnetic stirrer on which the cylinder was placed, was almost instantaneous and continuous dilution was accomplished. The reaction then proceeded in the usual manner through incubation and color development. The rate of dilution depended on the volume of fluid in the cylinder and the inner diameter of the diluent tube. The relationship may be expressed generally as follows:

$$[(a - n)/a]^t\, C = C' \tag{2}$$

where a is initial volume in milliliters, n equals flow in milliliters per minute, C is the initial, undiluted concentration of the substrate, and C' is the concentration of the substrate at a time, t, in minutes. The equation may be used to calculate the concentration of the substrate at the end of a stated period of time or, conversely, the time needed

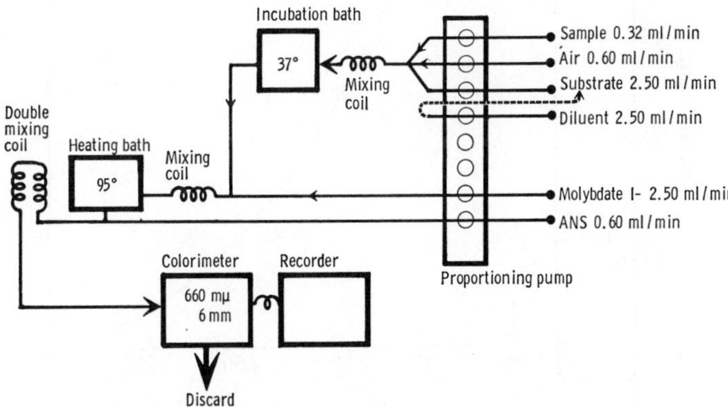

Fig. 7. Schematic outline of phosphatase studies by "equalized continuous-flow rate technique" (16). See text for details.

to reach a certain concentration. Thus, with a 10-ml volume and a 2 ml/min diluting line, the substrate would be diluted to 51% of its original concentration in 3 min. With the same size diluent line and 25 ml of fluid in the cylinder, the half-time for dilution would be 8 min.

This relationship was confirmed by diluting a known standard or a colored solution such as potassium ferricyanide. Dilution of 25 ml of a 20 μg/ml phosphorus standard with this manifold and subsequent development of the molybdenum blue color is shown in Figure 8. The spaces between the vertical grids represent 3-min intervals. The equalized continuous flow rate was 2.15 ml/min. The 20-μg standard, before dilution, had an optical density of 0.600. In 9 min (1.5 in. of chart paper) the optical density was exactly half or 0.300, representing 10 μg of phosphorus. In the next 9-min period, the optical density fell to 0.150, equal to 5 μg of the standard phosphorus.

The effect of varying concentrations of β-glycerophosphate on human bone phosphatase activity was studied in a similar fashion. The recorder tracing is shown in Figure 9. Thirty milliliters of

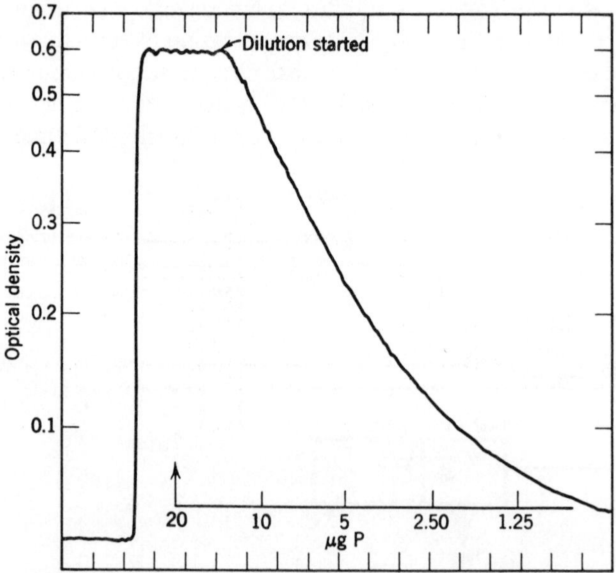

Fig. 8. Recorder tracing of continuous assay during dilution of a 20 μg/ml inorganic phosphorus standard (16). Each vertical grid represents 3 min.

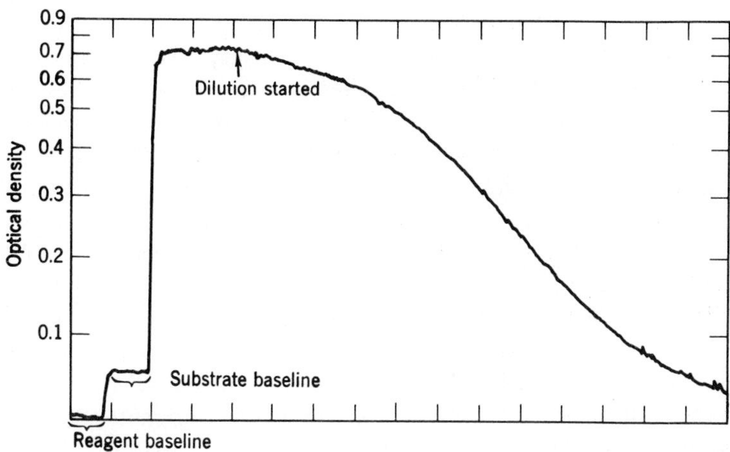

Fig. 9. Recorder tracing of alkaline phosphatase activity during dilution of β-glycerophosphate (16). See text for details.

β-glycerophosphate, 42m*M*, dissolved in 100m*M* Tris buffer, pH 9.5, containing 11m*M* Mg^{2+}, was placed in a 25-ml cylinder on a magnetic stirrer. Purified human bone phosphatase (29), diluted to contain 20 μg of protein per milliliter of cold water, was kept in a flask immersed in an ice bath. The delivery rate of the enzyme solution sample line was 0.34 ml/min, and the equalized continuous flow rate for the substrate was 2.14 ml/min. Concentrations of the reactants in the final reaction mixture were 86m*M* Tris buffer, 36.2m*M* β-glycerophosphate, 10m*M* Mg^{2+}, and 2.6 μg of enzyme protein per milliliter of reaction mixture. The reaction was allowed to proceed at this substrate concentration until the volume in the cylinder reached 25 ml. The inflowing solution consisting of 100m*M* Tris buffer, pH 9.5, and 11m*M* Mg^{2+}, but no substrate, was then started. In accordance with previous considerations [Eq. (2)], the substrate concentration was 36.2m*M* at the beginning of the inflow, 18.1m*M* 9 min later, 9.1m*M* at 18 min, and 4.6m*M* at 27 min. The substrate enzyme mixtures were incubated for 9.5 min at 37°C at all concentrations of substrate. The amount of inorganic phosphorus liberated during the 9.5-min incubation was within zero-order portion of the reaction at practically all concentrations of substrate.

Since it is known that maximum velocity is obtained with 36m*M* substrate, the Michaelis-Menten constant, K_m, is equal to the substrate concentration at one-half this velocity. This value was found to be 18mM. The Michaelis-Menten constant can also be determined from Figure 9 by correcting the absorbance for the substrate blank, calculating the velocities at exact substrate concentrations, and then plotting the data according to the Lineweaver-Burk formulation (30). The K_m value, 18mM, obtained by both methods is in the general range of the values previously reported for bone phosphatase (29).

Adenosine 5′-phosphate is also a substrate for human bone alkaline phosphatase. However, there is a zone of substrate inhibition when this compound is used. This phenomenon was studied with the manifold described in Figure 7. Adenosine 5′-phosphate, 23.6mM, dissolved in 100mM Tris buffer pH 9.60 and containing 11.4mM Mg^{2+} was diluted with 100mM Tris buffer containing 11.4mM Mg^{2+}. The continuous, equalized flow rate was 2.13 ml/min. After addition of enzyme, the concentration of reactants were: adenosine 5′-phosphate, 20.4mM; Tris buffer, 86mM; Mg^{2+}, 10mM; and enzyme, 2.74 μg of protein per milliliter of reaction mixture. Dilution was then begun. The chart recording is shown in Figure 10. Inhibition due to excess substrate ranged from 20.4mM substrate to a concentration of about 2mM. From this point on there was a progressive decrease in reaction velocity as the substrate concentra-

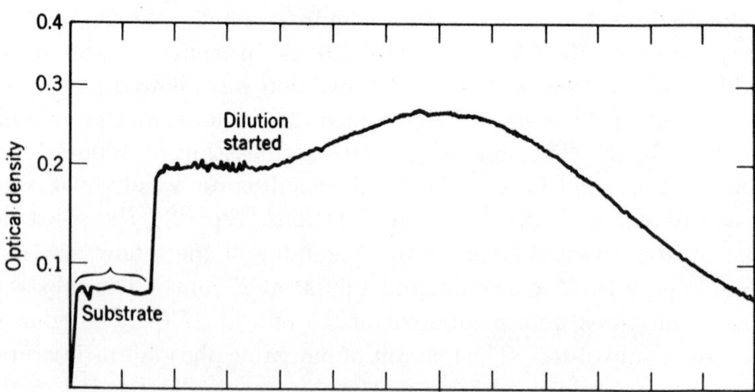

Fig. 10. Recorder tracing of alkaline phosphatase activity during dilution of adenosine 5′-phosphate (16). See text for details.

tion decreased. From this data it is possible to calculate both K_m, the Michaelis constant, and K_s', the equilibrium constant for the dissociation of the compound assumed to account for inhibition by high concentrations of substrate (30). In this example, the value for K_s' was 125mM and that for K_m was 0.7mM.

Equalized continuous-rate manifolds may be prepared for any enzyme reaction. In our laboratory similar investigations have been carried out with 5′-nucleotidase, phosphohexose isomerase, glutamic oxaloacetic transaminase, glucose-6-phosphate dehydrogenase, lactic dehydrogenase, and alkaline phosphatase with phenylphosphate as substrate. In Figure 11 is shown a manifold for the study of the effect of the concentration of sodium pyruvate as substrate on the velocity of lactic dehydrogenase. The method, based on an automated procedure for assay of enzymes utilizing NAD^+ or NADH (31), and determining change in absorbance at 340 mμ, is extremely sensitive with only a small fraction of the reaction mixture being passed through the phototube colorimeter. In the experiment to be described, the equalized continuous-flow rate was 1.90 ml/min. The sodium pyruvate, 12.4mM, was dissolved in 67mM phosphate buffer, pH 7.4, which also served as the diluent. Crystalline rabbit muscle lactic dehydrogenase (Worthington Biochemical Co.) was dissolved in 0.15% human serum albumin and kept in an ice bath

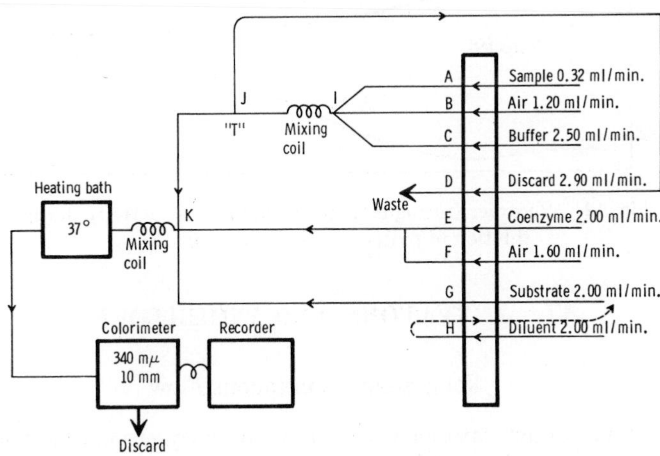

Fig. 11. Schematic outline for lactic dehydrogenase studies by "equalized continuous flow rate technique." See text for details.

during the experiment. The enzyme reaction produces a decrease in absorbance as NADH is converted to NAD^+ (Fig. 12). At first glance it might appear that inhibition of the lactic dehydrogenase activity occurs at higher concentrations of pyruvate (curve *a*). However, as is shown in the curve (curve *b*), pyruvate itself has a substantial absorption at 340 mμ; this curve was obtained when the pyruvate was diluted with buffer instead of NADH and enzyme. When corrections were made for the absorption by pyruvate the reaction velocity remained relatively constant at concentrations from 124m*M* until 3.1m*M* (curve *c*). From this point on there is a drop in activity with decreasing concentrations of substrate. The Michaelis-Menten constant can be calculated from the middle part of curve *c* and in this experiment was found to be 0.7m*M*.

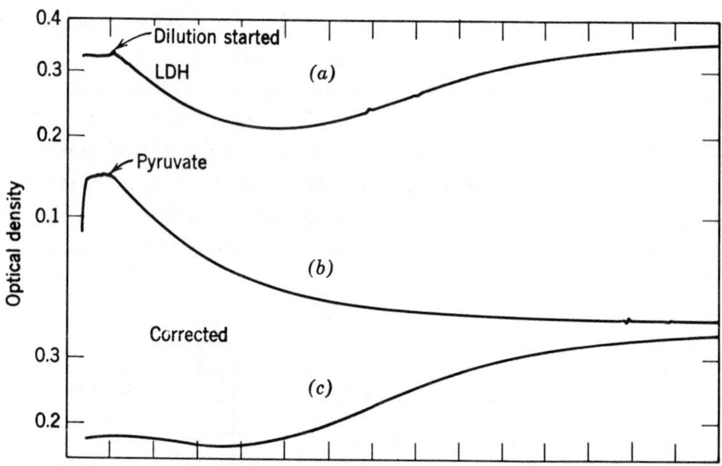

Fig. 12. Recorder tracing of lactic dehydrogenase activity during dilution of pyruvate. See text for details.

VI. ACTIVATORS AND INHIBITORS

1. Nongradient Continuous Flow

The effect of activators and inhibitors on enzymes can be studied by continuous-flow methods. Stein and his associates (32) have used the AutoAnalyzer in studies of the effect of cofactors and inhibitors on the hydrolysis of adenosine triphosphate (ATP) by rat brain microsomal adenosine triphosphatase (ATPase). A manifold for the assay of ATPase activity was devised (Fig. 13) with enzyme sample continuously aspirated from a beaker or cylinder kept at the suitable temperature, and the inhibiting or activating solutions placed in sample cups on the sampler plate. Sampling was set for 40 specimens/hr, and a "constant-volume" device aspirated 37.4m*M* Tris buffer, pH 7.4, containing 175m*M* sucrose between samples to maintain a constant volume of reaction mixture and to prevent air peaks from causing irregularities in the recording trace.

The final reaction mixture contained 20m*M* Tris buffer, pH 7.4, 2m*M* ATP, and 2m*M* Mg^{2+} in one series of experiments. To study the K^+, Na^+ activated ATPase, a second series of experiments were carried out in which the same concentrations of Tris, ATP, and Mg^{2+} were used and 2m*M* Na^+ and 2m*M* K^+ were also present. After incubation at 37°C for about 18 min, the liberated phosphate was dialyzed into a flowing stream of aminonaphtholsulfonic acid and then mixed with sulfuric acid–ammonium molybdate solution. The mixture was incubated at 37°C for about 15 min, and the intensity of the blue color that had developed was determined by passage through

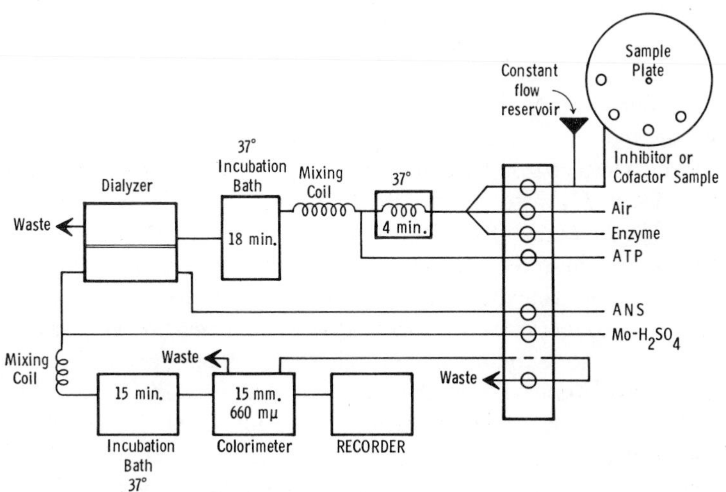

Fig. 13. Schematic outline for study of effect of inhibitors or activators on ATPase activity (32). See text for details.

a 15-mm tubular flow cell in a colorimeter equipped with 660-mμ filters.

N-Ethylmaleimide, fluorodinitrobenzene, iodoacetic acid, and iodacetamide exhibited no inhibitory effects on ATPase activity at concentrations as high as 0.1m*M*. The mercaptide-forming reagents, *p*-chloromercuribenzoate, *p*-chloromercuriphenol, and phenylmercuric acetate were strong inhibitors. The type of inhibition was not described. More detailed studies of the inhibitory effect of ouabain, a cardiac glycoside, on ATPase activity were reported. An enzyme preparation, containing 47 μg of protein per milliliter, was continuously aspirated. Ouabain was dissolved to yield final concentrations ranging from 0.00 to 0.0294m*M* in a series of buffered solutions containing 3.74m*M* Mg^{2+} and in another series containing 37.4m*M* Na^+ and 3.74m*M* K^+ in addition to the Mg^{2+}. These solutions were placed in alternate cups on the sample plate. When Mg^{2+}, Na^+, and K^+ were present, a concentration of 0.000294m*M* ouabain caused a 22% inhibition, and 0.0294m*M* led to an inhibition of 77% of the control ATPase activity. The inhibition by ouabain was found to be uncompetitive (32). In the absence of Na^+ or K^+, even though Mg^{2+} was present, essentially no inhibition was observed at these concentrations of ouabain.

2. Gradient Continuous Flow

The effect of Mg^{2+} on alkaline phosphatase activity has been investigated by using the equalized continuous flow rate technique (16) and the type of manifold shown in Figure 7. β-Glycerophosphate, 29.2m*M*, dissolved in 100m*M* Tris buffer, pH 9.5, containing 1.19*M* Mg^{2+}, was diluted with a solution containing the same concentrations of these constituents except for the magnesium. The delivery rate of the enzyme sample line was 0.36 ml/min. The concentrations of reactants after the inflow of enzyme and before dilution was begun were: human bone phosphatase, 0.6 μg of protein per milliliter; β-glycerophosphate, 25m*M*; Tris buffer, 86m*M*; and Mg^{2+}, 1.02*M*. The enzyme preparation was kept in an ice bath. The chart recording of this experiment is shown in Figure 14. At high concentrations of magnesium the phosphatase activity was inhibited. At 1.02*M* Mg^{2+}, only 23% of the optimal activity was observed. Decreasing inhibition occurred as the magnesium was diluted and

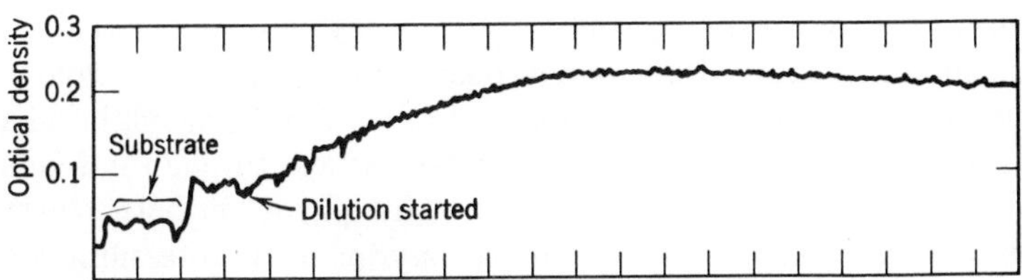

Fig. 14. Recorder tracing of alkaline phosphatase activity during dilution of Mg^{2+} (16). See text for details.

maximum activity was obtained at a magnesium concentration between 32 and 16m*M*. The enzyme activity decreased with further dilution of Mg^{2+} (not shown in Fig. 14). These results are similar to those previously reported for this enzyme from bone or from serum (29,33).

3. Specific Inhibitors

Hill and Sammons (34,35) have devised an automated method for the estimation of serum 5′-nucleotidase activity. The procedure was based on inhibition of the enzyme by Ni^{2+} (36). A set of specimens were run in an AutoAnalyzer manifold without the addition of Ni^{2+} to the reaction mixture. A duplicate set was then run with reaction mixtures containing a final concentration of 10m*M* Ni^{2+}. The reaction mixtures contained 1m*M* adenosine 5′-phosphate and 28m*M* Veronal buffer, pH 7.4, and inflowing enzyme. The temperature of the reaction was 37°C. Before the liberated inorganic phosphate was removed from the reaction mixture by dialysis into a flowing stream on aminonaphthosulfonic acid, the enzyme reaction was stopped by the introduction of the inhibitor EDTA in a concentration of 17m*M*. The nucleotidase activity was defined as the rate of liberation of inorganic phosphate in the absence of Ni^{2+}, subtracted from the rate in the presence of Ni^{2+}.

Fishman and his associates (37) have reported that L-phenylalanine inhibits alkaline phosphatase from human placenta, 79%; intestine, 77%; bone, 10%, liver, 8%; and kidney, 14%. D-Phenylalanine is without effect on any tissue phosphatase. An AutoAnalyzer technique based on the method of Marsh et al. (38) was devised to determine these inhibitory effects (39). The manifold split the enzyme sample into equal parts. One portion was mixed with buffered phenylphosphate containing D-phenylalanine and the other with similarly buffered phenylphosphate containing L-phenylalanine. The final concentrations were reported to be 18mM phenylphosphate and 5mM D- or L-phenylalanine. After passage through individual coils in a 37°C heating bath, the liberated phenol in both reaction mixtures was reacted with aminoantipyrine and potassium ferricyanide. The colored solutions were simultaneously passed through tubular flow cells in two colorimeters, each equipped with 505-mμ filters and the peaks recorded by a dual pen. Therefore, for each specimen two adjacent peaks were obtained, one indicated the total phosphatase activity (D-phenylalanine present) and the other the residual phosphatase activity after L-phenylalanine inhibition.

VII. ENZYME CONCENTRATION

Any of the gradient methods previously described (Sections IV and V) could be used for diluting enzyme and for automated studies of the effects of enzyme concentration on reaction velocity. If a proper measure of reaction velocity, such as the amount of substrate changed at a stated time during the initial zero-order portion of the reaction is used, then the velocity should be directly proportional to the enzyme concentration. A concave deviation from linearity indicates the presence of an activator in the enzyme preparation, while a convex curve indicates the presence of an inhibitor (40). Tappel and Beck (19) used their previously described system (Section IV) to study the effect of varying the concentration of rat liver lysosomal β-glucuronidase on reaction velocity when p-nitrophenyl-β-D-glucuronide was used as substrate. A linear relationship was obtained throughout the entire range of enzyme concentration studied, from 0 to 8×10^{-3} mg of protein per milliliter of reaction mixture. In their study of ferricyanide reduction by homogenates of yeast cells, Roodyn and Wilkie (27) used a gradient system (12) to increase the concentration of homogenate in a linear fashion during the assay.

VIII. MULTIPLE ENZYME ANALYSIS

1. Variable Sampling of Substrate

A number of workers have utilized automated procedures to study several enzymes in the same preparation. Tappel (41) has proposed a method for the simultaneous automated assays of acid phosphatase, β-glucuronidase, and sulfatase in rat-liver lysosomes. Chromogenic substrates, phenolphthalein phosphate, phenolphthalein glucuronide, and nitrocatechol sulfate, respectively, were used in these assays. The substrate solutions were placed in adjacent cups on an Auto-Analyzer sample plate and aspirated at a rate of 0.6 ml/min. The substrates were joined and mixed with a solution of enzyme (0.6 ml/min) and acetate buffer, pH 5.0 (0.6 ml/min). The total reaction mixture was segmented with air and incubated at 37°C for about 8 min. After incubation, 2.0*M* ammonium hydroxide, pH 10.7, joined the reaction mixture and served to stop the enzyme reaction, solubilize the protein, and develop the chromogen color. After heating at 50°C, the intensity of the developed yellow color was recorded by passage through a 15-mm tubular flow cell in a colorimeter equipped with 540-mμ filters. Thus, three adjacent peaks were obtained on the recorder—one representing phosphatase activity, the second β-glucuronidase, and the third sulfatase. This assay system can be applied to any group of enzymes which exhibit similar pH optima and whose activity can be determined with a group of chromogenic substrates which develop their color under the same conditions.

2. Continuous Time Course of Two Enzymes Acting on One Substrate

Frenkel (42) has employed automated methods to study the kinetics of the simultaneous action of hog-liver phosphohexose isomerase and glucose-6-phosphate dehydrogenase on glucose-6-phosphate; the product of the glucose-6-phosphate dehydrogenase reaction, 6-phosphogluconate, exerted an inhibitory effect on phosphohexose isomerase. The enzyme preparations were purified 16- to 18-fold, and each enzyme was free of the other as well as of 6-phosphogluconic dehydrogenase. Evaluation of this system with manual assays would be difficult since both the concentration of substrate and inhibitor were changing continuously and simultaneously.

In order to study these reactions, a single manifold was devised for measuring simultaneously the activity of the two enzymes in a single reaction mixture (Fig. 15). The reaction mixture, maintained at 37°C, contained glycylglycine buffer, pH 8.0, various concentrations of glucose-6-phosphate, and 0.1mM NADP. The reaction was started by the addition of the two enzymes. One aspirating tube (E, Fig. 15) withdrew reaction mixture for the measurement at 340 mμ of NADH formed during the glucose-6-phosphate dehydrogenase reaction and another tube (C, Fig. 15) simultaneously withdrew reaction mixture for the colorimetric determination of fructose-6-phosphate, the product of the phosphohexose isomerase reaction. Deflections representing enzyme activity appeared on the recorders 40 sec after the start of flow from the reaction vessel to the point (H, Fig. 15) where the isomerase reaction was stopped by the addition of 50% sulfuric acid and where the dehydrogenase reaction was stopped (I, Fig. 15) by the addition of 50% ethanol. The manifold can be used for the assay of the simultaneous reactions or for each of the enzymes by disconnecting one side. In Figure 16 is shown a drawing of the recorder tracing of the phosphohexose isomerase reaction during the simultaneous action of both enzymes on 14mM glucose-6-phosphate. Conditions were chosen so that the isomerase

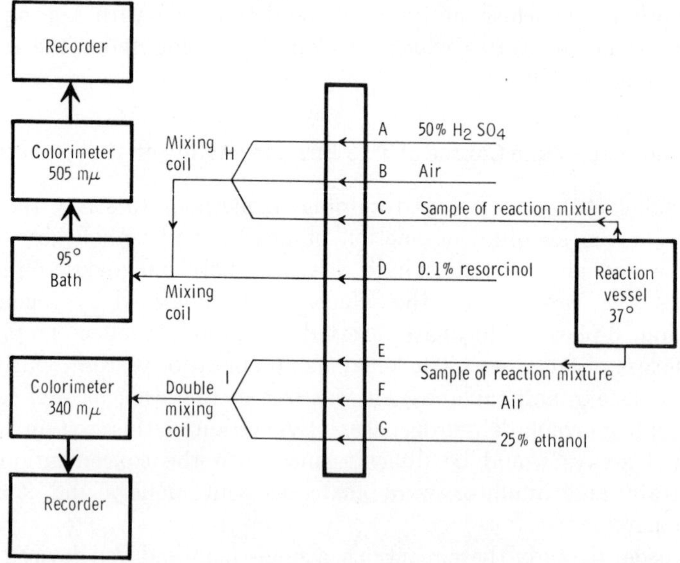

Fig. 15. Schematic outline for simultaneous automated assay of phosphohexose isomerase and glucose-6-phosphate dehydrogenase activity (42). See text for details.

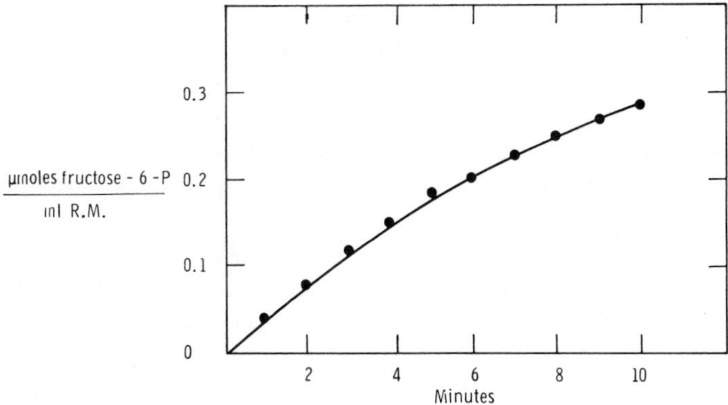

Fig. 16. Time course of action of phosphohexose isomerase on glucose-6-phosphate during simultaneous action of phosphohexose isomerase and glucose-6-phosphate dehydrogenase. The closed circles represent theoretical points (42).

reaction alone would be linear during this period. However, 6-phosphogluconate, formed in the glucose-6-phosphate dehydrogenase reaction, inhibits the activity and creates a deviation from linearity. A theoretical equation was derived to express the course of this reaction (42), and the closed circles in Figure 16 are theoretical points which agree very well with the experimental tracing.

3. Enzyme Induction

Dealy and Umbreit (43) applied automated methods to the study of multiple enzyme activity during enzyme synthesis in *Escherichia coli*. An AutoAnalyzer system was used, but details of the manifold were not given. Presumably, the suspension of bacteria continuously aspirated from the culture was split, and β-galactosidase, acid phosphatase, and alkaline phosphatase activity were determined simultaneously by use of separate but joined manifolds and the passage of the reaction mixtures through individual colorimeters. In addition to the enzyme assays, protein, DNA, RNA, and turbidity as indications of growth were also simultaneously determined by automated means.

In the case of β-galactosidase the aspirated cells were first mixed with the inducer, 10% lactose containing 0.2% Tween 20, and incubated at 37°C for 20 min. The cells were disrupted by mixing with a stream of toluene. The flowing suspension was passed through an extraction cell and the toluene layer discarded. The enzyme preparation was mixed with 1m*M* *o*-nitrophenyl-β-D-galactoside dissolved in 0.35*M* sodium phosphate buffer, pH 7.4, and incubated at 37°C for 20 min. After incubation 1.0*M* sodium carbonate was introduced and the intensity of formed yellow color recorded at 420 mμ.

Alkaline phosphatase activity was determined by mixing the aspirated bacterial suspension with 100m*M* Tris buffer, pH 8.8, containing 0.2% Tween 20. For the determination of acid phosphatase activity, the bacterial suspension was mixed with a solution containing 200m*M* acetate buffer, pH 4.0, and 0.2% Tween 20. After treatment with toluene the flowing streams were mixed with a solution of 1m*M* nitrophenylphosphate adjusted to pH 4.0 for determination of acid phosphatase activity and to pH 8.8 for determination of alkaline phosphatase activity. The reaction mixtures were incubated at 37°C for 20 min by passage through a suitable coil. As in the case of the β-galactosidase assay, 1.0*M* sodium carbonate was then added, and the intensity of the formed yellow color was recorded after passage through a tubular flow cell equipped with 420-mμ filters.

β-Galactosidase induction proceeded at an exponential rate during the early log phase of growth, but reached a stationary level just before the late log phase (after about 4 hr). These experiments were conducted with a strain of bacteria (*E. coli* B) which requires lactose for the induction of β-galactosidase. The studies indicate the ability of cells of various ages grown without lactose to induce enzyme when they were brought into contact with the inducer for 20 min. Under these conditions, β-galactosidase induction was very rapid and measurable amounts of enzyme were present within 2 min after the addition of lactose. When the experiments were conducted with a strain of bacteria (*E. coli* 15 ATU) which synthesizes β-galactosidase without the presence of an inducer, the rate of enzyme synthesis paralleled the growth rate as reflected by the measurement of turbidity, RNA, DNA, and protein.

A 3-hr lag period in the synthesis of alkaline phosphatase was observed when *E. coli* 15 ATU was grown in a low-phosphate medium. In a high-phosphate medium, synthesis of enzyme paralleled growth of bacteria, as reflected by turbidity measurements. In the low-phosphate medium, acid phosphatase synthesis paralleled growth until phosphate was depleted, but the synthesis of alkaline phosphatase continued.

IX. COMPUTER TECHNIQUES

In a previous volume of this series, the present authors (1) considered the use of computers in automated enzyme studies. Chance and his associates have applied computerization to studies of kinetics of multiple enzyme systems and of cellular intermediary metabolism. Subsequently, Chance et al. (44) reviewed the experimental methods used in studies of metabolic regulation and its representation by digital computer technique. A number of other workers have also described methods for computer calculation of enzyme reaction constants (45–48).

Cleland (45) has written FORTRAN language programs for computer handling of square fits or for equations used in enzyme kinetic studies. The computer presents a rapid calculation of the kinetic constants as well as the standard error. The programs were devised for use with an IBM 1620 or Control Data Corporation 1604 computer, but can be used with any computer that accepts FORTRAN language. Among the computer programs described by Cleland is one for analysis of substrate concentration–enzyme velocity data by the equation:

$$v = VS/(K + S) \tag{3}$$

where v is the reaction velocity, S the substrate concentration, V the velocity at maximal substrate concentration, and K the Michaelis constant. Cleland's procedure consisted in graphing manually $1/v$ vs. $1/S$, discarding points which clearly indicated effects of substrate inhibition or were "outside the expected range of experimental deviation." However, no criteria for making these judgements were presented. The data were then fed into the computer. The processing time for each set of data was 1–2 min with the IBM 1620 and 1 sec with the CDC 1604 computer. The values of K, V, K/V, V/K, and $1/V$, as well as the standard errors, were recorded. In addition, programs were written for a variety of situations such as the assumed combination of two molecules of substrate with one molecule of enzyme or for linear, parabolic, and hyberbolic enzyme–substrate–inhibitor reactions.

Higgins (46) has described a "hybrid" computer for investigation of chemical reactions (The Johnson Foundation Electronic Computer, Mark II). This instrument is capable of solving and analyzing ordinary nonlinear differential equations arising from the law of mass action. In 100 μsec it is capable of handling 14 bimolecular reactions and the possible effects on these reactions of 25 different compounds. The instrument has been used to analyze and solve the Michaelis-Menten nonlinear differential equations:

$$dx/dt = k_1 X(e - p) \tag{4}$$

and

$$dp/dt = k_1 X(e - p) - k_2 p \tag{5}$$

when dx/dt is the change in substrate in time and dp/dt is the simultaneous formation of an enzyme substrate complex; e is the concentration of enzyme and k_1 and k_2 are the rate constants. After feeding the data into the computer, the relationships noted are immediately visualized for specific conditions on a calibrated oscilloscope, and a camera is used to obtain a permanent record.

Kerson and his associates (48) have constructed a digital computer model for the pyruvate kinase reaction using a CDC 160-A digital computer and a compiler which, starting from a set of initial conditions, transforms chemical equations into instructions for solving the corresponding differential equations. The enzyme-simulated model included 31 related equations for the enzyme reaction without activation by a divalent cation and 64 equations for the reaction in the presence of Mg^{2+}. The reactions studied included inhibition by ATP, inorganic phosphate, pyruvate, and AMP under aerobic and anaerobic conditions. Experimental laboratory data using rabbit muscle pyruvate kinase confirmed the kinetic feasibility of the system.

X. CONCLUDING REMARKS

The present article has reviewed available methods for the automated study of enzyme kinetics. With these techniques it is possible to automate all aspects of enzyme kinetics: effects of varying the concentration of enzyme, substrate, hydrogen ion, activators, inhibitors, temperature, and reaction time. Consideration has also been given to simultaneous automated assay of several enzymes in the same preparation as well as computer calculation of kinetic constants. The advantages of these methods to the investigator have been pointed out. The eventual complete automation (third stage) (1) of these techniques will be achieved by combining automated enzyme methods with computer techniques to permit simultaneous automation of the reaction and analysis of the data.

References

1. M. K. Schwartz and O. Bodansky, *Methods of Biochemical Analysis,* Vol. 11 D. Glick, Ed., Interscience, New York, 1963, p. 211.
2. W. J. Blaedel and G. P. Hicks, in *Advances in Analytical Chemistry and Instrumentation,* Vol. 3, C. N. Reilly, Ed., Interscience, New York, 1964, p. 105.
3. G. G. Guilbault, *Anal. Chem., 38,* 527R (1966).
4. M. K. Schwartz, paper presented at Technicon European Symposium, Automation in Analytical Chemistry, Paris, 1966.
5. R. B. McComb, R. V. LaMotta, and H. J. Wetstone, *Clin. Chem., 11,* 645 (1965).
6. A. L. Babson, S. J. Greeley, C. M. Coleman, and G. E. Phillips, *Clin. Chem., 12,* 482 (1966).
7. T. O. Hausamen, R. Helger, W. Rick, and W. Gross, *Clin. Chem. Acta, 15,* 241 (1967).
8. G. N. Bowers, Jr. and R. M. McComb, *Clin. Chem., 12,* 70 (1966).
9. G. Szasz, *Am. J. Clin. Pathol., 47,* 607 (1967).
10. W. J. Blaedel and G. P. Hicks, *Anal. Biochem., 4,* 476 (1962).
11. H. C. Pitot, N. Pries, M. Poirier, and A. Cutler, in *Automation in Analytical Chemistry,* L. Skeggs, Ed., Mediad, New York, 1965, p. 555.
12. D. B. Roodyn, *Nature, 206,* 1226 (1965).
13. H. C. Pitot and N. Pries, *Anal. Biochem., 9,* 454 (1964).
14. J. Boy, in Internationales Technicon Symposium, *Automation in der Analytischen Chemie,* Technicon, Frankfurt, Germany, 1963, p. 105.
15. J. Lenard, S. L. Johnson, R. W. Hyman, and G. P. Hess, *Anal. Biochem., 11,* 30 (1965).
16. M. K. Schwartz and O. Bodansky, in *Automation in Analytical Chemistry,* Vol. 1, L. Skeggs, Ed., Mediad, New York, 1967, p. 489.
17. M. K. Schwartz, G. Kessler, and O. Bodansky, *Am. J. Clin. Pathol., 33,* 275 (1960).
18. P. M. Serrone, A. A. Stein, E. A. Menegaux, M. A. Gallo, and F. Coulston, in *Automation in Analytical Chemistry,* L. Skeggs, Ed., Mediad, New York, 1965, p. 586.
19. A. L. Tappel and C. Beck, in *Automation in Analytical Chemistry,* L. Skeggs, Ed., Mediad, New York, 1966, p. 559.
20. G. Brehmer, K. H. Holzer, and G. Binzus, *Klin. Wochschr., 41,* 40 (1963).

21. S. Posen, F. C. Neale, J. Brudenell-Woods, and D. J. Birkett, *Lancet, 1966-I*, 264.
22. P. N. Kitchener, F. C. Neale, S. Posen, and J. Brudenell-Woods, *Am. J. Clin. Path.*, *44*, 654 (1965).
23. P. E. Strandjord and K. J. Clayson, *J. Lab Clin. Med.*, *67*, 131 (1966).
24. J. S. Nisselbaum, D. E. Packer, and O. Bodansky, *J. Biol. Chem.*, *239*, 2830 (1964).
25. J. E. Taylor and S. L. Arora, *Public Health Rept.*, *81*, 466 (1966).
26. D. B. Roodyn, in *Automation in Analytical Chemistry*, L. Skeggs, Ed., Mediad, New York, 1966, p. 593.
27. D. B. Roodyn and D. Wilkie, *Biochem. J.*, *103*, 3c (1967).
28. J. S. Nisselbaum and O. Bodansky, *J. Biol. Chem.*, *238*, 969 (1963).
29. O. Bodansky and M. K. Schwartz, *J. Biol. Chem.*, *238*, 3420 (1963).
30. M. Dixon and E. C. Webb, *Enzymes*, Academic Press, New York, 1955.
31. M. K. Schwartz, G. Kessler, and O. Bodansky, *J. Biol. Chem.*, *236*, 1207 (1961).
32. H. H. Stein, A. J. Glassky, and W. R. Roderick, *Ann. N. Y. Acad. Sci.*, *130* 751 (1965).
33. M. K. Schwartz and O. Bodansky, *Am. J. Clin. Pathol.*, *42*, 572 (1964).
34. P. G. Hill and H. G. Sammons, *Clin. Chem. Acta*, *13*, 739 (1966).
35. P. G. Hill and H. G. Sammons, in *Automation in Analytical Chemistry*, L. Skeggs, Ed., Mediad, New York, 1965, p. 601.
36. D. M. Campbell, *Biochem. J.*, *82*, 34p (1962).
37. W. H. Fishman, N. I. Inglis, F. Sarke, and N. K. Ghosh, *Federation Proc.*, *25*, 748 (1966).
38. W. Marsh, B. Fingerhut, and E. Kirsch, *Clin. Chem.*, *5*, 119 (1959).
39. W. H. Fishman and N. K. Ghosh, *Advan. Clin. Chem.*, *10*, 255 (1968).
40. O. Bodansky, *Am. J. Med.*, *27*, 801 (1959).
41. A. L. Tappel, paper presented at 1964 Technicon International Symposium, New York, 1964.
42. R. Frenkel, dissertation, Cornell University Graduate School of Medical Sciences, 1964.
43. J. D. Dealy and W. W. Umbreit, *Ann. N. Y. Acad. Sci.*, *130*, 745 (1965).
44. B. Chance, A. Ghosh, J. J. Higgins, and P. K. Maitra, *Ann. N. Y. Acad. Sci.*, *115*, 1010 (1963).
45. W. W. Cleland, *Nature*, *198*, 463 (1963).
46. J. J. Higgins, *Ann. N. Y. Acad. Sci.*, *115*, 1025 (1964).
47. D. Garfinkel, S. W. Ching, M. Adelman, and P. Clark, *Ann. N. Y. Acad. Sci.*, *128*, 1054 (1966).
48. L. A. Kerson, D. Garfinkel, and A. S. Mildvan, *J. Biol. Chem.*, *242*, 2124 (1967).

METHODS OF BIOCHEMICAL ANALYSIS VOLUME 16

Enzymatic Synthesis and Hydrolysis of Cholesterol Esters

GEORGE V. VAHOUNY AND C. R. TREADWELL, *Department of Biochemistry, The George Washington University School of Medicine, Washington, D.C.*

I. INTRODUCTION

In the metabolism of cholesterol esters by the animal organism, extensive hydrolysis and reesterification occur, beginning with digestion and absorption and continuing through transport, deposition, and metabolic transformations. While cholesterol is not an essential dietary constituent, it is intimately involved in the intraluminar physicochemical organization of lipids and their digestion products. During digestion, a very active sterol ester hydrolase from pancreatic juice hydrolyzes cholesterol esters in the presence of trihydroxyl bile salts, producing free fatty acids and the unesterified sterol, the form required for absorption into the intestinal mucosal cell (103,137).

The intracellular transport of cholesterol is poorly understood, but at some point during transport cholesterol is reesterified with fatty acids by an intestinal sterol ester hydrolase which has several properties similar to those of the enzyme of pancreatic juice. Since there is little or no net accumulation of cholesterol esters in the mucosal cell during absorption, the esters are rapidly incorporated into chylomicrons and transported into the lymphatic system. Of the newly appearing cholesterol in lymph during absorption, between 85 and 95% exists in the esterified form, with the predominant ester as cholesterol oleate (46,132).

The primary metabolic fate of newly absorbed cholesterol esters appears to be a rather selective uptake by the liver (45), although recently it has been shown that several extrahepatic tissues are capable of removing and hydrolyzing lymph cholesterol esters (11). In the liver, as well as in extrahepatic tissues, there is slow hydrolysis of the chylomicron cholesterol esters by a low-energy sterol ester hydrolase, and subsequent resynthesis of esters occurs with a fatty acid distribution, dependent on tissue enzyme specificity and diet. The enzyme(s) mediating the formation of sterol esters in liver requires ATP and coenzyme A; the mechanism of the esterification reaction appears to proceed via the fatty acyl-coenzyme A intermediate. Following the release from the liver of free and esterified cholesterol in lipoprotein complexes, a plasma-specific transferase enzyme may be involved in the final regulation of the "normal" plasma cholesterol-ester fatty acid pattern which varies among species and with nutritional state.

In this presentation, three types of enzyme systems involved in synthesis and hydrolysis of cholesterol esters are described in some detail: the low-energy sterol ester hydrolases, a plasma-specific transferase enzyme(s), and ATP- and coenzyme A-requiring systems of liver and adrenal. For a comprehensive review of cholesterol absorption and cholesterol ester metabolism, the reader is referred to several recent communications (46,73,132,145).

II. TERMINOLOGY

Until recently, enzymes synthesizing and hydrolyzing cholesterol esters in various tissues were arbitrarily termed "cholesterol esterases," irrespective of the reaction mechanism or substrate specificities.

Very few of these reports contained any specificity studies with respect to the requirement for sterol esters as substrates (as opposed to esters of other alcohols, either primary, secondary, or tertiary). Thus, an enzyme hydrolyzing cholesterol acetate, but not cholesterol oleate, was referred to as cholesterol esterase, even though the hydrolysis of methyl esters of acetate, butyrate, oleate, or even glyceryl esters of these acids was not determined.

Earlier studies on distribution of "cholesterol esterases" often indicated the presence of either cholesterol ester synthesizing or hydrolyzing activities, or both, and such activities were referred to as synthetic or hydrolytic esterase activities. Since there are no definitive data indicating whether one or more enzymes are involved in these two activities in various tissues, the following names for enzymes synthesizing and/or hydrolyzing sterol esters are suggested, based on the Report of the Commission on Enzymes of the International Union of Biochemistry (25):

1. Sterol-ester hydrolase (E.C. 3.1.1.13), cholesterol esterase

$$\text{Cholesterol} + \text{fatty acid} \rightleftarrows \text{cholesterol ester}$$

2. Acyl-CoA:cholesterol *O*-acyltransferase (E.C. 2.3.1.)

$$\text{Acyl-CoA} + \text{cholesterol} \rightarrow \text{cholesterol ester} + \text{CoA}$$

3. Lecithin:cholesterol *O*-acyltransferase (E.C. 2.3.1.) (1,2-diacyl glycerylphosphorylcholine:cholesterol *O*-acyltransferase)

$$\text{1,2-Diacyl glycerylphosphorylcholine} + \text{cholesterol} \rightarrow \text{1-monoacyl glycerylphosphorylcholine} + \text{cholesterol ester}$$

III. GENERAL METHODS FOR DETERMINATION OF HYDROLYSIS AND ESTERIFICATION

1. Availability and Preparation of Substrates

Unlabeled cholesterol, fatty acids of all chain lengths, and corresponding cholesterol esters (except arachidonate) are commercially available in high purity. Similarly, carbon- or tritium-labeled cholesterol, fatty acids, and several long-chain fatty acid esters of cholesterol are commercially available; however, in certain instances, the purity of these radiochemicals should be checked and the compounds repurified if necessary. The most direct method for determination of radiopurity is by gas–liquid radiochromatography of sterols (117), fatty acids (e.g., 55,91), and sterol esters (e.g., 63,118).

Several methods are available for the synthesis of labeled cholesterol esters. Prior to preparation, the labeled sterol is purified by silicic acid column chromatography or, on a larger scale, can be purified as the dibromide (101). Fatty acids currently available are of sufficient purity ($> 99\%$). The most common method for large-scale synthesis of sterol esters involves the reaction of cholesterol with the fatty acyl chloride in the presence of pyridine (129). The acyl chloride is readily prepared by addition of excess thionyl chloride to the fatty acid; however, this procedure is unsuitable for preparation of unsaturated fatty acid esters of cholesterol. A reaction more suited to preparation of unsaturated esters involves an ester-interchange reaction (70) in which fresh sodium ethylate is added to cholesterol acetate and the desired fatty acid methyl ester.

Microscale synthesis can be accomplished by modification of the acyl chloride reaction (24), the ester-interchange method, or by an enzymatic procedure (130). This last method utilizes commercial pancreas powder for synthesis of saturated, unsaturated, and polyunsaturated fatty acid esters of cholesterol in amounts as low as 1 mg. Using the described method, high specific-activity esters are obtained in 70% yields with purities up to 99%. With respect to other radioactive sterols, there is a recent description of the preparation and purification of several sterol isomers by Smith (106).

2. Separation and Determination of Free and Esterified Cholesterol

The majority of methods reported for measuring enzymatic esterification and/or hydrolysis of sterol esters is based on determination of the loss or production of free or esterified sterol during the reaction. Several excellent reagents are available for quantitative sterol analysis, but all suffer the disadvantage that the free and esterified sterol must first be extracted and separated prior to spectrophotometric or isotope analysis. This is, by necessity, time consuming, and therefore makes all enzymatic assays lengthy and tedious. Various procedures for the extraction, isolation, and analysis of cholesterol are described in detail in Volume 10 of this series (54) and are only briefly summarized below.

A. EXTRACTION PROCEDURES

In the assay procedures the reaction is terminated and the free and esterified cholesterol are extracted by transferring aliquots of the reaction medium to 20 volumes ethanol–ether (3:1, v/v) or chloroform–methanol (2:1, v/v). With ethanol–ether, the extract is brought to a rolling boil, cooled, made to volume with solvent, and filtered to remove protein. The chloroform–methanol extract is allowed to stand 30 min before separation into two phases by addition of 0.4 volumes of water or salt solution (32). The lower chloroform phase contains cholesterol and its esters and is handled as described in subsequent sections.

B. PRECIPITATION OF STEROL DIGITONIDES

Precipitation of free cholesterol as the digitionide is accomplished from ethanol–ether extracts according to the procedure of Sperry and Webb (114) or by use of aluminum chloride as a gathering agent (135). The precipitated cholesterol digitonide is freed of contaminating lipids and digitonin by successive solvent washings (114,135). Total cholesterol (free + esterified) can be precipitated using the above procedure by initially hydrolyzing an aliquot of the ethanol–ether extract with alkali and neutralizing prior to addition of digitonin (114). Similarly, the supernatant from free cholesterol precipitation can be hydrolyzed and precipitated with digitonin to give esterified sterol values directly. The purified sterol digitonides are then dissolved in glacial acetic acid for spectrophotometric determination using Liebermann-Burchardt reagent (114), anthrone reagent (135), or ferric chloride reagent (147). Alternatively, the sterol digitonides are dissolved in 1–2 ml of methanol, and aliquots can be removed for simultaneous counting and spectrophotometric determination (137).

C. COLUMN CHROMATOGRAPHY

There are several excellent procedures for separation of lipid classes by silicic acid column chromatography (4,15,29,53), but these procedures should be standardized in each laboratory. A useful general procedure is as follows.

The alcohol–ether extract or chloroform phase is evaporated to dryness under N_2 and the sample is redissolved in 1–2 ml of hexane or petroleum ether. A convenient column is a 10-ml syringe barrel containing 4 g of silicic acid (100–200 mesh) prepared as a slurry in 3 ml of hexane or petroleum ether prior to loading. After applying the lipid sample (maximum load, 60 mg), cholesterol esters are eluted with 75 ml of 1% ethyl ether in petroleum ether; free cholesterol (including glycerides and free fatty acids) is eluted with 75 ml of 100% ethyl ether.

Another convenient and simple column separation of free and esterified cholesterol is described by Deykin and Goodman (24). The lipid sample in petroleum ether is placed on a small column containing 2 g of alumina, and sterol esters are eluted with 15 ml of benzene–petroleum ether (3:7, v/v). Free cholesterol (and other lipids) is subsequently eluted with 10 ml of acetone–ether (1:1, v/v).

After separation of free and esterified cholesterol, these can be determined spectrophotometrically or aliquots can be used for isotope determination.

D. THIN-LAYER SILICIC ACID CHROMATOGRAPHY (TLC)

This procedure is most commonly employed for separation of lipid classes in preparative or microgram amounts (45,71,115,116). A solvent system effective in separating lipid classes by TLC is hexane–ethyl ether–glacial acetic acid (83:16:1). The separated lipid classes are visualized in an iodine atmosphere or with 2′,7′-dichlorofluorescein. Silicic acid areas corresponding to free and esterified cholesterol are scraped and eluted with methanol prior to spectrophotometric or isotope analysis. A simple and rapid procedure for radioassay of free and esterified cholesterol is described by Vahouny et al. (136), in which microscope slides are spotted, developed, and equal halves of the silicic acid layer are scraped directly into scintillation vials. After addition of 1 ml of methanol, 10 ml of scintillation mixture [100 mg of 1,4-di-(2,5-phenyloxazolyl)benzene and 4 g of 2,5-diphenyloxazole per liter of toluene] is added. After shaking, the silicic acid is allowed to settle prior to liquid scintillation counting.

Further separation of cholesterol esters based on the degree of unsaturation of the fatty acid moiety is accomplished by silicic acid column chromatography (57) or thin-layer chromatography (75), using silicic acid impregnated with silver nitrate (5%, w/w). In this latter method, the cholesterol esters (up to 50 mg) are spotted on the plate and developed in diethyl ether–benzene (1:4, v/v), giving separation of cholesterol esters into groups of saturated, monounsaturated, diunsaturated, and polyunsaturated fatty acid esters. If desired, the polyunsaturated fatty acid esters can be further subfractionated on plates using diethyl ether as developing solvent (75).

E. GAS–LIQUID CHROMATOGRAPHY

Free fatty acids or cholesterol ester fatty acids can be determined quantitatively for mass and/or radioactivity by gas–liquid chromatography using any one of several suitable column packings and detectors (e.g., 55,91). Recently it has become possible to determine mass and/or radioactivity of cholesterol (as the trimethylsilyl ethers) with good quantitation (117). Also, gas–liquid radio chromatographic separation of intact cholesterol esters (both tritium- and carbon-labeled) has been achieved (118) with excellent resolution and quantitation.

IV. CHOLESTEROL ESTERASES (STEROL ESTER HYDROLASES)

1. Pancreas and Intestine

A. INTRODUCTION AND MECHANISM

Mueller (77) in 1915 demonstrated that cholesterol or cholesterol esters fed to dogs were readily absorbed into the thoracic lymph of the dog, and that regardless of the form in which the sterol was administered, absorption occurred with the same proportion of free and esterified cholesterol as was normally seen in lymph. After showing that both pancreatic juice and bile were intimately concerned with sterol absorption, Mueller (78) demonstrated that glycerol–water extracts of pancreas would promote esterification of substrate cholesterol but only in the presence of bile. However, contrary to an earlier report on pancreatic (1) hydrolysis of serum cholesterol esters, Mueller could not demonstrate pancreatic enzymatic hydrolysis of cholesterol oleate. Twenty years later, Nedswedski

(83–85) showed that a preparation of dried pancreas stimulated synthesis of cholesterol esters from colloidal substrates containing cholesterol and sodium salts of palmitic, stearic, or oleic acids. However, esterification was extremely slow if bile salts (cholate or its conjugates) were omitted. This bile salt effect was presumed due to enzyme stabilization (84). In 1938, Klein (58) showed hydrolysis of serum cholesterol esters and colloidal cholesterol palmitate and oleate by various bovine tissues, including pancreas and intestine, and subsequently (59) this worker confirmed the requirement for cholic acid and partially purified the enzyme.

In the 1940's, a French group (33,65,66) reported that dog and pig pancreas extracts effectively hydrolyzed substrate serum cholesterol esters at pH 7, while dog pancreatic juice had hydrolytic and synthetic activity at different pH's. Both lipase and cholesterol esterase activities of pancreatic juice were stimulated by addition of bile salts (65).

Nieft and Deuel (87) showed esterification of colloidal cholesterol substrates by preparations of rat intestine, but only after prefeeding cholesterol or lanolin for 2–3 weeks, and that hydrolysis of cholesterol esters was demonstrable only after dialysis of crude intestinal extracts. This latter activity was stimulated by addition of soy lecithin. Subsequently, Swell et al. (120), using aqueous emulsions of bile salts, cholesterol, and oleic acid, or cholesterol oleate, demonstrated the presence of hydrolyzing and esterifying activities in homogenates of rat intestine at pH 6.5 and 6.2, respectively. The enzyme required bile salt for activity and was substantially reduced in pancreatectomized animals, suggesting an intimate relationship of the enzymes from the two tissues.

Beginning in 1950, Swell et al. began an extensive series of studies on sterol-ester hydrolases of various tissues (119,121–123,127,129). These studies, which are described in later sections, included preparation of stable substrates, methods of enzyme extraction, optimal substrate concentrations, inactivation data, role of emulsifiers, optimal pH's, fatty acid and sterol specificities, comparative activation by different bile salts, and identity of the enzyme.

Recent investigations of the enzymes of pancreas using manometric (51,60,61), turbidometric (51), and mass analysis (80) have further elucidated the properties of pancreatic sterol-ester hydrolase and, to a lesser extent, the intestinal enzyme (8,52,80). In addition to reports of purification of the enzyme activities of pancreas (51,80), the properties (141) and purification of pancreatic juice sterol-ester hydrolase are considered in subsequent sections.

Based on available data, the following is the proposed mechanism for sterol-ester hydrolase of pancreas and intestinal mucosa:

$$\text{Steroi} + \text{unionized fatty acid} \underset{\substack{\text{enzyme–bile salt}\\ \text{complex}}}{\overset{}{\rightleftarrows}} \text{sterol ester}$$

There is no requirement for ATP and coenzyme A with either the purified pancreas enzyme (51,80) or with intestinal preparations (36,69), indicating that a high-energy acyl–coenzyme A intermediate is not involved in the esterification. Also, recent evidence (141) substantiates the earlier suggestion (51,80,122) that the bile salts act as specific cofactors for the enzyme protein. This aspect is more fully discussed in a later section.

B. ASSAY METHODS

a. Nephelometric Assay (51). When aqueous emulsions containing cholesterol, fatty acid, and an emulsifying agent such as bile salts are exposed to cholesterol esterases at pH 6.2, the formation of cholesterol ester is accompanied by an increase in turbidity of the assay mixture due to the less-soluble ester. Advantage was taken of this observation by Hernandez and Chaikoff (51), who describe a nephelometric procedure for determination of cholesterol esterification by rat pancreas preparations.

A clear emulsion of cholesterol (1mM) and oleic acid (2mM) is prepared by adding an ethereal solution of these lipid components to 224 ml of 0.154M phosphate buffer containing 2.4 g of sodium taurocholate. The ether is evaporated on a hot plate with rapid swirling, resulting in an aqueous emulsion showing a tyndall effect. Incubation of 7 ml of the substrate mixture with 0.5 ml of pancreatic enzyme is carried out for 30 min at 37°, measuring the increase in turbidity after 15 and 30 min, using a Klett photometer without a filter. The difference in turbidity between the 15 and 30-min readings is used as a measure of enzyme activity with a unit of activity defined as that amount of enzyme causing an increase in turbidity of 1 Klett unit/min during the second 15 min of incubation. However, the increase in turbidity during esterification is nonlinear, and constant and precise control is required in order to obtain accurate results.

The nephelometric procedure has a distinct advantage in terms of rapidity of assay but suffers many disadvantages peculiar to nephelometric methods. The use of different fatty acids will produce esters of differing solubilities, such that a direct comparison of formation of various cholesterol esters is arbitrary. The use of different emulsifiers, especially of different bile salts, can and does have a marked effect on the turbidity of the incubation mixture during esterification of cholesterol [and indeed on the solubilities of various esters in micellar media (140)]. Addition of various agents, such as heavy metal ions or other stabilizing agents, such as protein, influences the rate and extent of turbidity; the method cannot be used to measure hydrolysis of cholesterol esters since, in general, there is little if any decrease in the turbidity of ester substrates during enzymatic hydrolysis. In general, therefore, the nephelometric procedure has limited usefulness and at best gives empirical data.

b. Manometric Assay (61). During hydrolysis of cholesterol esters, CO_2 gas is released from an aqueous bicarbonate–CO_2 buffer and can be measured manometrically. Conversely, during esterification of cholesterol with fatty acids, CO_2 is absorbed into the buffer system. The method described below is taken from Korzenovsky et al. (61) and has been used by others (51).

The concentrations of components in the substrate mixture used for esterification studies are 0.04*M* cholesterol, 0.12*M* lithium oleate, and 0.02*M* cholic acid. These compounds are placed in distilled water and one-half volume of ethyl ether is added. The mixture is shaken until dissolution of the components and the ether is evaporated by warming (below 50°C). The mixture is adjusted to pH 6.6 with 0.1*M* H_2SO_4, ether is again added, and the mixture is reemulsified by shaking. After evaporation of the ether by stirring and warming, 0.8 ml of the colloidal substrate is used for assay. The total fluid volume during incubation should not exceed 2.2 ml in order to insure effective gas equilibration. Thus, in a final volume of 2.2 ml, which includes 0.3 ml of enzyme, the concentrations of components are: 0.0145*M* cholesterol, 0.044*M* lithium oleate, 7.3m*M* cholic acid, 0.014*M* $NaHCO_3$, and 0.09*M* NaCl or 0.045*M* $(NH_4)_2SO_4$. The complete reaction mixture is gassed with CO_2 and the final pH is 6.1–6.2.

A similar procedure is followed for hydrolysis, except that sterol ester, cholic acid, and lithium oleate are placed in 0.67% bovine serum albumin prior to ether addition. After stirring and evaporation of solvent, the mixture is made to volume and adjusted to pH 6.9–7.0. The final concentration of components in 2.2 ml are: 0.025*M* cholesterol oleate, 0.0127*M* lithium oleate, 5.1m*M* cholic acid, 0.24% albumin, 0.18*M* $NaHCO_3$, and 0.72*M* $(NH_4)_2SO_4$. After gassing with CO_2, the pH is 6.9–7.0. Reactions are conducted at 37°C in Warburg flasks with enzyme in the side arm and substrate in the flask. After temperature equilibration and gassing with CO_2, the enzyme is added to the substrate mixture. Initial readings are made after 3–6 min, and final readings are taken after 30 min of incubation.

Although the authors show convincing evidence concerning the validity of the described method (61), manometric assays require careful control and by necessity are subject to certain limitations, similar to those involving titrimetric assays. The presence of other buffers, such as occur in tissue homogenates, the formation of insoluble carbonates, and enzyme inactivation due to the rapid shaking needed for CO_2 equilibration are only a few of the limitations.

c. Spectrophotometric and Radioassay. (*1*) *Substrate Preparation and Conditions.* Basically, all substrate preparations used in recent studies of pancreatic or intestinal cholesterol ester hydrolase activities are similar, with slight modifications in concentration of substrates or volume of assay mixture. For spectrophotometric determination of cholesterol esterification (8), the substrate mixture contains 80 μmoles of cholesterol, 240 μmoles of oleic acid (or other fatty acids), 160 μmoles of sodium taurocholate, and 30.8 mg of bovine serum albumin fraction V, in 9 ml of 0.154*M* phosphate buffer, pH 6.2. For hydrolysis of cholesterol ester, oleic acid is omitted; cholesterol oleate (or other cholestereol esters) equivalent to 80 μmoles of free cholesterol and pH 6.6 phosphate buffer are used. The water-insoluble components are placed in ether in a Potter-Elevehjam homogenizing tube, the taurocholate, albumin, and buffer are added, and the mixture is homogenized. The ether is then evaporated at 37°C under N_2, and the mixture is rehomogenized (or sonicated). This substrate can be kept refrigerated and then rehomogenized before use. Nine milliliters of substrate mixture and 1 ml of enzyme preparation are incubated at 37°C with shaking; 1-ml aliquots are removed initially and at intervals for extraction and analysis of free and esterified cholesterol.

Recent studies (140,141) have indicated that the physical state of the substrate markedly influences the rate of cholesterol ester hydrolysis. Cholesterol ester dispersed in lecithin–taurocholate micelles are more rapidly hydrolyzed than when stabilized with albumin. However, due to the limited solubility of cholesterol esters in micellar media, the following modifications of the above procedure are suggested (141): Albumin-stabilized substrates are used for cholesterol esterification measurements using 9 mg of cholesterol, 19.8 mg of oleic acid, 26.2 mg of sodium taurocholate, and $0.154M$ phosphate buffer, pH 6.2, made up to 9 ml. For hydrolysis measurements using micellar media, 15 mg of cholesterol oleate is substituted for cholesterol and oleic acid, and 27 mg of lecithin is added to 9 ml of phosphate buffer, pH 6.6. The substrates are homogenized as above, ether is evaporated, and the mixture is sonicated for 30 min. The micellar media for hydrolysis studies are stable for weeks at 4°C and display no spontaneous hydrolysis (141).

The above substrate systems can conveniently be scaled down (138) for measurement of pancreatic or intestinal sterol-ester hydrolase activity by radioassay following microthin-layer (39,136) or column chromatography (24,53). The assay mixture for cholesterol ester synthesis contains 15.5 μmoles of cholesterol-4-^{14}C (specific activity 1×10^5 dpm/mg), 46.5 μmoles of oleic acid, 31 μmoles of sodium taurocholate, and 100 μmoles of ammonium chloride in a final volume of 1.5–2.5 ml of phosphate buffer, pH 6.2. Albumin (6 mg) may be added or omitted from the medium. For sterol ester hydrolysis, the medium contains 15.5 μmoles of cholesterol-4-^{14}C oleate (specific activity 1×10^5 dpm/mg), 31 μmoles of sodium taurocholate, and 18 mg of lecithin in a final volume of 6 ml of phosphate buffer, pH 6.6. The mixtures are homogenized, solvent evaporated, and the mixture rehomogenized or sonicated as described above. With these substrate preparations, 0.1–0.5 ml of enzyme are used, depending on activity and incubation time. Incubations are carried out at 37°C in a metabolic shaker. Initially and at 5-min intervals with pancreatic juice, or 15–30 min intervals with pancreas and intestine, 20-μl samples are withdrawn and placed in 20 μl of acetone ethanol (1:1, v/v). Incubations can vary from 5 min to several hours, depending on enzyme activity. After thin-layer silicic acid or column separation of the free and esterified cholesterol fractions, the appropriate aliquots can be counted as described previously.

A substrate system containing 10% ethanol has been described (80). Although comparable to those described earlier (51,127), this type of system may not be suitable for differentiating sterol ester hydrolase activity from other carboxylic ester hydrolases, notably aliesterases. A liquid crystalline cholesterol substrate has been described by Kelly and Newman (56) which involves preparation of myelin figures of cholesterol under N_2 in the presence of hydroquinone in 0.154*M* phosphate, pH 6.15. In the absence of nitrogen or hydroquinone, zero-order kinetics are not obtained and oxidation of the sterol occurs. Further studies with this substrate system will be of interest.

(*2*) *Enzyme Preparation.* Chilled fresh pancreas or intestine, thoroughly rinsed with cold physiological saline, is minced into a chilled homogenizing tube. Four volumes of either cold glycerol–water (1:1, v/v), 0.9% NaCl, or 0.154*M* phosphate buffer are added and the mixture is homogenized for 2 min in an ice bath. The homogenate is allowed to stand for 1 hr in ice with frequent shaking and is subsequently centrifuged at 5000*g* for 10 min in a refrigerated centrifuge. The supernatant fluid can be used or further diluted with the appropriate solution.

Starting with acetone powders (76) of pancreas or intestine, the material is extracted with 10 volumes of 0.154*M* phosphate buffer, 0.9% NaCl, or distilled water by stirring at 4°C for 30–60 min. The mixture is then centrifuged at 10,000*g* for 10 min and the supernatant is used for assay or further purification. The extracts are unstable and should be used immediately. In the authors' laboratory it has been found that extracts of acetone powders of rat pancreas have one-half or less the enzyme activity found in fresh tissue extracts prepared in the same manner.

The methods for obtaining fresh active rat pancreatic juice are described in detail elsewhere (8,138), and this source of enzyme has a specific activity about 10 times that of fresh pancreas.

C. ENZYME PURIFICATION

a. Pancreatic Juice. Rat pancreatic juice cholesterol ester hydrolase has recently been purified about 350-fold (138) to a specific activity of 2400 units/mg of protein using the following procedure

(Table I). Cold pancreatic juice is centrifuged at 12,000g for 10 min prior to use for removal of insoluble, nonenzyme protein. Ice cold acetone is slowly added while stirring to give a final acetone concentration of 35% (v/v), and the mixture is kept at 0°C for 10 min. The precipitate is separated by centrifugation at 23,500g for 10 min in a refrigerated centrifuge, and is redissolved in cold 0.05M phosphate buffer, pH 6.2. A purification of 15–20-fold is usually obtained.

DEAE cellulose (0.8 meq/g) is equilibrated for 24 hr with 0.05M phosphate buffer, pH 6.2, and is placed in a glass column (30 × 1 in.) under 2 psi nitrogen pressure. The phosphate buffer solution of the acetone precipitate is added, and the protein is eluted stepwise with 50–70-ml volumes of 0.05, 0.10, and 0.15M phosphate buffer, pH 6.2. The enzyme, eluted in the 0.15M buffer fraction, is dialyzed against distilled water overnight. This step usually gives an additional four- to fivefold purification.

Hydroxylapatite in 0.001M phosphate buffer, pH 6.2, is packed in a column as described above. The entire 50–70 ml of dialyzed DEAE-cellulose fraction is placed on the column, and after passage of the initial volume, the proteins are eluted with 40-ml volumes of 0.10, 0.15, and 0.20M phosphate buffer, pH 6.2, under nitrogen pressure to give a flow rate of 0.25 ml/min. The enzyme is eluted in the 0.20M buffer fraction with an additional four- to sixfold purification. The purified enzyme after DEAE-cellulose and hydroxylapatite fractionation is labile but can be lyophilized after addition of 25 mg of bovine serum albumin, fraction V. The addition of this amount of protein prior to hydroxylapatite chromatography increases stability and recovery of the enzyme during chromatography and the albumin is eluted with the enzyme in the 0.2M buffer fraction. The albumin can subsequently be removed by Sephadex G 200 chromatography.

b. Pancreas. Hernandez and Chaikoff (51) have reported the following method for 400-fold purification of pork pancreas cholesterol-ester hydrolase (Table II). Forty grams of an acetone powder of pork pancreas is extracted five times in the cold with 100-ml portions of 0.154M phosphate buffer, pH 6.2. Proteins are precipitated by addition of 22 g of ammonium sulfate per 100 ml of extract and the supernatant after centrifugation at 10,000g for 10 min is used for subsequent purification. One hundred grams of solid ammonium sulfate are added to the clear supernatant at room temperature, and the resulting precipitate is separated by centrifugation. The precipi-

TABLE I
Summary of Purification Procedure for
Rat Pancreatic Juice Cholesterol Ester Hydrolase (138)

Fraction	Volume, ml	Protein, mg	Units[a]	Specific activity[b]	Recovery, %
Original juice	24	384	2764	7.2	100
35% Acetone precipitate	12	14.4	1676	115	60
DEAE cellulose, 0.15M	70	2.4	1204	479	43
Hydroxylapatite,[c] 0.20M	40	0.06	168	2455	6

[a] A unit is 1 μmole of cholesterol esterified/hr under the conditions described in the procedure.

[b] Units/mg of protein.

[c] The recovery of enzyme by hydroxylapatite column fractionation is increased to 50% by stabilization of the enzyme with 25 mg of albumin prior to chromatography.

tate is dissolved in 100 ml of cold phosphate buffer (pH 6.2), and dialyzed against cold distilled water for 48 hr. The resulting white precipitates from 10 individual batches are dissolved in cold phosphate buffer (pH 6.2), combined, and made to 100 ml with buffer. To this solution is added 33 ml of cold 95% ethanol with stirring and the mixture is allowed to stand 10 min at 10°C, during which time a flocculent precipitate forms. This precipitate is separated by centrifugation, and discarded, and the supernatant is dialyzed against distilled water at 0°C for 24 hr. The dialysate is diluted to 225 ml, the pH is adjusted to 5.0 with 0.1N HCl, and 25 g of solid ammonium sulfate is added. The resulting precipitate is discarded, and to the supernatant is added 5 g of Whatman cellulose and 5 g of solid ammonium sulfate per 100 ml. This mixture is centrifuged and the protein–cellulose precipitate is suspended in 35% ammonium sulfate solution. This suspension is poured into a glass column and allowed to sediment. For gradient elution of proteins, 2 reservoirs are set up in series containing 35% ammonium sulfate in the lower reservoir and 20% ammonium sulfate in the upper reservoir. Fifteen-milliliter fractions are collected and assayed for protein and enzyme activity. The enzyme associated with the third protein peak is precipitated

TABLE II
Summary of Purification Procedure for
Pork Pancreas Cholesterol Ester Hydrolase (51)

Fraction	Volume, ml	Protein, mg	Units[a]	Specific activity[b]	Recovery, %
Acetone powder suspension	500	400,000	80,000	0.20	100
Phosphate extract	500	96,000	48,000	0.50	60
Ammonium sulfate, 30–60%	100	1,200	9,600	7.8	11
33% Ethanol supernatant	225	330	8,200	25	10
Second ammonium sulfate supernatant	225	85	3,700	44	4.6
Cellulose fractionation, third peak	2	6	550	89	0.7

[a] Based on nephlometric assay. One unit is the amount of enzyme producing an increase in turbidity of 1 Klett unit/min during the second 15 min of a 30-min incubation.
[b] Units per mg of protein.

by addition of ammonium sulfate (5 g/100 ml) and the precipitate is dissolved in 2 ml of distilled water for studies on properties and purity.

Murthy and Ganguly (80) have described a calcium phosphate gel purification procedure using extracts of rat pancreas acetone powder, and have reported differential purification of the cholesterol ester synthesizing (540-fold) and hydrolyzing (58-fold) activities. Acetone powder of pancreas is stirred in 10 volumes of cold water for 30 min and the residue is removed by centrifugation at 10,000*g* for 10 min. The water extraction is repeated once and combined with the first extract. Ten milliliters of enzyme extract is added to aged calcium phosphate gel (13 months at 0–5°C) in a protein-to-gel ratio of 1 : 3, and the mixture is allowed to stand 15 min in the cold. The gel is centrifuged and eluted twice with 4-ml portions of 0.025, 0.05, and 0.1*M* phosphate buffer, pH 7.4. The gel is then eluted twice with 4 ml of 0.1*M* phosphate buffer, pH 8.4, and then with 4 ml of 15% ammonium sulfate solution. The eluants are dialyzed against distilled water prior to determination of sterol ester hydrolase activity. The reported purification steps are shown in Table III. However, the authors state that the esterifying activity was recoverable from the gel in only 2 of 8 trials. Also, all assays were conducted using substrates in 10% ethanolic media.

c. Intestine. Thus far, no purification of intestinal cholesterol ester hydrolase has been reported.

D. PROPERTIES

a. pH. Despite a variety of enzyme preparations and assay techniques, there is general agreement on the optimum pH's for enzymatic synthesis and hydrolysis by pancreatic and intestinal sterol-ester hydrolases. In studies prior to 1940, it was apparently common to determine enzymatic activity at the pH of the tissue homogenate (1,78,83,85). Swell and Treadwell (129), using hog pancreas and substrates emulsified in solutions containing sodium taurocholate and egg albumin, reported optimal pH's of about 5 for formation of esters of acetate to palmitate, pH 5.5 for stearate, and pH 6.1 for oleate. The same optimum pH for cholesterol oleate formation (i.e., pH 6.1) was found by Korzenovsky et al. (60,61), using ox pancreas, Swell et al. (120), using rat intestine, and Murthy and Ganguly (80), using rat pancreas and intestine. Also, Fodor (31) reported an optimum pH of 5.2 using butyric acid and hog pancreas homogenates. Hernandez and Chaikoff (51) reported an optimum pH of 6.2 for esterification of cholesterol by a purified pork pancreas enzyme, irrespective of the fatty acid used (hexanoic to stearic acids).

Using a liquid crystalline cholesterol substrate and dog pancreatic juice, the optimal pH for esterification is 5.0–5.25 (56). With the same enzyme source and heated serum as substrate, LeBreton and Pantaleon (64,66) have reported a pH optimum of 7.5–8.0 for esterification. Thus, it is obvious that the optimal pH reported in many early and more recent studies is as much a reflection of the physical state of the substrate as of the enzyme requirement.

Swell and Treadwell (129), using a series of cholesterol esters from acetate to oleate, found an optimal pH of 6.6 for enzymatic hydrolysis by hog pancreas cholesterol esterase. A similar optimum pH was reported (120) using rat intestinal mucosa. Korzenovsky et al. (60, 61) reported an optimum pH of 7.0 for hydrolysis of cholesterol oleate by beef pancreas extracts. In a comparative study of albumin-stabilized substrates and ethanolic dispersions, Murthy and Ganguly (80) confirmed the earlier results for albumin-stabilized cholesterol oleate, but found an optimum pH of 8.6 for the hyrolytic process by both rat pancreas and intestinal enzymes when an ethanolic dispersion of cholesterol oleate was used. In general, then, the optimal pH for sterol esterification by the pancreatic and intestinal enzymes is 6.1–6.2, and for sterol ester hydrolysis it is 6.6–7.0 (8.6 for ethanolic suspensions).

b. Ions and Inhibitors. The activity of purified enzyme of pancreas, dialyzed prior to use, is enhanced 25–35% by $10^{-3}M$ NaCl, LiCl, KCl, or NH_4Cl and $5 \times 10^{-4}M$ $(NH_4)_2SO_4$ (51). Similar results have been observed with crude preparations of pancreas (60,61) and pancreatic juice (unpublished data, ref. 148). The cholesterol ester hydrolase activities of both pancreas and intestine are inhibited to varying extents by $10^{-3}M$ $HgCl_2$, NaF, Na_3AsO_3, $ZnCl_2$, and $CuSO_4$, while no effect is noted with $CaCl_2$ (51,80).

A requirement for a free sulfydryl group for enzyme activity is suggested by observations that marked inhibition of pancreatic enzyme activity results in the presence of $10^{-3}M$ *p*-chloromercurobenzoate (51,80) and that this inhibition is completely reversed by $10^{-2}M$ cysteine or glutathione (51). The possible involvement of a serine hydroxyl radical at the enzyme active site is suggested by the inhibitory effect of several organophosphorus compounds (80,82). The role of bile salts and detergents is discussed in Section IV-1-E.

c. Stability. Pretreatment of fresh pancreatic homogenate (31, 104,121), pancreatin extracts (127,146), pancreatic juice (unpublished data, ref. 148), and intestinal extracts (120) at 65°C for 15–30 min completely destroys enzymatic synthesis and hydrolysis of cholesterol esters. Dialysis of the crude or purified pancreatic preparations against salt solutions or phosphate buffer at 4°C does not affect enzyme activity (31,52), but when dialysis is carried out with distilled water, extensive inactivation results (51,61). However, activity of the dialyzed preparation is restored if monovalent cations are replaced (see Section b above).

Although acetone powders of pancreatic or intestinal tissue are stable for a year or more at −15°C (61), there are conflicting reports on the stability of fresh tissue or pancreatic juice preparations. These differences appear due to the extraction medium used; homogenates of pancreas in glycerol–water (1:1, v/v) or in phosphate buffer, pH 6.2, are stable for several days at 2°C (31,121). Fresh pancreatic juice (pH about 8.2), pancreas, or intestine in veronal or phosphate buffers, pH 8.6, are extremely labile at 37°C in the absence of substrate and bile salts. The pancreatic tissue or juice enzymes at pH 6.1 are also rapidly inactivated at 37°C (80,141) unless preincubated with glutathione (80), trihydroxycholanic acids (80,141) or trypsin inhibitor (141).

d. Sterol Specificity. Data from several laboratories (51,61,80,86, 123) on the sterol specificity for esterification by pancreatic sterol-ester hydrolase have been consistent and allow the following generalizations. The ring-saturated analog of cholesterol, cholestanol (dihydrocholesterol), is esterified with oleic acid at equal or somewhat greater rates than is cholesterol; β-sitosterol is esterified with oleic acid at about one-third to one-half the rate of cholesterol, while β-sitostanol is esterified equally as well as cholesterol; stigmasterol is 20% as effective as cholesterol. All other sterols tested, including ergosterol, epicholesterol, 7-dehydrocholesterol, coprostanol, and epicoprostanol are very poorly esterified, if at all, with oleic acid. From these data, the following general criteria are applicable to sterol specificity: (*1*) a free hydroxyl in the β configuration at carbon 3 of ring A is required, (*2*) fusion of rings A/B must be *trans*, (*3*) double bond at Δ 5:6 is not required, (*4*) double bond or hydroxyl at carbon 7 eliminates activity; and (*5*) alteration of side chain reduces activity.

Swell et al. (123) have investigated a number of sterol butyrate and oleate esters as substrates for hydrolysis by hog pancreas cholesterol esterase. In contrast to sterol specificity for the esterification reaction, the butyrate esters of cholesterol, dihydrocholesterol, β-sitosterol, mixed sitosterols, and stigmasterol were hydrolyzed equally well, while ergosterol butyrate was split to a lesser but still significant extent. With oleate esters, the relative extent of hydrolysis was cholesterol, 100; sitosterol, 92; stigmasterol, 76; and ergosterol, 31.

e. Fatty Acid Specificity. As pointed out in the previous section, the physical state of substrates and pH of the substrate mixture can influence the rate of enzymatic esterification. These factors should be kept in mind in comparing data from different laboratories. It has been reported that with hog pancreas extracts, oleic acid is most active in esterification of cholesterol; with saturated fatty acids from 8 to 18 carbons, the rate is about 60% of that with oleic, while with short-chain fatty acids up to caproate, there was essentially no esterification (129). With a 400-fold purified enzyme of hog pancreas, cholesterol was esterified with oleic and linoleic acids equally effectively, while stearic acid was utilized half as well (51). Palmitic and linolenic acids were poorly esterified and myristic acid was not esterified with cholesterol. However, these data were obtained by turbidometric assay, and aside from the different solubilities of the acids used in the substrate mixture, there is no assurance that different cholesterol esters produced the same degree of turbidity. As pointed out by Goodman (46), the use of substrates containing mixtures of fatty acids may provide additional information concerning fatty acid specificities and rates of ester formation.

Murthy et al. (81) have also reported that oleic acid was efficiently esterified to cholesterol by rat pancreas enzyme, while linoleic and linolenic acids were two-thirds and one-third as effective, respectively. Other fatty acids tested (C-2 to C-18) were essentially ineffective in cholesterol ester formation. On the other hand, these authors also used rat intestinal cholesterol esterase and ethanolic dispersion of substrates and found linolenate to be most efficiently esterified, while linoleic and oleic acids were progressively less effective. Lauric, myristic, palmitic, and stearic acids were used to about 10% of the extent of oleic acid, while fatty acids of chain length under C-8 were not esterified to cholesterol. Thus, there appear to be differences in fatty acid specificity between the pancreatic and intestinal enzymes using the same substrate dispersions.

With respect to cholesterol ester hydrolysis, short-chain fatty acid esters (C-2 to C-6) of cholesterol are efficiently hydrolyzed, while increasing fatty acid chain length results in lesser degrees of hydrolysis. Of the C-18 fatty acid esters, oleate is hydrolyzed three times as rapidly as sterate. These results were obtained on albumin–bile salt emulsions of the cholesterol esters (129). When cholesterol esters are solubilized in a micellar medium of taurocholate and lecithin, the hydrolysis of cholesterol oleate by pancreatic juice cholesterol esterase proceeds at 3–4 times the rate of hydrolysis when the ester is dispersed in an albumin–bile salt medium. All of the esters solubilized in micellar medium, from cholesterol propionate to cholesterol linoleate, are hydrolyzed at the same rate, indicating that only the sterol ester bond is required for maximum hydrolytic activity and that no specificity exists for fatty acid chain length (140). The earlier specificity data were probably a reflection of substrate solubility and accessibility to the enzyme.

f. Identity of the Enzyme. Nedswedski (85) in 1937 reported that pancreatic cholesterol esterase was unstable in water or 0.025*M* ammonium hydroxide, while the action of lipase under similar treatment was maintained for 10 days. However, in the presence of bile salts, both enzyme activities were maintained for 20 days. Differences between lipase, sterol-ester hydrolase, and aliesterase activities with respect to temperature inactivation and bile salt requirement have been reported (121). However, these data and those of others (31) were obtained on crude extracts using cholesterol butyrate as substrate. Data with this substrate may be less representative of cholesterol esterase activity and more of aliesterase activity than that with a long-chain fatty acid ester of cholesterol.

Meyers et al. (82) differentiated between rat pancreatic lipase, sterol ester hydrolase, and an esterase acting on salicyl esters by use of organophosphorus inhibitors. However, using only these inhibitors, the authors could not differentiate between enzymes splitting cholesterol palmitate or various phenyl esters. Mattson and Volpenhein (72) have reported on differences between rat pancreatic juice carboxylic ester hydrolases using as substrates methyl butyrate, triolein, cholesterol linolenate, and 2,3-dioleoyl butane. On the basis of effects of salt concentrations, temperature, pH optima, bile salt, and diethyl *p*-nitrophenylphosphate, it was concluded that sterol ester hydrolase was distinctly different from glycerol ester hydrolase (lipase) and methyl butyrase (aliesterase type). However, it was also suggested that sterol-ester hydrolase was active against water-insoluble esters of secondary alcohols and was capable of splitting primary alcohol esters (72). Recently (139,141) it has been found that pancreatic juice sterol-ester hydrolase is rapidly inactivated by added trypsin or chymotrypsin, whereas lipase is inactivated at a considerably slower rate. The strongest evidence for separate enzymes is that taurocholate completely protects sterol ester hydrolase activity against proteolytic inactivation while the inactivation of lipase is unaffected by the presence or absence of bile salts (unpublished data, ref. 148). Final decision as to the identity and specificity of these hydrolases awaits their separation and purification.

g. Intracellular Distribution. Gallo and Treadwell (36) reported that 85–95% of the mucosal sterol ester hydrolase activity was in the supernatant fraction of the cell, as opposed to the brush border;

95% of the invertase activity, which is used as a measure of separation (74), was associated with washed brush border. Of the subcellular fractions, 65% of the sterol ester hydrolase activity (both synthetic and hydrolytic) was cytoplasmic with small levels associated with the particulate fractions. However, David et al. (19), using similar methods, reported completely different results, finding 90% of the hydrolytic and 32% of the esterifying activity associated with brush border. No explanation for these differences is evident.

E. ROLE OF BILE SALTS

The stimulatory effect of bile on the esterifying activity of pancreatic sterol ester hydrolase was first demonstrated by Mueller (78) in 1916, and since that time many attempts to define the mechanism of this stimulatory effect have been reported (51,59,61,80,83,84,120, 122,123,127,129,141). The requirement of bile salt for pancreatic or intestinal sterol ester hydrolase activity appears to be unique and absolute. Of the recent reports, only in studies using the nephelometric and manometric assays (51,61) has there been any indication of even low levels of enzyme activity in the absence of bile salt. However, the nephelometric assay is initially standardized in the presence of bile salt, and in the manometric study, crude pancreas extracts were used. Except for an apparent stimulation of pancreatic hydrolysis of cholesterol butyrate and caproate (129), use of Tween 20 or Tween 80 as detergents in place of bile salts has not only been ineffective in supporting enzyme activity (129), but has actually inhibited the stimulatory effects of bile salts (51,80,127).

Despite the use of different assay systems, the data on the effects of various bile salts have been generally consistent (51,61,80,122,141). The trihydroxycholanic acid, cholic acid, and its taurine and glycine conjugates, are the only bile acids [except for taurochenodeoxycholate (51)] stimulating either enzymatic synthesis or hydrolysis of cholesterol esters. All other bile salts, containing either none, one, or two hydroxyl groups on the cyclopentanophenanthrene ring, have little if any activity (51,141). This specificity suggests a cofactor function for trihydroxycholanic acids in addition to the long-known detergent properties of most bile salts (107,143). Recently, evidence has been presented for a cofactor function of bile salts in sterol ester hydrolase activity. Taurocholate is effective in protection of the pancreatic enzyme against pH (80) and proteolytic inactivation (141). Deoxycholate, however, is ineffective in protection of the enzyme against pH inactivation and actually inhibits enzyme activity in the presence of taurocholate, suggesting competition for an active or allosteric enzyme site (80). Even more suggestive data for binding of bile salt and enzyme is the recent finding that although only trihydroxy-cholanic acids (3α,7α,12α-trihydroxy) activate sterol ester hydrolase, these compounds along with deoxycholate (3α,12α-dihydroxy) will completely protect the enzyme against proteolytic inactivation (unpublished observations, ref. 148). Partial protection is also provided by chenodeoxycholate (3α,7α-dihydroxy) and lithocholate (3α-monohydroxy), suggesting binding of the bile salts via the hydroxyl groups to an allosteric site on the enzyme. Presumably, only the trihydroxy compounds bind completely and alter the conformation of the protein to produce the active enzyme.

F. RELATIONSHIP OF ESTERIFYING AND HYDROLYTIC ENZYME ACTIVITIES

There is a variance of opinion concerning the number of proteins involved in the sterol ester synthesizing and hydrolyzing activities

TABLE III
Partial Purification of Pancreas Sterol Ester Hydrolases (80)

			Esterifying activity		Hydrolytic activity	
Fraction	Volume, ml	Protein, mg	Total[a]	Relative specific activity	Total[a]	Relative specific activity
Acetone powder	—	100	43.2	1	26.9	1
Water extract	10	25	41.2	3.8	26.4	3.7
Initial calcium phosphate gel supernatant	10	7	2.5	0.8	1.1	0.6
0.025M phosphate, pH 7.4	8	9.6	4.6	1.0	3.0	1.0
0.05M phosphate, pH 7.4	8	3.2	6.5	4.6	—	—
0.10M phosphate, pH 7.4	8	0.96	24.1	58.1	—	—
0.10M phosphate, pH 8.1	8	0.48	2.5	12.1	—	—
15% $(NH_4)_2SO_4$	4	0.12	2.0	38.8	15.5	540

[a] Total = μmoles of cholesterol esterified or liberated.

of pancreas and intestine. Due to differences in substrate solubilities and assay media, this problem has not yet been satisfactorily resolved. During extensive purification of acetone powder pancreatic enzyme, the ratio of synthesizing and hydrolyzing capacity remained constant, indicating the two activities were the property of a single enzyme (51). However, even with 400-fold purification, there remained at least two protein components. In contrast, Murthy and Ganguly (80), using 10% ethanolic dispersion of substrates, reported differential purification of the two activities of rat pancreas by calcium phosphate gel chromatography (see Table III). In addition, it was shown that the hydrolytic activity of pancreas was more labile during incubation at 37°C and more sensitive to Tween 20 inhibition than was esterifying activity. However, both activities showed only slight differences with respect to organophosphorus and heavy metal inhibition and pH inactivation.

2. Liver

A. INTRODUCTION

Prior to 1938, there were scattered and conflicting reports in the literature concerning the hydrolysis of cholesterol esters by liver extracts or homogenates (78,88,100,104,109). In 1938, Klein (58) reported that saline extracts of liver hydrolyzed up to 85% of substrate serum cholesterol esters in 15 hr at pH 5.3. Using a similar system, Sperry and Brand (110) reported the occurrence in the liver of various species, of a cholesterol esterifying system at pH 4–6. The same preparations were active in hydrolysis of cholesterol esters at pH 5.3. Nieft and Deuel (87) confirmed these findings using saline extracts of rat liver as enzyme and cholesterol palmitate stabilized with lecithin and dioxane as substrate.

Byron et al. (12) reported on the activity of various preparations of rat liver on the synthesis and hydrolysis of several fatty acid esters of cholesterol using emulsified substrates. There was considerable hydrolysis of short-chain esters of cholesterol, such as butyrate, at pH 6.6. Synthesis of cholesterol esters was negligible using long-chain fatty acids. Hydrolysis of cholesterol acetate by chicken (62) and rat (99) liver microsomes has also been reported, but the identity of this enzymatic activity is uncertain.

Recently, Deykin and Goodman (24) have described an extensive study of liver cholesterol ester hydrolytic enzymes. The subcellular distribution of enzymes, relative activity against different sterol esters, properties, and partial purification are included in this communication, which is discussed in detail in the following sections.

B. ASSAY METHODS

a. Substrate Preparation and Incubation Conditions (24). Labeled (^{14}C or ^{3}H) cholesterol oleate (1–2 mμmoles; 0.2 μc) in 50 μl of acetone is forcefully added via microsyringe beneath the surface of 2 ml of enzyme preparation containing 200 μmoles of phosphate buffer, pH 7.4. Attempts to add the labeled substrate as a dispersion in saline solutions of 5% Tween 20, 15m*M* bile salt, or in a mixture of ethanol, acetone, and albumin, all resulted in activities of one-half or less. Incubations are carried out for 1 hr in 25-ml Ehrlenmeyer flasks at 37°C in a metabolic shaker. Aliquots of 0.1 ml are withdrawn at 10–15-min intervals for extraction, separation of free and esterified cholesterol, and radioassay.

b. Enzyme Preparation (24). After sacrifice and exsanguination of the animal, the liver is rinsed with 0.1*M* phosphate buffer (pH 7.4), blotted, and weighed. The tissue is then minced into 2.5 volumes of phosphate buffer in a Potter-Elvejham homogenizing tube, and homogenized in ice with a loose-fitting Teflon pestle. Nuclei and cell debris are removed by centrifugation at 2000*g* for 30 min in a refrigerated centrifuge. This preparation may be used for studies of sterol ester hydrolysis; however, the 2000*g* supernatant contains approximately 0.36 mg/ml of cholesterol, of which about one-third is esterified. This endogenous cholesterol ester will interfere with studies of hydrolysis of labeled sterol esters and yield results which are low (24). A better preparation for studying hydrolytic activity without interference of endogenous sterol ester is obtained by centrifugation at 100,000*g* for 1 hr. The resulting supernatant is separated from the layer of floating fat and contains no sterol ester.

C. PURIFICATION (24)

The initial steps of purification of soluble sterol ester hydrolase are essentially as described above. After removal of the floating fat layer from the 100,000*g* supernatant fraction, solid ammonium sulfate is added to 30% saturation at 0°C (2.12 g/10 ml). The mixture is stirred in ice for 30 min and the precipitate is sedimented by centrifugation for 30 min at 10,000*g*. The precipitate is dissolved in a small volume of water and separated from ammonium sulfate by Sephadex G-25 chromatography. The eluted protein is diluted to 5 mg/ml with 0.01*M* phosphate buffer, pH 7.4. This fraction contains approximately one-half the original activity with about two-fold purification over that in the 100,000*g* soluble fraction. Further purification involves mixing the enzyme solution with calcium phosphate gel (100 mg protein/28 mg gel) at ice temperature, agitation of the suspension for 1 min, and allowing to stand at least 15 min in ice. The gel is removed by centrifugation, and the supernatant is adjusted to pH 6.5 with dilute acetic acid. Additional calcium phosphate gel is added (100 mg protein/100 mg gel) and, after centrifugation, the gel is eluted with 0.1 and 0.5*M* phosphate buffer, pH 7.4, using 1 ml of eluant per 10 mg of gel. The 0.5*M* phosphate eluant is diluted to 0.1*M*, and this fraction contains 25% of the activity of the ammonium sulfate fraction and is purified an additional fourfold.

A typical purification sequence is shown in Table IV. As mentioned previously, the 2000 and 10,000*g* supernatants contain significant amounts of endogenous cholesterol esters, and since the assay involves hydrolysis of added labeled cholesterol ester, the values for total enzyme activity in these fractions are low. In addition to the

TABLE IV
Partial Purification of Liver Sterol Ester Hydrolase (24)

Fraction	Volume, ml	Protein, mg	Specific activity[a]	Total activity[b]	Relative purification
Supernatant,					
2,000*g*	190	7,600	19.3	146,680	1
10,000*g*	168	6,132	30.1	184,573	1.5
100,000*g*	137	3,699	160.0	591,840	8.2
Ammonium sulfate ppt. 30% saturation	164	820	338.0	277,160	17.5
Calcium phosphate gel; eluant V	171	51	1,321	67,371	68.4

[a] % Hydrolysis × 5000 cpm/mg protein.

[b] Specific activity × total protein; the values for the 2,000*g*, and 10,000*g* supernatant are relatively low due to the presence of endogenous substrate.

enzyme in the 100,000*g* soluble fraction, the mitochondria and microsomes also show low levels of sterol ester hydrolytic activity. It was estimated that the microsomes contained 11–32% of the activity of the original homogenate. The microsomal activity has not been purified, but has a similar cholesterol ester specificity and sensitivity to sulfhydryl inhibitors (24).

D. PROPERTIES

a. pH (24). Using the partially purified supernatant enzyme and labeled cholesterol palmitate substrate, there is a broad pH optimum for sterol ester hydrolysis between 6.5 and 7.5, with a sharp drop in activity below pH 6.2 and above pH 8.0. In contrast, microsomal activity has a sharper optimum pH at 6.1.

b. Ions and Inhibitors (24). Both soluble fraction (partially purified) and microsomal hydrolytic activities are higher in Tris-HCl buffer than in phosphate buffer. The presence of Ca^{2+}, Mn^{2+}, Co^{2+}, or Ni^{2+} (0.5 and 3.0 m*M*) slightly enhances activity; Mg^{2+} up to 5.0m*M* has no effect; Zn^{2+} and Cu^{2+} (0.5m*M*) inhibit to the extent of 85 and 93%, respectively. Both activities are strongly inhibited by sulfhydryl agents; *N*-ethyl maleimide, 95% inhibition at 6 μmoles; iodoacetamide, 33% at 6 μmoles; *p*-hydroxymercuribenzoate, 95% at 2 μmoles. These inhibitions are reversed by a 3–5-fold excess of GSH. The activity of both fractions is also strongly inhibited by 0.10 μmole of DFP and 1 μmole of eserine sulfate.

c. Effects of Bile Salts and Emulsifiers. While there is evidence for a cofactor function of bile salt for pancreatic (and intestinal) sterol ester hydrolase, the effect of bile salts on the liver cholesterol ester hydrolase is uncertain.

With a 20% glycerol–water extract of liver and a cholesterol butyrate emulsion as substrate, there is no ester hydrolysis in 6 hr in the absence of sodium taurocholate. Addition of the bile salt in concentrations up to 10% increased hydrolysis to 71% of the added ester, while further addition to 20% caused progressive inhibition. The addition of Tween 20 alone in amounts up to 15% resulted in about 14% hydrolysis; however, Tween 20 inhibited the stimulatory effect of bile salt (12).

Nieft and Deuel (87) reported that addition of soy lecithin stimulated the hydrolysis of colloidal cholesterol palmitate by liver extracts. In partially purified fractions, no activity was observed in the absence of lecithin. However, Deykin and Goodman (24) have reported that addition of Tween 20, bile salt, albumin, human β-lipoprotein, or polyglycerolphosphatide to incubations before addition of labeled cholesterol ester (in acetone) all inhibit ester hydrolysis.

d. Stability. There is a rapid loss of enzyme activity upon dialysis or lyophilization of the 30% ammonium sulfate fraction of soluble sterol ester hydrolase (24). However, it appears that dialysis of the original homogenate does not affect activity (12). The enzyme eluted from calcium phosphate gel in the last step of the described purification (24) is unstable; 50% or more activity is lost overnight when stored in the cold or at −18°C. Liver hydrolase is inactivated by heating at 65°C for 15 min (12).

e. Identity of the Enzyme (24). Hydrolyses of cholesterol palmitate (sterol ester hydrolase), *p*-nitrophenylacetate (esterase), *p*-nitrophenylphosphate (phosphatase), ethyl palmitate (esterase), and tripalmitin (lipase) have been determined using various liver fractions and purified preparations. The purified sterol ester hydrolase (see Table IV) was completely separated from esterase activity whose major site was microsomal, and from phosphatase activity which was eluted from calcium phosphate gel in an earlier fraction than was sterol ester hydrolase. Tripalmitin was poorly hydrolyzed by the soluble fraction (5.3% hydrolysis of 10–20 μmoles/hr); however, both the soluble and microsomal fractions were active with respect to ethyl palmitate hydrolysis. These results leave some uncertainty as to the specific identity of the liver enzyme or enzymes involved in cholesterol ester hydrolysis.

f. Substrate Specificity (24). The fatty acid specificities of the partially purified soluble enzyme and the microsomal enzyme have been tested using individual labeled sterol esters and equimolar mixtures of esters of which only one was labeled. The latter study was performed to eliminate apparent differences in specificity due to differences in substrate solubility. The observed order of hydrolysis of individual cholesterol esters or equimolar mixtures was oleate = linoleate > acetate > palmitate > stearate. With the microsomal enzyme, the order of hydrolysis with individual cholesterol esters was acetate > linoleate > oleate > palmitate = stearate. However, due to the insolubility of cholesterol esters in general, and the necessity of addition of substrates in acetone, the data are still equivocal with respect to both rates of reaction and fatty acid specificities.

g. Kinetics (24). The rate of sterol ester hydrolysis is linear for approximately 40 min; at 120 min incubation, linearity was decreased by 20% and thereafter velocity declined rapidly. Lineweaver-Burk plots indicate the K_m for cholesterol acetate, oleate, and linoleate = 14.3×10^{-6}, and for cholesterol palmitate = 4.8×10^{-6}.

3. Adrenal

A. INTRODUCTION

Although Dailey et al. (16) originally failed to demonstrate hydrolysis of labeled cholesterol ester by hog adrenal homogenates during steroid synthesis, these workers subsequently showed (17,18) that dog adrenal homogenates could hydrolyze labeled cholesterol esters under conditions optimal for adrenal steroid synthesis. Brot et al. (10) reported that homogenates or acetone powders of rat, guinea pig, and beef adrenals all had the capacity to esterify labeled free cholesterol at pH's between 4 and 6, while no reaction occurred above pH 7. Using phosphate buffer (pH 6.1) extracts of hog adrenal acetone powder, and rat serum β-lipoprotein labeled with free cholesterol-4-^{14}C or cholesterol-7α-^{3}H oleate, it was found that 7% of the labeled free cholesterol was esterified in 3 hr at pH 5.7 without addition of high-energy cofactors; at pH 7.4, 10–12% of the available labeled cholesterol oleate was hydrolyzed in the same time period. Further studies showed a sharp optimal pH for esterification between pH 4–5.

Since others (17,68) have reported that rat and dog adrenal homogenates appear to require ATP and CoA for esterification of cholesterol (Section VI-5), the report of Brot et al. (10) was of interest with respect to the possible presence of a non-high-energy esterifying system in adrenal gland. This system due to the low pH optimum would not be observed in studies employing incubations above pH 7. Recently this same laboratory (105) has demonstrated enzymatic hydrolysis of cholesterol oleate by bovine adrenal particulate fractions at pH 7.5; the cell sap contained an enzyme system capable of esterifying added labeled cholesterol with a pH optimum of 5 and required no addition of ATP, CoA, and Mg^{2+}. Furthermore, no stimulation of esterification by these cofactors occurred even after dialysis of the 100,000*g* supernatant, indicating the presence of a low-energy esterifying system similar to that found in pancreas and intestine.

B. ASSAY METHOD

a. Substrate Preparation. Labeled free or esterified cholesterol dissolved in a small volume of acetone or complexed to rat plasma β-lipoprotein is employed as substrate. For the former, between 10^6 and 10^7 cpm substrate are dissolved in 1 ml of acetone and 0.1 ml of the solution is added/milliliter (100 mg) of adrenal homogenate in a total incubation volume of 4–5 ml. For the latter, rat serum β-lipoprotein is prepared by dextran sulfate precipitation (13). After reprecipitation, the β-lipoprotein is dissolved in a volume of 0.9% NaCl equivalent to the original volume of serum, and then incubated for 20 hr with labeled free or esterified cholesterol dispersed on Celite as described by Avigan (3). After incubation, the Celite is removed by centrifugation, β-lipoproteins are again purified by dextran sulfate precipitation, and subsequently are redissolved in 0.9% NaCl. This method yields a protein preparation containing 50% of the added free cholesterol or 20% of the added sterol ester (105). One-half milliliter of this preparation is used per milliliter of adrenal homogenate in a total incubation volume of 4–5 ml.

b. Enzyme Preparation. Fresh adrenals are homogenized in three volumes of ice cold 0.9% NaCl or 0.25*M* sucrose solution essentially as described by Frantz and Bucher (34). The homogenate is centrifuged at 350*g* for 15 min to remove cell debris and can be used for measurement of total enzyme activity. This supernatant is centrifuged at 10,000*g* for 10 min, yielding a mitochondrial pellet and a supernatant which is subsequently centrifuged at 100,000*g* for 45 min to obtain a microsomal pellet and supernatant cell sap. The pellets are resuspended by gentle homogenization in a volume of saline equivalent to the original supernatant.

c. Conditions (105). To 1-ml aliquots of the homogenate or cell fractions is added 0.1 ml of the substrate ester in acetone or 0.5 ml of the β-lipoprotein preparation, and the mixture is diluted to 5 ml with phosphate buffer, pH 7.4, for studies on cholesterol ester hydrolysis. Similarly, labeled free cholesterol in acetone or complexed to β-lipoprotein is added to 1 ml of the enzyme preparation and diluted to 5 ml with acetate buffer, pH 5.0, for esterification studies. When the effect of cofactors is studied, they are dissolved in the dilutant buffer. Incubations are carried out for 2–3 hr under 95% O_2–5% CO_2 at 37°C with continuous shaking. Termination of reaction, extraction, and analysis are as described in previous sections.

C. PURIFICATION

No purification of adrenal sterol ester hydrolase has been reported.

D. PROPERTIES

a. pH. The optimum pH for sterol ester hydrolysis by dog, hog, and bovine adrenal homogenates is 7.5 (10,18,105) and a marked reduction of activity results at pH 6 or 8. For enzymatic esterification of free cholesterol by bovine adrenal particulate fractions, without addition of cofactors, there is a sharp optimum at pH 5.0 (105).

b. Cofactors. No cofactors are required for enzymatic hydrolysis of cholesterol esters by adrenal homogenates (10,17,18,105). For esterification of free cholesterol, there appear to be two different systems. The first of these is microsomal (105), requires ATP, CoA, and Mg^{2+} (17,68), and the reaction presumably proceeds via the classical acyl-CoA thiol ester. This enzyme system is discussed in Section VI-E. The second system requires no high-energy cofactor addition (10,105), and occurs in the soluble fraction of the cell (105). This system is active at pH 5.0 and dialysis of the soluble fraction (to eliminate endogenous cofactors) does not affect the extent of enzymatic esterification of cholesterol (105).

c. Specificity. In a comparative study (18), the hydrolyses of labeled cholesterol palmitate, oleate, linoleate, and arachidonate were compared using acetone addition of the esters to crude dog adrenal homogenates. The three unsaturated fatty acid esters were hydrolyzed to the same extent at pH 7.4, while the hydrolysis of cholesterol palmitate proceeded to about one-half the extent.

4. Distribution in Tissues

The early work of Nomura (89), Shope (104), and Klein (58,59) indicated that substrate serum cholesterol esters were hydrolyzed by extracts of spleen, kidney, muscle, lung, and brain of various species. However, Swell et al. (119), using stabilized substrates of cholesterol and oleic acid, or cholesterol oleate, demonstrated the presence of sterol-ester hydrolase activity in pancreas, intestine, and liver with little or no activity in any other rat tissue measured. More recently it has been reported that brain homogenates and brain microsomes from newborn to 9 month-old rats hydrolyze cholesterol linoleate at pH 6.6–7.6 (94). Rabbit macrophage preparations have the capacity to hydrolyze cholesterol acetate at pH 7.3, and to esterify cholesterol with palmitic or oleic acids at pH 6.0 (20,21). With normal or atherosclerotic rabbit aorta sections, hydrolysis of cholesterol oleate is demonstrable. Synthesis of sterol esters did not occur in either preparation of aorta (22).

Recently, Brot et al. (11) incubated preparations of various rat tissues with a very low-density lipoprotein fraction of rat chyle which contained cholesterol-4-^{14}C primarily in the esterified form (70% or more). With slices, sections, or muscle strips, it was found that adipose tissue, adrenal, intestinal mucosa, liver, kidney, and to a small extent, muscle, were capable of hydrolyzing the labeled esters.

V. LECITHIN–CHOLESTEROL ACYLTRANSFERASE

1. Introduction and Mechanism

The first conclusive evidence for the presence of cholesterol esterifying activity in plasma was reported in 1935 by Sperry (108), who found that incubation of human serum for three days at 37–40°C resulted in a net increase in serum cholesterol esters without change in the total serum cholesterol. In subsequent studies by Sperry and Stoyanoff (111–113), it was shown that the cholesterol-esterifying activities in human and dog were inhibited by small concentrations of bile salts. However, in dog serum, the addition of bile salts also promoted rapid hydrolysis of cholesterol esters with the reaction half completed within 5 min.

In 1944, LeBreton and Pantaleon (65) provided direct evidence for the earlier suggestion (108) that phospholipid might be the fatty acid donor for cholesterol esterification in serum. A fall in serum lipid phosphorus during sterol esterification led to the suggestion that phospholipase cleavage of lecithin released fatty acids necessary for "cholesterol esterase" activity in a sequential coupled reaction (65). These findings were subsequently confirmed by Etienne and Polonovski (26–28), who reported that of the serum phospholipids, only lecithin was degraded significantly during 37°C incubation, giving rise first to lysolecithin and subsequently to glycerylphosphorylcholine with the release of 2 moles of fatty acid. However, the actual formation of fatty acid esters of cholesterol was not demonstrated. Swell and Treadwell (128) incubated sera from various species with aqueous emulsions containing cholesterol and oleic acid, or cholesterol oleate, in the presence and absence of sodium taurocholate. Dog serum (but not human, rat, rabbit, or guinea pig serum) contained cholesterol ester hydrolyzing activity which was demonstrable only in the presence of bile salts, and which had temperature and pH characteristics similar to the enzyme system of pancreas (127).

Elucidation of the mechanism of the plasma cholesterol esterifying system, and partial purification of the enzyme have been reported since 1960. During incubation of rat or human plasma containing labeled cholesterol-4-^{14}C, both saturated and unsaturated fatty acids became esterified to the labeled cholesterol in proportions approximating those present in the preexisting cholesterol esters (41). Incubation of plasma with labeled free fatty acids did not result in incorporation of labeled fatty acids into any esterified lipid fractions, including sterol esters. However, when lecithin, containing labeled linoleic acid, was incubated with rat or human plasma labeled cholesterol linoleate was formed (in addition to labeled free fatty acids, glycerides, and cephalins). Also when labeled tripalmitin was incubated with serum, labeling was found throughout the serum lipid fractions, suggesting the presence of transesterification reactions with both neutral and acidic lipids. In a subsequent study (37), it was shown that the molar decrease in free cholesterol in human plasma incubated at 37°C was closely paralleled by the molar decrease in plasma lecithin. Furthermore, the fatty acids of the newly synthesized sterol esters closely resembled the fatty acids in the β position of plasma lecithin suggesting transesterification between the β position of lecithin and the free hydroxyl of cholesterol.

Further evidence for the involvement of lecithin as a primary fatty acid donor for plasma cholesterol-esterifying enzyme is that addition of dimyristoyl glycerylphosphorylcholine to human serum caused a stimulation of cholesterol esterification during incubation. However, similar results were not found with the distearyl- or dioleyl esters (142). Also incubation of extracts of acetone powders of rat plasma with lecithin containing labeled fatty acids in the β position resulted in formation of labeled cholesterol esters (102). It was estimated that up to 13% of the labeled acyl groups in the 2 position of lecithin were transesterified to cholesterol during incubation, while less than 1% of triglyceride fatty acids participated in sterol ester formation (102). Preincubation of these extracts with phospholipase A resulted in degradation of the labeled lecithin and prevented the esterification of cholesterol via transesterification. A similar finding has been reported by Rowen (97).

The reaction mechanism proposed for esterification of cholesterol during incubation of plasma is as follows (37).

$$\text{Diacylglycerylphosphorylcholine} + \text{cholesterol} \xrightarrow[\text{transferase}]{\text{plasma}} \text{cholesterol ester} + \alpha\text{-monoacylglycerylphosphorylcholine}$$

The involvement of lipoprotein lipids in the transferase reaction has been investigated by Glomset (38). Incubation of human plasma for 24 hr followed by ultracentrifugal separation into very low density, low density, high density, and very high-density lipoprotein fractions showed that (*a*) the greatest decrease of free cholesterol occurred in the low density fraction (although free cholesterol decreased in all fractions); (*b*) the greatest loss of lecithin occurred in the high density fraction; and (*c*) the major increase in lysolecithin was in the very high density fraction. When individual lipoprotein fractions were incubated, a greater proportion of the high density lipoprotein cholesterol was esterified, presumably because proportionally more of the high-density lecithin was available for acyl transfer to the cholesterol. Since the lysolecithin formed from transacylation was now associated with a physically different lipoprotein fraction, it was suggested that the net formation of cholesterol esters in plasma might be dependent on dissociation or removal of the lysolecithin from the site of sterol ester synthesis (38). In an extension of these studies, Glomset et al. (39) showed that although the greatest increases in esterified cholesterol occurred in the plasma low density lipoprotein fraction, the initial rate of esterification was highest in the high density lipoprotein fraction. The possible physiological importance of these findings is discussed in a subsequent section.

2. Assay Method

A. SUBSTRATE PREPARATIONS

a. Unlabeled Substrates. (*1*) *Common Substrate-Enzyme Preparation.* Whole blood is collected into heparinized syringes or into citrate, and plasma is harvested by centrifugation at 2°C. EDTA is added to the plasma in a final concentration of 5–10m*M* to prevent autooxidation. Initial rate determinations are made from 0.5 to 2.5 hr of incubation at 37°C in a metabolic shaker by measuring the loss in free cholesterol. Determination of changes in whole plasma or in lipoprotein fractions in respect to lecithin, lysolecithin, esterified cholesterol, and the fatty acid distribution in these fractions are most conveniently carried out on samples which have been incubated for 24 hr.

(*2*) *Heat-Inactivated Plasma Substrate* (38). A large common pool of substrate plasma can be prepared by inactivating the cholesterol-esterifying activity at 60°C for 30 min. This provides a constant lipoprotein–cholesterol substrate source which can be frozen in small aliquots.

b. Radioactive Cholesterol Substrates. Blood plasma or serum can be labeled with 4-^{14}C-cholesterol or 7α-^{3}H-cholesterol in several ways, as described below. The first two procedures have been used extensively and are the methods of choice.

(*1*) *Labeled Cholesterol–Albumin Mixture.* The method is essentially as described by Glomset and Wright (42) after the procedure of Porte and Havel (92). A 5% solution of human albumin, fraction V, in 0.15*M* NaCl is heated to 60°C, centrifuged, and complexed with labeled (carbon or tritium) cholesterol to give a final mixture containing about 1×10^6 dpm sterol per ml of albumin solution. The complexing is as follows.

Five microcuries of isotopic cholesterol are dissolved in 1 ml of ethanol and this solution is injected with a syringe and 23-gauge needle beneath the surface of 10 ml of the albumin solution at 25°C. After 20 min the ethanol is evaporated under nitrogen; the volume of the solution is maintained by addition of distilled water. The substrate for lecithin–cholesterol transferase measurements is prepared by adding 1 volume of the albumin solution to 8 volumes of heat-inactivated (60°C, 30 min) plasma and allowing it to stand for 1 hr.

(*2*) *Labeled Cholesterol-Coated Celite Particles.* This method has been used to prepare labeled substrate for transferase studies (41,93), and is essentially the method of Avigan (3) except that tracer levels of cholesterol are used rather than the amounts originally described. Higher levels of cholesterol apparently inhibit the transferase reaction in plasma (9).

^{14}C- or ^{3}H-labeled cholesterol in pentane is added to Celite 545 in a ratio of 50 μg of cholesterol (5 μc)/g Celite, and the solvent is evaporated under N_2. The cholesterol-coated Celite is incubated with heat-inactivated plasma at 37°C for 3 hr in a proportion of 50 mg of Celite/milliliter of plasma. The suspension is subsequently passed through a Swinny filter to remove the Celite. Two milliliters of the labeled, heat-inactivated plasma containing 5 μg of added cholesterol (0.5 μc) is incubated with 0.1–0.5 ml of enzyme plasma in a 25-ml Erlenmeyer flask at 37°C for period up to 3 hr (93). At intervals during the incubation, 0.05-ml aliquots are removed for extraction and determination of labeled free and esterified sterol.

(*3*) *Labeled Cholesterol in Tween 20.* This preparation has been described by Whearet and Staple (144). A benzene solution containing 1–3 μg of cholesterol (0.5 μc) is taken to dryness under N_2, and 0.1–0.2 ml of a 4% solution of Tween 20 in methanol is added to dissolve the cholesterol. The methanol is removed at 40–50°C under N_2 and the warmed solution of cholesterol is diluted with an equal volume of 0.15*M* NaCl (see ref. 102). Up to 0.1 ml of the Tween 20–cholesterol preparation is incubated with 1–10 ml of heat-inactivated plasma at 37°C for 2 hr prior to use as substrate. However, during the incubation, considerable esterification of the labeled sterol occurs.

(*4*) *Labeled Cholesterol in Acetone.* Glomset and Wright (42) have described the direct addition of labeled cholesterol in acetone to dialyzed, heat-inactivated plasma. The acetone is subsequently evaporated under N_2. However, no characteristics of this substrate preparation have been described.

c. Other Labeled Lipids. In order to study the incorporation or effect of fatty acid sources, and to define the mechanism of action of the plasma transferase reaction, labeled free fatty acids, or triglycerides and phospholipids labeled with fatty acids, can be added to plasma or serum prior to incubation. There is, however, uncertainty regarding the availability of these added compounds due to the form in which they are added and their relative water insolubility. Labeled free fatty acids are most effectively added to plasma or serum complexed to fraction V albumin. Labeled triglycerides or labeled phosphoglycerides can be added in small volumes of Tween 20. Approximately 0.1 μmoles of labeled lipid (0.1 μc) are dissolved in 2 ml of 0.5% Tween 20 in ethanol, and the solvent is evaporated on a steam bath under N_2. The lipid in Tween 20 is suspended in 2 ml of 0.9% NaCl, and 0.1–0.2 ml of the saline suspension is incubated with 2 ml of plasma (102). The addition of triglyceride to plasma has also been accomplished by dispersion of the lipid on Celite according to the method of Avigan (3), and subsequent incubation of the lipid-coated Celite with plasma or β-globulin (41).

B. ENZYME PREPARATION

Plasma or serum is prepared from freshly drawn whole blood using as anticoagulant either heparin, citrate, or sodium EDTA (5 mμmole/ml). There appears to be no effect of the anticoagulant on the rate or extent of enzyme activity.

For measurement of tissue enzyme activity, it is first necessary to remove all blood from the tissue prior to homogenization. This has been done by perfusion of the whole animal via the thoracic aorta using cold 0.25*M* sucrose (40). The organ(s) is then removed, minced in cold 0.25*M* sucrose, blotted, and weighed. Aliquots of the tissue are then homogenized in 4 volumes of either heat-inactivated or nonheated plasma/0.15*M* Na Cl (1:1, v/v) for simultaneous control and experimental values. The computations for determining "corrected tissue activity" and tissue activity relative to plasma are described in detail by Glomset and Kaplan (40).

C. ANALYSIS AND CHARACTERIZATION OF PRODUCTS

Only examples of those methods specifically for lecithin–cholesterol transferase activity are described in this section, since other general methods are described in Section III.

Following extraction of the incubation mixture, lipids are fractionated into major classes by silicic acid column chromatography. Phospholipids can subsequently be subfractionated (2,49,50,90,123) by silicic acid column or thin-layer chromatography.

An aliquot from the lecithin fraction after column separation, or the silicic acid area corresponding to lecithin on the thin-layer plate, is placed in a digestion tube and digested with perchloric acid at 220°C for 2 hr or at 100°C overnight. Subsequently, phosphorus is determined by the method of Fiske and Subbarow (30) or by the micromodification of Bartlett (5). Lysolecithin, after separation from other phospholipids by column or thin-layer silicic acid chromatography, is determined in a similar manner.

In the estimation of the fatty acid composition in the C-2 position of lecithin, phospholipase A of snake venom or pancreas is used for specific hydrolysis of the fatty acid in the C-2 position, yielding free fatty acid and lysolecithin. A quantitative procedure is described by Hanahan (48; see also refs. 96,98). Following extraction of lipids and fractionation of the major neutral lipid classes, the phospholipids are fractionally eluted from silicic acid. The lecithin fraction is rechromatographed on an alumina column to remove contaminants (95). The eluant containing lecithin is evaporated under N_2 and the residue is redissolved in ethyl ether and transferred to a 5-ml volumetric flask with a total of 2 ml of ether. Dry *Crotalus adamanteris* snake venom (Ross Allan Reptile Institute, Silver Springs, Fla., or Sigma Chemical Co., St. Louis, Mo.) is dissolved in 0.1*M* Tris buffer, pH 7.5, containing $2 \times 10^{-3}M$ $CaCl_2$ to give a final concentration of 0.5 mg/0.1 ml. Enough enzyme is added to the ethereal solution of lecithin to give a turbid solution after 9 min. The incubation mixture is shaken at intervals for 6 hr and the reaction is stopped by addition of 3 ml of ethanol. Solvents are evaporated under N_2 and the residue is dissolved in 5 ml of chloroform. This is placed on a silicic acid column (12–15 mg lipid/g of silicic acid) and free fatty acids are eluted with 50 ml of chloroform. The free fatty acids are subsequently methylated and the distribution determined by gas–liquid chromatography. The fatty acids remaining on lysolecithin can also be determined by transmethylation and gas–liquid chromatography.

3. Purification (42)

Human plasma (825 ml) is dialyzed against 0.01*M* Tris-HCl (pH 7.4) for 24 hr and subsequently applied to a column containing 165 g DEAE-cellulose (exchange capacity = 0.64 equiv/g), preequilibrated with 0.01*M* Tris-HCl (pH 7.4). The first fraction is eluted with 0.01*M* Tris-HCl buffer, pH 7.4, until a total fraction volume of

1100 ml is collected; this is discarded. Fraction II is eluted with 880 ml of 0.01*M* Tris-HCl buffer containing 0.1*M* NaCl, and contains approximately 24% of the original enzyme activity. Fraction III is obtained by elution with 48.5 ml of 1*M* NaCl in 0.01*M* Tris-HCl. This fraction contains about 10% of the original protein and between 60–90% of the total enzyme activity. This preparation, representing a two- to tenfold purification, has been used to determine several of the properties of the transferase system. Fraction III is concentrated by ultrafiltration and dialyzed against 0.01*M* Tris-HCl (pH 7.4). A solution of ammonium sulfate saturated at 6°C (pH 7.4) and containing 0.01*M* EDTA is used for subsequent purification.

Precipitates are obtained at 40, 66, and 90% ammonium sulfate saturation. About 80% of the activity is recovered in the 66% ammonium sulfate precipitate with about twofold purification.

The 66% ammonium sulfate fraction, obtained as described above, was contaminated with ceruloplasmin as evidenced by a blue-green color in the solution. For further purification and removal of ceruloplasmin, the precipitate from 66% ammonium sulfate saturation was dissolved in 125 ml 0.07*M* potassium phosphate, pH 6.8, and dialyzed against the same buffer. This solution was filtered through a column (3 by 15 cm) of hydroxylapatite, and the column

TABLE V

Partial Purification of Human Plasma Lecithin–Cholesterol Acyl Transferase (42)

Fraction	Volume, ml	Protein absorbancy at 280 mu/ml	Total activity[a]	Recovery, %	Relative purification
Dialyzed blood bank plasma	825	55.1	70,475	—	1
DEAE Fraction III	48.5[b]	72.8	35,987	52.6	6.8
Ammonium sulfate, 66% saturation precipitate	125	9.7	22,875	33.4	12.6
Hydroxylapatite 0.7*M* phosphate eluant	24[b]	12.5	14,496	21.2	32.2

[a] Micromoles of cholesterol esterified $\times 10^{-4}$/ml/hr.

[b] Concentrated by ultrafiltration.

was eluted at 6°C with 0.07*M* phosphate buffer, pH 6.8. The eluant was dialyzed against 0.01*M* phosphate buffer (pH 7.4) and contained about 63% of the applied enzyme activity with an additional 2.5-fold purification.

As shown in Table V, the final enzyme preparation contained 21% of the original activity and was purified 32-fold. The preparation was still contaminated with albumin and lipoprotein ($d > 1.21$).

4. Properties

A. pH

The optimal pH for the transferase reaction has been determined in a 6-hr incubation using unfractionated human plasma labeled with ^{3}H-cholesterol (42,93). There is a rapid increase in activity between pH 6 and 7 with a broad optimum between 7.0–8.5. These results confirm the earlier findings of Sperry using 3-day incubations of serum (108). In dog serum, which contains a bile salt-dependent esterification system, there is a broad optimum between pH 5.9 and 6.5, with an apparent peak at 6.1.

B. IONS AND INHIBITORS

Sodium chloride at physiological concentrations stimulates the transferase reaction, while $CaCl_2$ and Na_2HPO_4 have no effect (42). Preincubation of a tenfold purified enzyme with 4*M* urea for 1 hr at 37°C results in almost complete inhibition of enzyme activity which is not restored by dialysis. However, exposure to urea at concentrations of 2*M* or less, followed by dialysis, does not reduce enzyme activity (42). Bile salt (sodium taurocholate) completely inhibits human (42,108) and rat (42) plasma cholesterol esterification, but this inhibition is reversible, since dialysis of the bile salt-treated enzyme results in complete restoration of activity (42). In contrast, both the synthesis and hydrolysis of cholesterol esters by dog plasma are bile-salt dependent. Addition of bile salt to either human or dog plasma or mixtures of the two results in net hydrolysis of substrate cholesterol esters (113,128).

Human serum or plasma cholesterol-esterifying activity is blocked by various sulfhydryl or alkylating agents including monobromoacetate (37,67), monoiodoacetate, monoiodoacetamine, and *N*-ethylmaleimide in concentrations of $10^{-2}M$ (37). *p*-Hydroxymercuribenzoate ($10^{-1}M$) also completely blocks enzymatic esterification of plasma cholesterol, and this inhibition is reversed by $10^{-2}M$ mercaptoethanol (37). Since control experiments showed that the effect of the sulfhydryl reagents was not on the lipoprotein substrate, the data indicated the possible importance of a SH group at the enzyme site and suggested the involvement of an acyl-*S*-enzyme intermediate in the transferase reaction.

C. STABILITY

It appears that plasma or serum enzyme is reasonably stable at 2–3°C (40,42,128) and retains 81% of its original activity after two days at 0°C (93). A partially purified (sevenfold) preparation of human plasma can be stored frozen (42). Heating plasma at 56–60° for 30–60 min completely inactivates the enzyme (40,113); however, this does not preclude the use of this heated preparation as a convenient substrate for fresh plasma enzyme (42,93).

D. FATTY ACID SPECIFICITY

Glomset (37) has compared the fatty acid distribution in cholesterol esters before and after 20 hr of incubation of plasma with the initial fatty acid distribution in plasma triglyceride and in the β-position of plasma lecithin. Of the newly synthesized cholesterol esters, the unsaturated fatty acid esters ranged between 86–97% of the total fatty acids transesterified to cholesterol, and it was concluded that these fatty acids were derived primarily from the β-position of lecithin. However, in a similar study by Portman and Sugano (93), the only apparent changes in phospholipid fatty acids were a decrease in palmitic acid and an increase in linoleic acid relative to the control samples.

The incubation of heat-inactivated human plasma containing labeled cholesterol with active rat plasma will result in the formation of labeled cholesterol esters characteristic of rat plasma; conversely, active human plasma will form labeled esters characteristic of human plasma when the substrate is ^{14}C-cholesterol heat-inactivated rat plasma (93). Thus, while the fatty acids may be derived from the β-position of lecithin, it appears that the fatty acid distribution in the newly formed esters is determined in part by the specificity of the enzyme. In one study, plasma from rats on fat-free diets, which is deficient in linoleic and arachidonic acids, was incubated with heat-inactivated plasma from fat-fed rats. The newly formed cholesterol esters were rich in polyunsaturated fatty acids, indicating the importance of the substrate compositions on the distribution in the cholesterol esters (93).

E. LIPOPROTEIN SPECIFICITY

Glomset and co-workers have shown that the initial rates of esterification of cholesterol are 3–5 times greater in the high-density plasma lipoproteins (HDL) than in the low-density fraction (LDL) (39). However, the largest increases in cholesterol ester after incubation of plasma at 37°C for 24 hr occurs in the LDL fraction. A point in question with these studies is that free cholesterol equilibrates rapidly between lipoprotein fractions (39), and such equilibration during incubation of plasma would continuously alter the percentages of free and esterified cholesterol between lipoprotein fractions. It has been suggested that initially there is less free cholesterol in the HDL fraction; during incubation at 37°C the initial rate of esterification in the HDL is greater, resulting in a greater deficit of free cholesterol and a redistribution of free cholesterol between the LDL and HDL (39).

F. KINETICS

The rate of esterification has been determined using 37°C incubations of human plasma by measuring the decrease in free cholesterol over a period of 48 hr. During this period, about two-thirds of the preexisting free cholesterol is esterified, but the rate is linear for only the first few hours (37). Initial rates, determined on plasma from 12 human donors, indicate an average value of 110 μmoles of cholesterol esterified/hr/liter of plasma (37). Assuming a plasma volume of 3 liters, this would account for the esterification of 3 g of cholestrol per day. Portman and Sugano have reported a mean value of serum cholesterol-esterifying activity (10 male subjects) to be approximately 120 μmoles of cholesterol esterified/hr/liter (93). The absolute rates for any species can be greatly influenced by the substrate (inactivated) plasma source. For example, values ranging from 19.8–62.3 μg of cholesterol esterified/hr/ml were obtained on a single human plasma sample when the heat-inactivated substrate source was blood bank, human, rat, or monkey plasmas (93).

The initial reaction rates for cholesterol esterification in high- and low-density lipoprotein fractions have been determined using radioactive assay methods (39). Initial concentrations of free and esterified cholesterol in the HDL were 1.11 and 0.40 μmoles, respectively, and for the LDL, were 1.99 and 0.70 μmoles. The initial rate for the HDL was 0.08 μmole of cholesterol esterified/ml/hr and for the LDL was 0.015 μmole/ml/hr.

G. OTHER PROPERTIES

Fresh human serum has been subject to starch block and poly-(vinyl chloride)–poly(vinyl acetate) copolymer zone electrophoresis using 0.1*M* barbital, pH 8.6 (42). Enzyme activity migrates closely with the α-globulin peak, but tailing of the enzyme activity was evident, and 40–50% of the activity was lost, suggesting hetereogeneity and lability of the total plasma cholesterol esterifying activity. Fresh serum dialyzed for 24 hr against 0.01*M* Tris-HCl, pH 7.4, and subjected to DEAE-cellulose chromatography gives consistent recovery of two cholesterol-esterifying enzyme peaks, which also suggests heterogeneity in total plasma cholesterol esterifying activity (42).

5. Source and Distribution

Friedman and Byers (35) in 1955 reported that partially hepatectomized or eviscerated rats lost their ability to restore normal levels of plasma cholesterol esters, and suggested that the liver was the major source of plasma cholesterol esters. Based on this study, Brot et al. (9) determined the effect of hepatectomy evisceration in rats on the capacity of plasma to esterify free cholesterol *in vitro* (presumably via the transferase reaction). It was found that plasma from sham-operated rats esterified 44.5% of added labeled sterol, while plasma from operated rats averaged 5.4% esterification. Other investigators (7,14,133,134) have reported that plasma of patients with liver disturbances, i.e., insufficiency, hepatitis, and cirrhosis, all have depressed capacity for sterol esterification. These data suggest that the plasma transferase is derived from the liver; however, attempts to actually demonstrate transferase activity in liver have been unsuccessful (6,40). In a study on the distribution of plasma transferaselike cholesterol-esterifying activity, Glomset and Kaplan

(40) reported that such activity was either completely absent or present in extremely low concentrations in all tissues studied.

VI. LIVER ACYL CoA-CHOLESTEROL *O*-ACYLTRANSFERASE

1. Introduction and Mechanism

As discussed under Section IV-1-B above, there were a number of reports of cholesterol esterifying activity in liver prior to 1958. However, these described very low levels of esterifying activity and the activity was either attributed to cholesterol ester hydrolase or the experimental conditions were such that the mechanism of the reaction could not be specified. In 1958 Mukerjee et al. (79) reported that ATP and CoA were required for the esterification of cholesterol with palmitic acid by rat liver homogenates. Since palmityl CoA could replace the ATP and CoA requirement, they suggested that the esterification proceeded in two steps: (*a*) The activation of the fatty acid by ATP to form the acyl-CoA derivative, and (*b*) a transesterification reaction in which the fatty acid exchanges its CoA group for the hydroxyl of cholesterol to form the cholesterol ester. These reactions are shown schematically below.

$$\text{Fatty acid} + \text{coenzyme A} \xrightarrow[\text{Mg}^{2+}]{\text{ATP}} \text{acyl-CoA}$$

$$\text{Acyl-CoA} + \text{cholesterol} \longrightarrow \text{cholesterol ester}$$

This reaction is distinct from that in serum, catalyzed by lecithin–cholesterol acyl *O*-transferase, since attempts to demonstrate cholesterol esterification with labeled fatty acids derived from the β position of lecithin have been unsuccessful (6,23).

Recent studies (23,47,125,126,131) have described the subcellular sites of the reaction and have defined in more detail the characteristics of the reaction. The enzyme occurs in microsomes, mitochondria, and possibly in the soluble fraction. Goodman et al. (47) reported that the enzymes mediating cholesterol esterification in rat liver were exclusively particulate, while Swell and co-workers (131) have described similar activity in the soluble fraction. Recently, Swell and Law (124) have reported (in the absence of added fatty acid sources) liver microsomes and soluble fraction from the fed rat synthesize cholesterol esters having a similar fatty acid composition. Microsomes from the fasted rat synthesize the same pattern of esters as those of the fed rat (principally saturated and monosaturated fatty acid esters). However, the soluble fraction synthesizes esters in which arachidonate is a major component.

2. Assay Method

The general assay method allows the study of several aspects of the esterification reaction, but there are some serious limitations to the procedure. The degree of esterification of radioactive cholesterol by the enzyme in the presence of acyl-CoA is determined by isolation of radioactive cholesterol ester by column or thin-layer chromatography and subsequent radioactive counting. Possible variations are: (*a*) Substitution of ATP + CoA + fatty acid for the acyl-CoA, and (*b*) use of radioactive acyl-CoA or fatty acids. The assay method is essentially that described by Goodman et al. (47). A comparable system has been described by Swell and co-workers (126).

A. ENZYME PREPARATION

Rats, weighing 125–250 g, are sacrificed by decapitation and the livers are perfused *in situ* via the superior vena cava with ice cold 0.85% NaCl solution. The livers are then removed, passed through a tissue press, and the pulp homogenized at 0°C in a Potter-Elvehjem homogenizer with a loose-fitting pestle; 1 g of liver is added to 2.5 ml of 0.1*M* phosphate buffer, pH 7.4. The homogenate is serially centrifuged at 2000*g*, 10,000*g*, and 104,000*g* to prepare mitochondrial, microsomal, and soluble supernatant fractions. The mitochondrial and microsomal fractions are resuspended in a large volume of buffer followed by ultracentrifugation as above. The washed particles are then suspended in a volume of buffer equal to 1/10 volume or to the same volume of homogenate from which the particles have been separated.

B. SUBSTRATES

Fatty acyl-CoA esters are synthesized chemically by the method of Goldman and Vagelos (43). Solutions of the potassium salts of fatty acids are prepared by dissolving the fatty acid in a small volume

(1 or 2 ml) of methanol followed by addition of that amount of KOH solution required to neutralize the fatty acid. After evaporation of the solvent, the potassium salt is dissolved in hot water and the solution diluted to the desired concentration. Free fatty acid-deficient serum albumin is prepared by extracting dry serum albumin with 5% glacial acetic acid in isooctane (44). The ATP and coenzyme A solutions are adjusted to pH 7 with NaOH before use or are prepared in the phosphate buffer medium. The radioactive cholesterol is dissolved in acetone prior to addition to the incubation system.

The enzyme preparations used in the assay contain both endogenous cholesterol and free fatty acids so that it must be assumed that there is an exchange or equilibration between the radioactive substrate added to the medium and that in the microsomes or mitochondria. In the case of microsomes, it has been shown that after incubation, reisolated microsomes contain all of the newly formed radioactive cholesterol esters. Also, it has been demonstrated that the equilibrium must be rapidly attained, since after one hour incubation calculation of the synthesis of cholesterol esters, based on the calculated specific activity of the free cholesterol pool in the microsomes at zero time, agrees well with the chemically determined increase in cholesterol esters (126).

C. CONDITIONS

The basic incubation system contains in 3 ml: washed microsomes, usually 0.5 mg of protein; 300 μmoles of potassium phosphate buffer, pH 7.1; 3 mg of fatty acid-deficient serum albumin; 75 mμmoles of fatty acyl-CoA, and 20 mμmoles of radioactive cholesterol (added in 100 μl of acetone via a 100-μl syringe). The incubation is carried out in 25-ml Erlenmeyer flasks in a metabolic shaker at 37°C with air as the gas phase. Duration of incubation is from 30 to 60 min.

A variation of this system is the substitution of 6 μmoles of ATP, 0.3 μmoles of CoA, and 75 mμmoles of potassium oleate for the fatty acyl-CoA. A further variation described by both Goodman et al. (47) and Swell and co-workers (126) is the use of radioactive fatty acids or fatty acyl-CoA. In this case no cholesterol need be added to the system since there is adequate endogenous cholesterol in the microsomes. This latter system has been used to determine the free fatty acid pools and their relative availability for cholesterol esterification. The systems involving the addition of ATP, CoA, and fatty acid salt are actually an assay for a combination of two reactions, namely the formation of the fatty acyl-CoA and its utilization in the esterification of cholesterol.

The reaction is terminated by transferring the incubation mixture to chloroform–methanol (2:1, v/v) and washing with chloroform methanol to a final volume of 50 ml. The mixture is separated into two phases by the addition of 7 ml of 1:2000 H_2SO_4, and the entire lower phase, except 0.5 ml, is withdrawn and evaporated to dryness under a stream of N_2. The lipid residue is taken up in an appropriate solvent and analyzed, as described in Section III above.

3. Purification

No purification of liver acyl CoA-cholesterol *O*-acyltransferase has been reported.

4. Properties

A. COFACTORS

In the basic assay system containing microsomes (enzyme) and radioactive cholesterol as the substrate, addition of both ATP and CoA is required for cholesterol esterification. There is no requirement for added free fatty acid, since this is present in adequate amounts in microsomes. If fatty acyl-CoA is included as the second substrate, then ATP and CoA are not required. There is no requirement for added divalent cations (47).

The addition of free fatty acid bound to albumin (with ATP plus CoA) or fatty acyl CoA produces some stimulation of esterification, and the added fatty acid makes up a major fraction of the acids in the newly synthesized cholesterol esters. This is particularly true with oleic, palmitic, and linoleic acids (118). This effect may be due to shifts in the sizes of the endogenous fatty acid pools, relative specificity of the enzyme, or both. According to Goodman et al. (47), the omission of fatty acid-poor albumin when either free fatty acids or fatty acyl-CoA compounds are added, results in a marked inhibition of cholesterol esterification. They suggest that the albumin binds the added free fatty acids or those liberated from the fatty acyl-CoA compounds. In the system described by Swell and co-workers (126), they were unable to demonstrate the protective effect of added serum albumin.

B. pH

The optimum pH appears to lie in a broad range between 6.5 and 7.4, probably between 6.9 and 7.1 (47,126).

C. INHIBITORS

Sodium taurocholate and glycocholate, when added at a concentration of 1 mg/ml produce more than 70% inhibition of cholesterol esterification. In systems containing either ATP + CoA + potassium oleate or oleyl-CoA, increasing the bile salt concentration to 4 mg/ml gives more than 97% inhibition (47).

D. KINETICS

Since the enzyme source (microsomes or mitochondria) contains both substrates, i.e., free cholesterol and fatty acids, a study of the detailed kinetics of the reaction catalyzed by liver acyl-CoA-cholesterol *O*-acyltransferase are not available. In the incubation system, described above, the rate of esterification is constant for 20–30 min with less than 30% decrease in linearity at one hour. Goodman et al. (47) observed that maximum esterification of cholesterol occurred with small amounts of microsomes (less than 1 mg protein), while larger amounts of microsomes inhibited esterification. Swell et al. (126) showed that increasing amounts of ATP and CoA stimulated cholesterol esterification; the optimum level was 5 μmoles of ATP and 0.65 μmoles of CoA in their incubation system.

E. FATTY ACID SPECIFICITY

The rat liver microsomal enzyme has a high relative specificity for the formation of cholesterol oleate in comparison to other cholesterol esters. Only oleate added as the potassium salt or as the CoA derivative gives a marked stimulation of esterification. This is true whether the acids or CoA derivatives (oleic, palmitic, stearic, and linoleic) are added singly or as equimolar mixtures of the four acids (47). The relative rates of esterification with either fatty acids or CoA derivatives are: oleic > palmitic > stearic > linoleic.

5. Occurrence of Similar Enzyme Systems

The only other tissue in which an enzyme system comparable to the liver acyl CoA cholesterol *O*-acyltransferase has been reported is adrenal. Dailey and co-workers (17), Longcope and Williams (68), and Brot et al. (10) have described a cholesterol esterification reaction in dog and rat adrenals which requires ATP and CoA or fatty acyl-CoA. Like the liver system, there is activity in microsomes, mitochondria, and possibly in the soluble fraction. In general, the requirements and characteristics of the reaction are like that in liver except for a requirement for the divalent cations, Mg^{2+} or Ca^{2+} (68).

Acknowledgments

The authors are indebted to DeWitt S. Goodman and John A. Glomset for providing unpublished manuscripts and other information. Mrs. M. Pogue gave valuable assistance in the preparation and checking of the manuscript. Investigations reported from the authors' laboratory were supported by grants from the National Heart Institute, National Institutes of Health, and the Life Insurance Medical Research Fund. The valuable technical assistance of Mr. H. Kothari and Miss E. Herm is gratefully acknowledged.

References

1. Abderhalden, E., *Biochem. Handlexikon, 3,* 178 (1911).
2. Ansell, G. B., and J. N. Hawthorne, Eds., *Phospholipids,* Elsevier, Amsterdam, 1964, pp. 46, 89.
3. Avigan, J., *J. Biol. Chem., 234,* 787 (1959).
4. Barron, E. J., and D. J. Hanahan, *J. Biol. Chem., 231,* 493 (1958).
5. Bartlett, G. R., *J. Biol. Chem., 234,* 466 (1959).
6. Bennett, J. R., *Biochim. Biophys. Acta, 70,* 465 (1963).
7. Bertolini, A., C. Guardamagna, and N. Massori, *Acta Vitaminol., 13,* 3 (1959).
8. Borja, C. R., G. V. Vahouny, and C. R. Treadwell, *Am. J. Physiol., 206,* 223 (1964).
9. Brot, N., W. J. Lossow, and I. L. Chaikoff, *J. Lipid Res., 3,* 413 (1962).
10. Brot, N., W. J. Lossow, and I. L. Chaikoff, *Proc. Soc. Exptl. Biol. Med., 114,* 786 (1963).
11. Brot, N., W. J. Lossow, and I. L. Chaikoff, *J. Lipid Res., 5,* 63 (1964).
12. Byron, J. E., W. A. Wood, and C. R. Treadwell, *J. Biol. Chem., 205,* 483 (1953).
13. Castaigne, A., and A. Amselem, *Ann. Biol. Clin. (Paris), 17,* 336 (1959).
14. Castro, M. H., and J. C. Jennings, *Bull. Inst. Med. Res. (Madrid), 2,* 81 (1949).
15. Creech, B. G., and B. Sewell, *Anal. Biochem., 3,* 119 (1962).
16. Dailey, R. E., L. Swell, and C. R. Treadwell, *Proc. Soc. Exptl. Biol. Med., 110,* 571 (1962).

17. Dailey, R. E., L. Swell, and C. R. Treadwell, *Arch. Biochem. Biophys.*, *99*, 334 (1962).
18. Dailey, R. E., L. Swell, and C. R. Treadwell, *Arch. Biochem. Biophys.*, *100*, 360 (1963).
19. David, J. S. K., P. Malathi, and J. Ganguly, *Biochem. J.*, *98*, 662 (1966).
20. Day, A. J., *Quart. J. Exptl. Physiol.*, *45*, 55 (1960).
21. Day, A. J., and P. R. S. Gould-Hurst, *Quart. J. Exptl. Physiol.*, *46*, 376 (1961).
22. Day, A. J., and P. R. S. Gould-Hurst, *Biochim. Biophys. Acta*, *116*, 169 (1966).
23. Deykin, D., and D. S. Goodman, *Biochem. Biophys. Res. Commun.*, *8*, 411 (1962).
24. Deykin, D., and D. S. Goodman, *J. Biol. Chem.*, *237*, 3649 (1962).
25. *Enzyme Nomenclature*, Elsevier, Amsterdam, 1965, p. 34.
26. Etienne, J., and J. Polonovski, *Bull. Soc. Chem. Biol.*, *41*, 321 (1959).
27. Etienne, J., and J. Polonovski, *Bull. Soc. Chem. Biol.*, *41*, 813 (1959).
28. Etienne, J., and J. Polonovski, *Bull. Soc. Chem. Biol.*, *42*, 857 (1960).
29. Fillerup, D. L., and J. F. Mead, *Proc. Soc. Exptl. Biol. Med.*, *83*, 574 (1953).
30. Fiske, C. H., and Y. Subbarrow, *J. Biol. Chem.*, *66*, 375 (1925).
31. Fodor, P. J., *Arch. Biochem.*, *26*, 331 (1950).
32. Folch, J., M. Lees, and G. H. S. Stanley, *J. Biol. Chem.*, *226*, 497 (1957).
33. Fontaine, T., E. LeBreton, and J. Pantaleon, *Compt. Rend. Soc. Biol.*, *137*, 611 (1943).
34. Frantz, I. D., Jr., and N. L. R. Bucher, *J. Biol. Chem.*, *206*, 471 (1954).
35. Friedman, M., and S. O. Byers, *J. Clin. Invest.*, *34*, 1369 (1955).
36. Gallo, L., and C. R. Treadwell, *Proc. Soc. Exptl. Biol. Med.*, *114*, 69 (1963).
37. Glomset, J. A., *Biochim. Biophys. Acta*, *65*, 128 (1962).
38. Glomset, J. A., *Biochim. Biophys. Acta*, *70*, 389 (1963).
39. Glomset, J. A., E. Janssen, R. Kennedy, and J. Dobbins, *J. Lipid Res.*, *7*, 639 (1966).
40. Glomset, J. A., and D. M. Kaplan, *Biochim. Biophys. Acta*, *98*, 41 (1965).
41. Glomset, J. A., F. Parker, M. Tyadan, and R. H. Williams, *Biochim. Biophys. Acta*, *58*, 398 (1962).
42. Glomset, J. A., and J. L. Wright, *Biochim. Biophys. Acta*, *89*, 266 (1964).
43. Goldman, P., and P. R. Vagelos, *J. Biol. Chem.*, *236*, 2620 (1961).
44. Goodman, D. S., *J. Am. Chem. Soc.*, *80*, 3887 (1958).
45. Goodman, D. S., *J. Clin. Invest.*, *41*, 1886 (1962).
46. Goodman, D. S., *Physiol. Rev.*, *45*, 747 (1965).
47. Goodman, D. S., D. Deykin, and T. Shiratori, *J. Biol. Chem.*, *239*, 1335 (1964).
48. Hanahan, D. J., *J. Biol. Chem.*, *195*, 199 (1952).
49. Hanahan, D. J., *Lipid Chemistry*, Wiley, New York, 1960.
50. Hanahan, D. J., J. C. Dittmer, and L. Warashena, *J. Biol. Chem.*, *228*, 665 (1957).
51. Hernandez, H. H., and I. L. Chaikoff, *J. Biol. Chem.*, *228*, 447 (1957).

52. Hernandez, H. H., I. L. Chaikoff, and J. Y. Kiyasu, *Am. J. Physiol.*, *181*, 523 (1955).
53. Hirsch, J., and E. H. Ahrens, Jr., *J. Biol. Chem.*, *233*, 311 (1958).
54. Kabara, J. J., in *Methods of Biochemical Analysis*, Vol. 10, D. Glick, Ed., Interscience, New York, 1962, p. 263.
55. Karmen, A., I. McCaffrey, and P. L. Bowman, *J. Lipid Res.*, *3*, 372 (1962).
56. Kelly, L. A., and H. A. I. Newman, *Federation Proc.*, *24*, 210 (1965).
57. Klein, P. D., and E. T. Janssen, *J. Biol. Chem.*, *234*, 1417 (1959).
58. Klein, W., *Z. Physiol. Chem.*, *254*, 1 (1938).
59. Klein, W., *Z. Physiol. Chem.*, *259*, 268 (1939).
60. Korzenovsky, M., C. P. Walters, O. A. Harvey, and E. R. Diller, *Proc. Soc. Exptl. Biol. Med.*, *105*, 303 (1960).
61. Korzenovsky, M., E. R. Diller, A. C. Marshall, and B. M. Auda, *Biochem. J.*, *76*, 238 (1960).
62. Krishnamurthy, S., P. S. Sastry, and J. Ganguly, *Arch. Biochem. Biophys.*, *75*, 6 (1958).
63. Kuksis, A., *Can. J. Biochem.*, *112*, 407 (1964).
64. LeBreton, E., and J. Panteleon, *Compt. Rend. Soc. Biol.*, *138*, 20 (1944).
65. LeBreton, E., and J. Pantaleon, *Compt. Rend. Soc. Biol.*, *138*, 38 (1944).
66. LeBreton, E., and J. Pantaleon, *Arch. Sci. Physiol.*, *1*, 63 (1947).
67. LeBreton, E., and J. Pantaleon, *Exposes Ann. Biochem. Med.*, *7*, 111 (1947).
68. Longcope, C., and R. H. Williams, *Endocrinology*, *72*, 735 (1963).
69. Lossow, W. J., R. H. Migliorino, N. Brot, and I. L. Chaikoff, *J. Lipid Res.*, *5*, 198 (1964).
70. Mahadevan, V., and W. O. Lundberg, *J. Lipid Res.*, *3*, 106 (1962).
71. Mangold, H. K., *J. Am. Oil Chem. Soc.*, *38*, 708 (1961).
72. Mattson, F., and R. A. Volpenhein, *J. Lipid Res.*, *7*, 536 (1966).
73. Mead, J. F., and D. R. Howton, *Radioisotope Studies of Fatty Acid Metabolism*, Pergamon Press, New York, 1960, p. 88.
74. Miller, D., and R. K. Crane, *Anal. Biochem.*, *2*, 1 (1961).
75. Morris, L. J., *J. Lipid Res.*, *4*, 357 (1963).
76. Morton, R. K., in *Methods of Enzymology*, Vol. 1, S. P. Colowick and N. O. Kaplan, Eds., Academic Press, New York, 1955, p. 34.
77. Mueller, J. H., *J. Biol. Chem.*, *22*, 1 (1915).
78. Mueller, J. H., *J. Biol. Chem.*, *25*, 561 (1916).
79. Mukherjee, S., G. Kunitake, and R. B. Alfin-Slater, *J. Biol. Chem.*, *230*, 91 (1958).
80. Murthy, S. K., and J. Ganguly, *Biochem. J.*, *83*, 460 (1962).
81. Murthy, S. K., S. Mahadevan, P. S. Sastry, and J. Ganguly, *Nature*, *189*, 482 (1961).
82. Myers, D. K., A. Schotte, H. Boer, and H. Borsje-Bakker, *Biochem. J.*, *61*, 521 (1955).
83. Nedswedski, S. W., *Z. Physiol. Chem.*, *236*, 69 (1935).
84. Nedswedski, S. W., *Z. Physiol. Chem.*, *239*, 165 (1936).
85. Nedswedski, S. W., *Biokhimiya*, *2*, 758 (1937).
86. Nedswedski, S. W., *Biokhimiya*, *11*, 323 (1946).

87. Nieft, M. L., and H. J. Deuel, Jr., *J. Biol. Chem.*, *177*, 143 (1949).
88. Nomura, T., *Tohoku J. Exptl. Med.*, *4*, 281 (1924).
89. Nomura, T., *Tohoku J. Exptl. Med.*, *4*, 677 (1924).
90. Phillips, G. B., *Biochim. Biophys. Acta*, *29*, 594 (1958).
91. Popjak, G., A. E. Lowe, and D. Moore, *J. Lipid Res.*, *3*, 364 (1962).
92. Porte, D., Jr., and R. J. Havel, *J. Lipid Res.*, *2*, 357 (1961).
93. Portman, O. W., and M. Sugano, *Arch. Biochem. Biophys.*, *105*, 532 (1964).
94. Pritchard, E. T., and N. E. Nichol, *Biochim. Biophys. Acta*, *84*, 781 (1964).
95. Rhodes, D. N., and C. H. Lea, *Biochem. J.*, *65*, 526 (1957).
96. Robertson, A. F., and W. E. M. Lands, *Biochemistry*, *1*, 804 (1962).
97. Rowen, R., *Biochim. Biophys. Acta*, *84*, 761 (1964).
98. Saito, K., and D. J. Hanahan, *Biochemistry*, *1*, 521 (1962).
99. Schotz, M. C., L. I. Rice, and R. B. Alfin-Slater, *J. Biol. Chem.*, *207*, 665 (1955).
100. Schultz, J. H., *Biochim. Z.*, *42*, 255 (1912).
101. Schwenk, E., and N. T. Werthessen, *Arch. Biochem. Biophys.*, *40*, 334 (1952).
102. Shah, S. N., W. J. Lossow, and I. L. Chaikoff, *Biochim. Biophys. Acta*, *84*, 176 (1964).
103. Shiratori, T., and D. S. Goodman, *Biochim. Biophys. Acta*, *106*, 625 (1965).
104. Shope, R. E., *J. Biol. Chem.*, *80*, 127 (1928).
105. Shyamala, G., W. J. Lossow, and I. L. Chaikoff, *Proc. Soc. Exptl. Biol. Med.*, *118*, 138 (1965).
106. Smith, P. F., *J. Lipid Res.*, *5*, 121 (1964).
107. Spanner, G. O., and L. Bauman, *J. Biol. Chem.*, *98*, 181 (1932).
108. Sperry, W. M., *J. Biol. Chem.*, *111*, 467 (1935).
109. Sperry, W. M., *J. Biol. Chem.*, *113*, 599 (1936).
110. Sperry, W. M., and F. C. Brand, *J. Biol. Chem.*, *137*, 377 (1941).
111. Sperry, W. M., and V. A. Stoyanoff, *J. Biol. Chem.*, *117*, 525 (1937).
112. Sperry, W. M., and V. A. Stoyanoff, *J. Biol. Chem.*, *121*, 101 (1937).
113. Sperry, W. M., and V. A. Stoyanoff, *J. Biol. Chem.*, *126*, 77 (1938).
114. Sperry, W. M., and M. Webb, *J. Biol. Chem.*, *187*, 97 (1950).
115. Stahl, E. G., *Thin-Layer Chromatography*, Academic Press, New York, 1965.
116. Stahl, E. G., G. Schroter, G. Kraft, and R. Renz, *Pharmazie*, *11*, 633 (1956).
117. Swell, L., *Anal. Biochem.*, *16*, 70 (1966).
118. Swell, L., *Proc. Soc. Exptl. Biol. Med.*, *121*, 1290 (1966).
119. Swell, L., T. A. Boiter, H. Field, Jr., and C. R. Treadwell, *Am. J. Physiol.*, *181*, 193 (1955).
120. Swell, L., J. E. Byron, and C. R. Treadwell, *J. Biol. Chem.*, *186*, 543 (1950).
121. Swell, L., R. E. Dailey, H. Field, Jr., and C. R. Treadwell, *Arch. Biochem. Biophys.*, *59*, 393 (1955).
122. Swell, L., H. Field, Jr., and C. R. Treadwell, *Proc. Soc. Exptl. Biol. Med.*, *84*, 417 (1953).
123. Swell, L., H. Field, Jr., and C. R. Treadwell, *Proc. Soc. Exptl. Biol. Med.*, *87*, 216 (1954).
124. Swell, L., and M. D. Law, *Biochem. Biophys. Res. Commun.*, *26*, 206 (1967).
125. Swell, L., M. D. Law, and C. R. Treadwell, *Proc. Soc. Exptl. Biol. Med.*, *109*, 176 (1962).

126. Swell, L., M. D. Law, and C. R. Treadwell, *Arch. Biochem. Biophys.*, *104*, 128 (1964).
127. Swell, L., and C. R. Treadwell, *J. Biol. Chem.*, *182*, 479 (1950).
128. Swell, L., and C. R. Treadwell, *J. Biol. Chem.*, *185*, 349 (1950).
129. Swell, L., and C. R. Treadwell, *J. Biol. Chem.*, *212*, 141 (1955).
130. Swell, L., and C. R. Treadwell, *Anal. Biochem.*, *4*, 335 (1962).
131. Swell, L., and C. R. Treadwell, *Proc. Soc. Exptl. Biol. Med.*, *110*, 55 (1962).
132. Treadwell, C. R., and G. V. Vahouny, *Handbook of Physiology*, in press.
133. Turner, K. B., G. H. McCormack, Jr., and A. Richards, *J. Clin. Invest.*, *32*, 801 (1953).
134. Turner, K. B., and V. Pratt, *Proc. Soc. Exptl. Biol. Med.*, *71*, 633 (1949).
135. Vahouny, G. V., C. R. Borja, R. M. Mayer, and C. R. Treadwell, *Anal. Biochem.*, *1*, 371 (1960).
136. Vahouny, G. V., C. R. Borja, and S. Weersing, *Anal. Biochem.*, *6*, 555 (1963).
137. Vahouny, G. V., and C. R. Treadwell, *Proc. Soc. Exptl. Biol. Med.*, *116*, 496 (1964).
138. Vahouny, G. V., and C. R. Treadwell, *Methods of Enzymology*, S. P. Colowick and N. O. Kaplan, Eds., Academic Press, New York, in press.
139. Vahouny, G. V., S. Weersing, and C. R. Treadwell, *Biochem. Biophys. Res. Commun.*, *15*, 224 (1964).
140. Vahouny, G. V., S. Weersing, and C. R. Treadwell, *Arch. Biochem. Biophys.*, *107*, 7 (1964).
141. Vahouny, G. V., S. Weersing, and C. R. Treadwell, *Biochim. Biophys. Acta*, *98*, 607 (1965).
142. Wagner, A., *Circulation Res.*, *7*, 818 (1959).
143. Weiland, H., and H. Sorge, *Z. Physiol. Chem.*, *97*, 1 (1916).
144. Whearet, A. F., and E. Staple, *Arch. Biochem. Biophys.*, *90*, 224 (1950).
145. Wiseman, G., *Absorption from the Intestine*, Academic Press, New York, 1964, p. 148.
146. Yamamoto, R. S., N. P. Goldstein, and C. R. Treadwell, *J. Biol. Chem.*, *180*, 615 (1949).
147. Zlatkis, A., B. Zak, and A. J. Boyle, *J. Lab. Clin. Med.*, *41*, 486 (1953).
148. Authors' unpublished results.

METHODS OF BIOCHEMICAL ANALYSIS VOLUME 16

Determination of Histidine Decarboxylase Activity*

RICHARD W. SCHAYER, *Research Center, Rockland State Hospital, Orangeburg, New York*

* Supported by U. S. Public Health Service Grant AM 10155.

I. INTRODUCTION

The first demonstrations of enzymatic decarboxylation of L-histidine by extracts of mammalian tissues were made independently in the laboratories of Werle (1) and Holtz (2). Later studies using isotopic L-histidine revealed the presence in rat stomach of a histidine-decarboxylating enzyme with different characteristics (3).

The originally discovered enzyme, which has a low affinity for L-histidine and catalyzes decarboxylation of a number of amino acids, has been found to be identical with dopa decarboxylase (4,5). The other histidine-decarboxylating enzyme, which has a high affinity for substrate molecules and, as far as is known, is specific for L-histidine, is probably the major *in vivo* catalyst of histamine formation (4,5). Werle now agrees that the nonspecific and specific histidine-decarboxylating enzymes are different enzymes and not two different centers on the same protein (6).

A recent review (7) covers the historical development in this field, the sources and properties of histidine decarboxylase, its relationship to histamine formation *in vivo*, the mechanism of action, and inhibitor studies.

Since the nonspecific histidine-decarboxylating enzyme seems to be identical with dopa decarboxylase, methods for its determination will not be included in this paper. Certain bacteria have a powerful specific histidine decarboxylase; methods have been published in Volume 4 of this series (8). Methods for determination of histidine decarboxylase activity in plants have been reported (9,10).

II. GENERAL PRINCIPLES

Histidine decarboxylase in the presence of the cofactor pyridoxal phosphate catalyzes the conversion of L-histidine to histamine plus carbon dioxide. Methods of enzyme assay are based on the amount of histamine or carbon dioxide formed. Disappearance of L-histidine cannot ordinarily be used as a measure of histidine decarboxylase activity since histamine formation is a minor metabolic pathway in mammalian tissues.

Since the amount of histamine formed may be extremely small, and since relatively large quantities of endogenous histamine may be present, isotopic methods are particularly useful.

The methods to be described are of three types: (*a*) nonisotopic; newly formed histamine is measured, (*b*) isotopic; newly formed labeled histamine is measured, and (*c*) isotopic; $^{14}CO_2$ is measured. The merits and disadvantages of each will be discussed in Section VI.

III. PURIFICATION OF HISTIDINE DECARBOXYLASE

1. Method of Hakanson (11)

Pregnant rats are decapitated 15–20 days after mating and the litters removed and pooled. In frozen condition the fetal rat tissue maintained its enzyme activity for several weeks. At least 50 g of whole fetus are taken for each preparation.

Tissue is homogenized in 2 volumes of 0.1*N* NaAc buffer, pH 4.5, centrifuged at 20,000*g* for 20 min at 0° and the precipitate discarded. The supernatant (I) is warmed at 55° for 5 min, centrifuged at 20,000*g* for 10 min at 0°, and the precipitate discarded. The supernatant (II) contains more than 90% of the initial enzyme activity.

The enzymic material is precipitated with $(NH_4)_2SO_4$ at 25, 40, and 60% saturation. The precipitates are spun down, redissolved in 10–20 ml 0.1*M* phosphate buffer, pH 7.0, and dialyzed against redistilled water at 4° overnight. Sediments develop on dialysis and the extracts are centrifuged before the assay of activity. The fraction which precipitates between 25 and 40% saturation contains most of the enzyme activity. This fraction is again treated with $(NH_4)_2SO_4$ and the material which now precipitates between 28 and 42% saturation contains all enzyme activity. This extract is dialyzed against water for at least 6 hr. This treatment results in a 200-fold purification of the enzyme as compared to supernatant II. More than 50% of the enzyme activity of the initial homogenate is preserved in this extract, which is diluted to contain 0.4% protein before use. In this state the enzyme is stable for several days.

2. Method of Aures

Dr. Dorothea Aures kindly provided her method prior to publication; preliminary results have been reported (12,13). A special

TABLE I

D. Aures, Purification of Histidine Decarboxylase

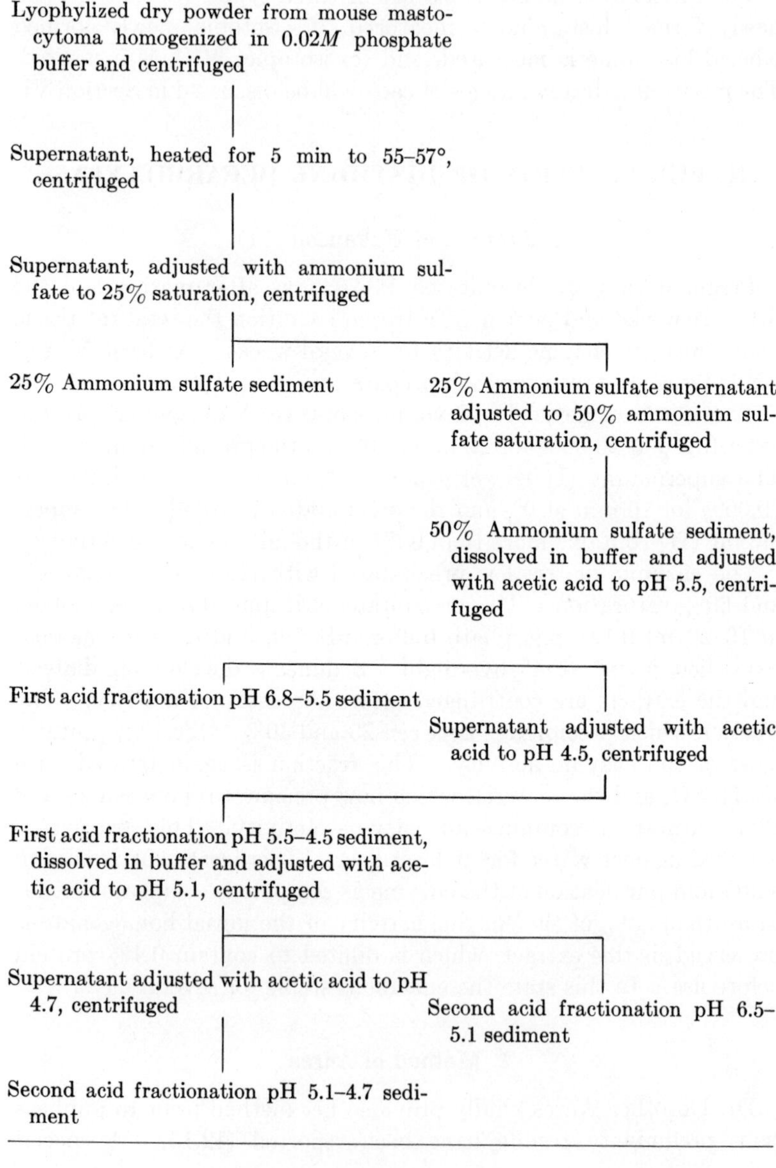
Lyophylized dry powder from mouse mastocytoma homogenized in 0.02*M* phosphate buffer and centrifuged
Supernatant, heated for 5 min to 55–57°, centrifuged
Supernatant, adjusted with ammonium sulfate to 25% saturation, centrifuged
25% Ammonium sulfate sediment
25% Ammonium sulfate supernatant adjusted to 50% ammonium sulfate saturation, centrifuged
50% Ammonium sulfate sediment, dissolved in buffer and adjusted with acetic acid to pH 5.5, centrifuged
First acid fractionation pH 6.8–5.5 sediment
Supernatant adjusted with acetic acid to pH 4.5, centrifuged
First acid fractionation pH 5.5–4.5 sediment, dissolved in buffer and adjusted with acetic acid to pH 5.1, centrifuged
Supernatant adjusted with acetic acid to pH 4.7, centrifuged
Second acid fractionation pH 6.5–5.1 sediment
Second acid fractionation pH 5.1–4.7 sediment

transplantable mouse mast-cell tumor, extremely rich in histidine-decarboxylating enzyme, is used. Its properties are similar to those of fetal rat-liver histidine decarboxylase but the tumor is a more convenient source. Dopa decarboxylase is also present. The purification steps are shown in Table I and the results in Table II.

TABLE II

Purification of Histidine Decarboxylase

Fraction	μmoles CO_2/g protein/hr[a]		Increase in specific activity over the starting material	
	Histidine	Dopa	Histidine	Dopa
Dry powder homogenate	1.64	23.28	—	—
25% Ammonium sulfate sediment	15.06	17.82	9.2	—
50% Ammonium sulfate sediment	6.36	29.26	3.8	1.26
First pH fractionation				
6.8–5.5	3.76	1.74	2.3	—
5.5–4.5	18.90	40.00	12.0	1.72
Second pH fractionation				
6.5–5.1	88.42	2.64	56.5	—
5.1–4.7	12,982.00	1,242.00	7,910.0	53.5

[a] Values obtained from initial rates, 30-min incubation time.

The heat treatment does not increase the specific activity of either dopa or histidine decarboxylase under the conditions employed, but decreases both. However, the ammonium sulfate fractionation is less successful if the heat treatment is omitted. The sediment from the 25% saturated ammonium sulfate solution is 20 times as active as the starting material, but the yield is low and the bulk of the decarboxylases appears in the sediment of the 50% saturated ammonium sulfate solution. The latter therefore was used for repeated pH fractionations, and the proteins were simply separated by lowering the pH with diluted acetic acid. A fraction which precipitated between pH 5.5 and 4.5 was refractionated. The most active material was obtained in the fraction between pH 5.1 and 4.7. Under the incubation conditions employed, 12,982 μmoles CO_2/g protein/hr was formed, measured as initial rates.

This fraction had a great increase in specific activity for histidine; its decarboxylating rate was 150 times as high as that for dopa. The acid treatment also has an activating effect, since quantitatively more activity units were obtained in the sediment than in the starting material. This highly active fraction was stable for a few days, then about 10% of its activity remained for several weeks.

IV. PURIFICATION OF L-HISTIDINE

Commercial L-histidine, isotopic and nonisotopic, is contaminated with a variable amount of histamine. It is important to realize that drastic treatment of histidine at any stage of the assay may cause a serious error through nonenzymatic histamine formation.

1. Method of Schayer (unpublished)

Dissolve the ^{14}C L-histidine in 1.3% sodium chloride solution equal to about one-half the desired final volume. Then add sufficient 1*N* NaOH solution so that after neutralization and dilution, the solution will be in approximately isotonic saline. ^{14}C-histamine is removed by extracting the alkaline solution six times with butanol–chloroform, 3:1 (volume about that of the aqueous fraction), centrifuging each time if the layers do not separate completely. The organic layer (upper) is discarded. The volume of the aqueous layer (lower), which contains the ^{14}C-L-histidine, is kept constant by addition of water as necessary.

Residual butanol and chloroform are removed by extracting the aqueous layer three times with ether. Residual ether is blown off in a stream of nitrogen, the aqueous solution neutralized with 0.1*N* HCl, and water is added to give the desired final volume. Portions of the ^{14}C-L-histidine solution are transferred to several suitable containers and stored in a freezer.

2. Method of Kahlson et al. (14,15)

The ^{14}C-histidine, not exceeding 4 mg, is dissolved in 1 ml of 0.1*M* sodium phosphate buffer, pH 6.5. A column 30 × 4 mm (0.20 g) of Dowex 50 W-X4 (100–200 mesh), buffered in advance with sodium acetate buffer (pH 6.0), is prepared in a glass tube. The histidine solution is passed through the column, followed by 4 ml of the phosphate buffer. In the collected effluent > 96% of the ^{14}C-histidine and < 2% of the ^{14}C-histamine are present, compared with the original solution. It should be mentioned that the purchased ^{14}C-histidine never contained > 0.1% of the ^{14}C-histamine before purification, and that the size of the column and volume of the buffer must be increased should the histamine content be greater.

In testing this method, by adding radioactive histamine to non-radioactive histidine, a higher degree of purification than that mentioned above was regularly seen, i.e., < 0.05% of the ^{14}C-histamine was found in the effluent. It would thus appear that some ^{14}C-histamine was formed on purification of radioactive histidine. However, the procedure, as employed, reduces contamination to such an extent that the activity of blank samples is negligible.

3. Method of Mackay and Shepherd (16)

Nonisotopic L-histidine zwitterion is prepared from commercial L-histidine monohydrochloride monohydrate to eliminate traces of histamine. The hydrochloric acid is removed by adding excess silver carbonate to a warm, concentrated, aqueous solution of the monohydrochloride. After filtration, silver ions are removed from the filtrate by passing in hydrogen sulphide. Filtration and subsequent concentration of the filtrate, under reduced pressure, give a saturated solution of L-histidine zwitterion from which the latter may be precipitated by addition of dehydrated alcohol.

V. METHODS FOR HISTIDINE DECARBOXYLASE ASSAY

1. Isotopic Methods: $^{14}CO_2$ Measured

A. METHOD OF KOBAYASHI (17)

Kobayashi has modified his original procedure (17) and currently uses the following method.

Histidine decarboxylase is prepared by excising rat tissues, washing quickly with physiological saline, and mincing with 3 volumes (1 g tissue/3 ml buffer) of 0.1*M* phosphate buffer, pH 7.4. The tissue is homogenized in an Omnimixer for 1 min at maximum speed, then frozen and thawed three times. The entire brei is centrifuged at approximately 20,000g for 30 min at 4° and the cell-free solution decanted. The pH of the cell-free preparation is adjusted to 6.9 with 0.01N HCl.

Analyses are made in a single-arm Warburg vessel incubated in a Dubnoff metabolic shaker. The side arm is sealed with a small multiple-dose bottle rubber stopper, and the main chamber sealed with a solid ground-glass stopper. In a typical assay, the flasks are prepared as follows: 2 ml of rat enzyme, 30 μg of pyridoxal phosphate, and 5 μg of ^{14}C-L-histidine (carboxyl-labeled) in the side arm. Hyamine hydroxide (Packard Instrument Co.), used to absorb carbon dioxide, is placed in the center well in one of two ways: 0.2 ml in the center well or 0.01–0.02 ml placed on a strip of Whatman No. 3 filter paper, 7.5 × 25 mm. The filter paper is edged with a thin strip of paraffin wax to prevent the Hyamine from touching the walls of the vessel. The final volume of each flask is usually 2.60 ml. The reaction time is 2 hr at 37°, after which the reaction is stopped by introducing 0.3 ml of 1M citric acid through the rubber stopper with a syringe. The flasks are shaken for an additional hour to ensure quantitative absorption of the carbon dioxide by Hyamine hydroxide. Longer absorption times do not result in increased recovery of carbon dioxide as measured by the amount of radioactivity found.

When Hyamine hydroxide is placed directly into the center well, the Hyamine solution is washed out quantitatively into a counting vial with water such that Hyamine plus the wash water weighs 1.5 g on a triple beam balance; 15 ml of counting solution is then added. The counting solution consists of 6 parts dioxane, 1 part anisole, and 1 part dimethoxyethane: 500 ml of this solvent mixture contained 3.5 g of diphenyloxazole (Pilot Chemical Co.) and 25 mg of *p*-bis(*O*-methylstyryl) benzene (Pilot Chemical Co.).

When Hyamine hydroxide on paper is used, it is placed in 10 ml of scintillation solvent consisting of 3 parts ethanol, 7 parts toluene, and 0.4% diphenyl-oxazole. It is essential that the caps of the counting vials be tightly sealed. Samples are counted in a liquid scintillation spectrometer.

More recently Kobayashi has found that the methanol in the Hyamine hydroxide used directly does not inhibit histidine decarboxylase activity. Therefore, the preparation of aqueous Hyamine hydroxide has been abandoned (personal communication).

B. METHOD OF AURES AND CLARK (18)

The incubation mixture consists of 0.1–0.4 ml of enzyme preparation, 0.0102 μmole of pyridoxal 5-phosphate, 0.03 μmole of aminoguanidine, and 0.2 μc DL-histidine-1-^{14}C (specific activity 1.05–1.1 μc/μmole), which is adjusted to a total of 0.5 μmole of the L-isomer with unlabeled L-histidine. The inhibitors are dissolved in Teorell-Stenhagen buffer (19) or Carbowax-300, a liquid ethylene glycol polymer of 300-cp viscosity, and added in different concentrations. The final mixture is adjusted to pH 6.8 and a volume of 0.5 ml with Teorell-Stenhagen buffer. The blanks are identical except that the enzyme is denatured by heating at 100° for 10 min. The incubations are done in a 5-ml Erlenmeyer flask which was connected with a sleeve-type rubber connector to a short-neck, 20-ml ampule (shown in Fig. 1 of ref. 18).

After a 2-hr shaking in a Dubnoff shaking incubator, the reaction is stopped by injecting 0.4 ml of 1*M* citric acid with a hypodermic syringe and 23-gauge needle through the rubber connector into the incubation mixture. The whole apparatus is then fastened with rubber bands in a horizontal position on a rotating wheel, as shown in (18), Figure 2, and cooled to 0°.

Two milliliters of a phenethylamine, methanol, and toluene–POPOP–PPO mixture (20) (27 ml of phenethylamine, 27 ml of absolute methanol, 0.5 g of PPO, 0.01 g of POPOP, and a sufficient quantity of toluene to make up 100 ml) is injected into the empty 20-ml ampule, and 0.1 ml of 1*M* sodium bicarbonate solution is injected into the Erlenmeyer flask. The diffusion chamber is then rotated on the wheel at a slow speed in a refrigerated ethylene glycol bath at 0°. After 30 min the sodium bicarbonate injection is repeated, and after a total rotation time of 60 min a third sodium bicarbonate addition is made and the rotation continued for a final 30 min. The phenylethylamine is rinsed quantitatively into a counting vial with 10 ml of toluene–methanol scintillator mixture (27 ml of absolute methanol, 0.5 mg of PPO, 0.01 g of POPOP, and a sufficient quantity of toluene to make a total volume of 100 ml). The counting is done in a liquid scintillation spectrometer.

C. METHOD OF LEVINE AND WATTS (21)

Specific histidine decarboxylase is prepared from whole fetal rats (19–20 days gestation) using the method of Hakanson (see Section III-1) modified by performing the initial homogenization in $0.1M$ sodium acetate buffer at pH 5.5 rather than 4.5. When the lower pH is used there is occasional complete loss of enzyme activity.

To prepare incubation vessels, a straightened wire paper clip with a narrow loop on one end is forced through the center of the cap of a 25-ml polyethylene vial (Packard Instrument Co.) so that the loop end projects from the lower part of the cap. A rectangular piece of Whatman 3MM filter paper measuring 1×3 cm is rolled into a cylinder and clamped firmly into the loop. Next, the paper is dipped into hydroxide of Hyamine and the excess allowed to drip off. It is essential that the loop be so positioned that when the cap is screwed onto the vial the filter paper is entirely in the upper third of the vial and does not touch the edges of the vial. The incubation is carried out at 37° in a Dubnoff metabolic shaker with the vial cap screwed on tightly. Components of the incubation mixture are listed below.

Component	Final concentration
Enzyme preparation	4–10 mg of protein/2 ml
Pyridoxal-5′-phosphate	$3.7 \times 10^{-5}M$
Streptomycin sulfate	$1.0 \times 10^{-4}M$
Sodium phosphate buffer, pH 6.8	$12.5 \times 10^{-2}M$
L-Histidine	$2.5 \times 10^{-4}M$
^{14}C-DL-Histidine	0.25 μc/ml
Water	To final volume of 2.0 ml

Streptomycin is added to suppress the growth of bacteria that may decarboxylate histidine; it does not affect histidine decarboxylase activity. The protein content varies with the activity of the enzyme preparation.

Blanks are prepared by adding 4-bromo-3-hydroxybenzyloxamine (NSD-1055), final concentration $10^{-4}M$, to incubation mixtures which are duplicates of experimental incubates; under these conditions histidine decarboxylase activity is totally inhibited.

Substrate is prepared in advance as follows. In $10^{-4}N$ HCl, nonradioactive L-histidine and ^{14}C-DL-histidine are dissolved to final concentrations of $5 \times 10^{-3}M$ (concentration is calculated to accommodate contribution of radioactive material to total L-histidine) 5 μc/ml (2.5 μc/ml of ^{14}C-L-histidine) respectively. Thus, addition of 0.1 ml of this solution to the incubation mixture results in the desired concentration of substrate.

The entire mixture, with the exception of substrate, is preincubated for 10 min. To start the reaction, substrate is added, and the cap of the vial screwed on tightly. $^{14}CO_2$ formation is linear for 2.5 hr. Thirty or 60-min incubations are used for routine assays.

To terminate the reaction, 2 ml of 6*N* HCl is injected through the side of the vial through a 23-gauge disposable needle; the hole in the vial is sealed promptly with adhesive tape. Incubation is continued for 30 min after acidification to permit quantitative adsorption of CO_2 into the filter paper. Then the vial is opened, the filter paper transferred to another vial containing 10 ml of a scintillation fluorophor solution (22), and the radioactivity determined in a scintillation spectrometer. Recovery of $^{14}CO_2$ is virtually quantitative. Under the conditions used by the authors, with enzyme preparations free from endogenous L-histidine, the recovery of 655 counts/min represents the decarboxylation of 1.0 mμmole of L-histidine.

Smith and Code have also devised a modified method in which $^{14}CO_2$ is measured (23).

2. Isotopic Methods: Radioactive Histamine Measured

A. METHOD OF SCHAYER

The method is a modification of isotope dilution methods previously described (7,24). It involves incubation of tissue extracts with minute quantities of radioactive L-histidine (diluted by endogenous free L-histidine), and determination of the ^{14}C-histamine formed by addition of carrier which is subsequently extracted, converted to the benzenesulfonyl derivative (BSH), and counted.

Soft tissues are homogenized by hand in cold phosphate buffer, 0.1*M*, pH 7.2–7.4, containing 0.2% glucose. Enzyme is extracted from tough tissues by repeated freezing and thawing. Either homogenates or cell-free extracts may be used.

If the tissue is significantly active with respect to either dopa decarboxylase, diamine oxidase (histaminase), or histamine methylation, additives are required. Dopa decarboxylase and diamine oxidase can be inhibited by addition of α-methyl dopa and aminoguanidine, respectively. No potent inhibitor of the histamine-methylating enzyme is available. If the tissue contains significant quantities of endogenous histamine there is no significant error; methylation of endogenous histamine occurs in the early phases depleting the supply of the methyl donor, *S*-adenosylmethionine; ^{14}C-histamine formed during the course of the anaerobic incubation is unaffected. If a tissue is low in endogenous histamine, it is advisable to add histamine to the buffer to give a concentration of about 10 μg/ml.

Incubations are done in 20-ml beakers to which are added 1.8 ml of crude enzyme preparation, 0.10 ml (50 μg) of pyridoxal phosphate, and 0.10 ml of purified ^{14}C-L-histidine solution. Currently, we use about 0.10 μc of ^{14}C-L-histidine, weighing approximately 0.4 μg; it is, of course, diluted by the endogenous free L-histidine in the tissue extract unless the latter is removed by dialysis or some other procedure. Blanks are prepared either by substituting 0.10 ml of 0.01*M* hydroxylamine for pyridoxal phosphate, or by using heat-denatured enzyme preparations. Incubations are carried out in a Dubnoff metabolic shaker for 3 hr at 37° under nitrogen. The reaction is stopped by adding with mixing 1 ml of carrier solution containing 66.4 mg of histamine dihydrochloride (40 mg of histamine base) plus 50 mg of L-histidine monohydrochloride (the latter is used to dilute the substrate) and 3 ml of 0.6*M* perchloric acid. If the protein precipitate is bulky, it may be removed by centrifugation and washed once with 1 ml of 0.2*M* perchloric acid. If only a small amount of protein is present it need not be removed.

Each incubate–carrier–perchloric acid mixture is transferred through a funnel to an extraction tube of suitable size using 1 ml of 0.2*M* perchloric acid for rinsing. Then are added through the same funnel, 1 ml of 5*N* NaOH solution and 20 ml of butanol–chloroform 3:1 (freshly mixed). Through a powder funnel is added approximately 3 g of solid NaCl, the tubes are closed tightly with a neoprene stopper, and are shaken mechanically for about 10 min. The solution is centrifuged and most of the organic layer (upper) is transferred to a clean tube (avoid contamination by traces of the lower aqueous layer) and washed by shaking with 5 ml of 0.1*N* NaOH solution saturated with NaCl for about 5 min. Most of the organic layer is transferred to a third tube and shaken with 10 ml of 0.1*N* HCl for about 5 min and most of the acid (which contains ^{14}C-histamine plus carrier) is transferred to a 25-ml Erlenmeyer flask, and evaporated to dryness in a stream of warm air.

The carrier histamine dihydrochloride samples are dissolved in 2 ml of water, and then are added 3 ml of a solution of benzenesulfonylchloride in dioxane (60 mg of benzenesulfonylchloride/ml of dioxane) and about 0.3–0.4 g of sodium bicarbonate. The reaction is allowed to proceed at 37° for 30 min with gentle shaking. Water is added in small quantities over a period of 1–2 hr to a total of about 20 ml. The sides of the flask are scratched to help start crystallization. The crude dibenzenesulfonyl histamine (BSH) is collected and washed with cold 25% ethanol in water. Recrystallization is effected by dissolving in a few milliliters of warm acetone, treating with charcoal, filtering through a 2-ml coarse sintered-glass funnel containing filter-aid, and slow addition of small amounts of warm water, avoiding turbidity. The crystals of BSH are collected, dried, and accurately weighed in vials suitable for liquid scintillation counting.

Since this is an isotope-dilution procedure, losses during the various steps need not be kept uniform for all samples and blanks.

BSH samples are dissolved in 15 ml of a suitable phosphor and counted in a liquid scintillation spectrometer. After subtracting the background count, samples and blanks are corrected for weight by calculating radioactivity per 100 mg of BSH. After subtracting blank values, samples are reported as counts per minute per 100 mg of BSH. Since approximately equal weights of a purified compound are counted, all samples and blanks show almost identical counting efficiency.

B. METHOD OF KAHLSON, ROSENGREN, AND THUNBERG, CURRENT MODIFICATION (15)

Minced tissue, usually 0.2 g, is incubated for 3 hr at 37° under nitrogen, in beakers containing 40 μg of ^{14}C-L-histidine (base), labeled in the 2 position of the imidazole ring, $10^{-4}M$ aminoguanidine sulphate, $10^{-5}M$ pyridoxal-5-phosphate, and $10^{-1}M$ sodium phosphate buffer (pH 7.4) containing 0.2% glucose, made up to a final volume of 3.2 ml. After incubation, nonradioactive histamine dihydrochloride equivalent to 40 mg of the base is added as carrier, followed by perchloric acid, giving a final concentration of $0.4M$.

After mixing, the sample is allowed to stand for 30 min and is then filtered. The pH of the filtrate is adjusted to approximately 6.5 with sodium hydroxide. The sample is transferred to a 50 × 10 mm (2.5 g) column of Dowex 50W-X_4 (100–200 mesh), and buffered in advance with 1N sodium acetate buffer, pH 6. The resin is then rinsed with 200 ml of $10^{-1}M$ sodium phosphate buffer, pH 6.5, at a flow rate of 15 ml/hr and finally with 10 ml 1N hydrochloric acid. The histamine is then eluted from the column with 10 ml of 10N hydrochloric acid and the eluate is evaporated to dryness on a steam bath. The residue, dissolved in 5 ml of water, is treated with activated charcoal and filtered. The histamine in the filtrate is allowed to react with pipsyl chloride (*p*-iodobenzenesulfonyl chloride) by adding 0.5 g of sodium bicarbonate and 5 ml of acetone, and after mixing, 0.22 g of pipsyl chloride in 4 ml of acetone. Crystals of pipsyl histamine precipitate when distilled water is added slowly. They are collected on a glass filter, rinsed with 25% alcohol, and dissolved in hot acetone. Charcoal is added, and, after filtration, the pipsyl histamine recrystallizes on the gradual addition of water. The crystals are again collected on a filter, rinsed with alcohol, transferred to a counting plate, and counted after drying. The samples are repeatedly recrystallized from acetone until they display constant radioactivity.

Measurement of the radioactivity is made at infinite thickness in a flow counter until at least 2000 counts are obtained. Values are corrected for background radiation, usually about 20 counts/min and for blanks. Traces of ^{14}C-histamine present as contamination are removed before use (see Section IV-2). Activity of blank samples is a few counts per minute above the background. A fuller description of the various procedures is given by Kahlson, Rosengren, and Thunberg (14).

3. Isotopic Methods: Radioactivity Extractible by Butanol-Chloroform Measured

A. METHOD OF SCHAYER (UNPUBLISHED)

The incubation is similar to that described in Section V-2-A. At the end of the incubation period the incubate is transferred to a tube suitable for extraction (with glass or neoprene stopper), and the beaker is rinsed with 1 ml of water. Two milliliters of saturated sodium chloride solution, 1.0 ml of 5N sodium hydroxide, 12 ml of butanol–chloroform, 3:1, and approximately 1.5 g of sodium chloride is then added. This is shaken mechanically and centrifuged. Ten milliliters of the organic (upper) layer is transferred to another tube and washed by shaking with 5 ml of 0.1N sodium hydroxide saturated with sodium chloride and centrifuged. Eight milliters of the organic layer is transferred to another tube, and shaken with 5 ml of 0.1N hydrochloric acid. Four milliliters of the acid layer is transferred to a counting vial, evaporated to dryness in a stream of warm air, the residue is dissolved in 0.50 ml of water, a water-miscible phosphor is added and the solution is counted.

4. Nonisotopic Methods: Newly Formed Histamine Determined

Only those methods not included in a previous review (7) will be included in this section.

A. METHOD OF WERLE AND LORENZ (6)

This method was used for thyroid and thymus; modifications may be required for other tissues (25,26). Assays are run in a Warburg apparatus at 37°. To the reaction vessel are added 2.0 ml of crude homogenate (equivalent to 0.35 g of fresh tissue), 0.10 ml of aminoguanidine (final concentration $5 \times 10^{-4}M$), 0.10 ml of tetracycline (20 g/ml of incubate), and 0.3 ml of 0.2M phosphate buffer, pH 7.0. For the nonspecific decarboxylase the pH is 8.0, and 20 mg of benzene are added. Additives are dissolved in 0.05M phosphate buffer, pH 7.45, and the pH readjusted with 1N HCl or 1N NaOH.

To the side arm is added 0.5 ml of L-histidine (to produce a final concentration of $10^{-2}M$). The final volume is 3.0 ml and the final buffer concentration is 0.06M.

Blanks containing acid-inactivated enzyme plus substrate, and blanks containing active enzyme but no substrate, are included.

Assays are gassed for 10 min with pyrogallol-purified nitrogen prior to adding substrate. After 3 hr the reaction is stopped by addition of 0.5 ml of 3N perchloric acid. After removal of protein, histamine is extracted, reacted with *o*-phthalaldehyde, and the product assayed spectrofluorimetrically by Burkhalter's modification (27) of the method of Shore et al. (28).

B. METHOD OF KIM AND GLICK FOR MICROGRAM SAMPLES OF TISSUE (29)

This method was devised to measure histidine decarboxylase activity in microtome sections of rat stomach and in other minute quantities of tissue.

Tissue samples, 50 mg, are homogenized in 5 ml of cold distilled water, centrifuged for 15 min at 2000*g*, and the supernate assayed. The preparation of microtome sections is described in the original paper (29).

Reagents are $3.34 \times 10^{-2}M$ histidine (base), $8.27 \times 10^{-5}M$ puridoxal phosphate, $6.67 \times 10^{-4}M$ aminoguanidine sulfate, $0.667M$ phosphate buffer, pH 6.4. Equal volumes of the four solutions are mixed just before use. The assay procedure is as follows:

1. Forty microliters of distilled water is added to each reaction tube containing a tissue section and mixed by vibration to break up the tissue, or 40-μl samples of supernatant fluid are pipetted from tissue homogenate into separate reaction tubes. Tubes are 27 mm long and 4 mm bore.

2. To measure the preformed histamine in the sample, 20 μl of $2.4N$ perchloric acid is added, mixed, 60 μl of reagent mixture is added, centrifuged at 1000*g* for 4 min, and 100 μl of the clear supernatant liquid is used for the histamine analysis.

3. To measure the enzyme, 60 μl of reagent mixture is added to each tube (after step *1*), mixed, stoppered, and placed in a water bath at 37°C for 3 hr. The reaction is stopped by adding 20 μl of $2.4N$ perchloric acid, mixing, centrifuging, and using 100 μl for histamine analysis as in step *2*.

4. The enzyme activity is expressed as the difference in the values for preformed histamine and that obtained after the incubation.

Histamine is determined by the fluorometric procedure of von Redlich and Glick (30) with the following two changes.

a. 20 μl of perchloric acid is used in place of 120 μl, and the concentration of the perchloric acid is $2.4N$ in place of $0.4N$.

b. A 3:2 mixture (by volume) of *n*-butanol and chloroform (both redistilled) is used in place of butanol alone.

VI. COMMENTS

Nonisotopic methods can be recommended only for tissues high in histidine decarboxylase activity. These methods are insensitive and there are a number of instances in which nonisotopic procedures have failed to detect activity readily measured by isotopic methods (7,15).

Methods in which $^{14}CO_2$ is measured (see Section V-1) are rapid. For relatively active tissues such as rat stomach, rat fetus, and mast-cell tumors, the results are precise. However, the $^{14}CO_2$ method has not as yet been shown to be capable of measuring histidine decarboxylase in tissues with low activity and hence has not been proven generally useful. Formation of $^{14}CO_2$ from ^{14}C-L-histidine may not always be a satisfactory criterion of histidine decarboxylase activity; in metabolically active tissues such as liver, it is possible that histidine is catabolized by the glutamic acid pathway, and $^{14}CO_2$ released only after further degradation. Blocking of $^{14}CO_2$ liberation by inhibitors of histidine decarboxylase does not ensure that a direct decarboxylation is involved. If compounds inhibit one pathway of histidine metabolism they may also inhibit others. Histidine decarboxylase activity of several tissues has been determined simultaneously by the $^{14}CO_2$ method and a nonisotope method; the results agreed reasonably well (31).

Obviously, the $^{14}CO_2$ method can not be used to study *in vivo* decarboxylation of histidine.

Methods requiring isotope-dilution techniques (Section V-2) are tedious. Nevertheless, they are, in the author's opinion, the only available means for evaluating the physiological significance of histamine through the ability of tissues to produce it. Radioactive histamine formed during incubation may be destroyed to some extent, however, errors can be minimized by proper precautions. The isotope dilution methods are used to measure histamine, the only specific product of histidine decarboxylase action, and the substance of biological interest. When purified ^{14}C-L-histidine is used, blanks are low; blanks in the $^{14}CO_2$ method are said to be high. The isotope-dilution methods have consistently produced results which are consonant with *in vivo* findings (7,15). Finally, their value is enhanced by the fact that the same procedure can be used to measure histamine formation *in vivo*.

The short isotopic method (Section V-3) does not identify the product as histamine. Nevertheless, it is useful for specialized studies such as testing effects of inhibitors on a single sample of histidine decarboxylase. It is not reliable for complex *in vitro* experiments or for measuring radioactive histamine formed *in vivo*.

Relative to units for expressing histidine decarboxylase activity, there is no standardization. Some laboratories describe enzyme activity as the "amount histamine formed per unit time per unit weight of tissue." This practice might lead uncritical readers to make the unwarranted assumption that similar rates of histamine formation occur *in vivo* (7). For this reason the author prefers to use a unit which shows relative rather than absolute rates of histamine formation.

References

1. E. Werle, *Biochem. Z.*, *288*, 292 (1936).
2. P. Holtz and R. Heise, *Naturwiss.*, *25*, 201 (1937).
3. R. W. Schayer, *Am. J. Physiol.*, *189*, 533 (1957).
4. P. O. Ganrot, A. M. Rosengren, and E. Rosengren, *Experientia*, *17*, 263 (1961).
5. W. H. Lovenberg, H. Weissbach, and S. Udenfriend, *J. Biol. Chem.*, *237*, 89 (1962).
6. E. Werle and W. Lorenz, *Biochem. Pharmacol.*, *15*, 1059 (1966).
7. R. W. Schayer, in *Handbook of Experimental Pharmacology*, Vol. XVIII, Part 1, M. Rocha e Silva, Ed., Springer, Berlin, 1966, p. 688.
8. E. F. Gale, in *Methods of Biochemical Analysis*, Vol. 4, D. Glick, Ed., Interscience, New York, 1957, p. 285.
9. G. R. Lloyd and P. J. Nicholls, *Nature*, *206*, 298 (1965).
10. U. von Haartmann, G. Kahlson, and C. Steinhardt, *Life Sci.*, *5*, 1 (1966).
11. R. Hakanson, *Biochem. Pharmacol.*, *12*, 1289 (1963).
12. D. Aures, W. J. Hartman, and W. G. Clark, *Federation Proc.*, *21*, 269 (1962).
13. D. Aures, W. J. Hartman, and W. G. Clark, *Proc. Intern. Congr. Physiol. Sci.*, *22nd*, *Leiden*, *2*, 694 (1962).
14. G. Kahlson, E. Rosengren, and R. Thunberg, *J. Physiol.*, *169*, 467 (1963).
15. G. Kahlson and E. Rosengren, *Physiol. Rev.*, in press.
16. D. Mackay and D. M. Shepherd, *Brit J. Pharmacol.*, *15*, 552 (1960).
17. Y. Kobayashi, *Anal. Biochem.*, *5*, 284 (1963).
18. D. Aures and W. G. Clark, *Anal. Biochem.*, *9*, 35 (1964).
19. T. Teorell and E. Stenhagen, *Biochem. Z.*, *299*, 416 (1938).
20. F. H. Woeller, *Anal. Biochem.*, *2*, 508 (1961).
21. R. J. Levine and D. E. Watts, *Biochem. Pharmacol.*, *15*, 841 (1966).
22. G. A. Bray, *Anal. Biochem.*, *1*, 279 (1960).
23. R. D. Smith and C. F. Code, *Mayo Clin. Proc.*, *42*, 105 (1967).
24. R. W. Schayer, Z. Rothschild, and P. Bizony, *Am. J. Physiol.*, *196*, 295 (1959).
25. W. Lorenz and E. Werle, *Z. Physiol. Chem.*, *348*, 468 (1967).
26. W. Lorenz, C. Pfleger, and E. Werle, *Arch. Exptl. Pathol. Pharmakol.*, in press.

27. A. Burkhalter, *Biochem. Pharmacol.*, *11*, 315 (1962).
28. P. A. Shore, A. Burkhalter, and V. H. Cohn, *J. Pharmacol. Exptl. Therap.*, *127*, 182 (1959).
29. Y. S. Kim and D. Glick, *J. Histochem. Cytochem.*, *15*, 347 (1967).
30. D. von Redlich and D. Glick, *Anal. Biochem.*, *10*, 459 (1965).
31. B. A. Callingham, Y. Kobayashi, D. V. Maudsley, and G. B. West, *J. Physiol.*, *179*, 44P (1965).

The Estimation of Total (Free + Conjugated) Catecholamines and Some Catecholamine Metabolites in Human Urine

H. WEIL-MALHERBE, *Division of Special Mental Health Research Programs, National Institute of Mental Health, St. Elizabeths Hospital, Washington, D. C.*

I. INTRODUCTION

It has been said that there are as many modifications of the trihydroxyindole method as there are laboratories engaged in catecholamine analysis. No doubt preferences for one technique or another will vary according to the material analyzed, the information desired, and the time, manpower, and facilities available. What is good for one analyst may not suit another, and there is, therefore, justification for an opening statement in the first person singular: The methods to be described or discussed in this article are those of which I have some personal experience, and the methods which are recommended are those which are now in use in my laboratory, at the time of writing. Their principal field of application is the analysis of human urine.

Although the isolation of the urinary catecholamines by a single adsorption–elution cycle, with alumina as adsorbent, appears to be satisfactory when unhydrolyzed urine is analyzed, it cannot be emphasized too strongly that acid hydrolysis makes a further purification of the urinary catecholamine fraction an inescapable necessity. Whether urine should or should not be hydrolyzed before analysis is for the analyst to decide. In the opinion of some (1), the free fraction of catecholamines provides a sufficient, if not a superior, parameter of stress responses. This may be true, but it is also a fact that there is practically no published evidence in support of this assertion. It may well be a rationalization due to the added complications of the hydrolysis procedure. The alumina eluates of acid-hydrolyzed urines are frequently yellow, their "faded blanks" are high, quenching is serious, and the fluorescence obtained includes unspecific components. A second purification step, consisting in the adsorption of the amines on a cation exchange column (2,3), largely eliminates interfering impurities.

Changes in the conjugated fraction of urinary catecholamines are today practically unexplored. Observations such as those of Henkin and Bartter (4), who found a decrease of conjugated epinephrine excretion in familial dysautonomia and also described a case of pheochromocytoma with an increased excretion of conjugated norepinephrine and a normal excretion of free norepinephrine, suggest that this fraction deserves greater attention.

The procedures to be described are the following: (*1*) hydrolysis of urine at pH 1.5 and 100°, (*2*) adsorption of catecholamines on a column of alumina at pH 8.4 and elution with 0.2*M* acetic acid, (*3*) adsorption of catecholamines on a column of a carboxylic cation exchange resin at pH 6 and elution with 1*M* acetic acid, and (*4*) fluorimetric estimation of the three catecholamines, epinephrine, norepinephrine, and 3,4-dihydroxyphenylethylamine (dopamine) in different aliquots of the eluate.

The basic metabolites, metanephrine and normetanephrine, are estimated in the filtrate from the alumina column after desalting by electrodialysis, adsorption on a column of carboxylic cation-exchange resin at pH 6 and elution with 1*M* formic acid. The fluorimetric estimation is carried out with two aliquots, one for the estimation of metanephrine and one for the estimation of metanephrine plus normetanephrine.

The two principal acid metabolites of epinephrine and norepinephrine, 3-methoxy-4-hydroxymandelic acid (vanillylmandelic acid, VMA), and 3,4-dihydroxymandelic acid DHMA), are estimated in two separate portions of unhydrolyzed urine.

II. GENERAL DIRECTIONS

Because of their great sensitivity and their susceptibility to various interfering factors, fluorimetric methods require a more than ordinary degree of care and cleanliness. Two common sources of contamination with fluorescent impurities are lubricants and detergents; whenever they are used, the possibility of extraneous fluorescence being introduced into the sample must be avoided. The water used for the preparation of the reagents must be of high quality; in our laboratory, distilled water is redistilled over alkaline potassium permanganate in an all-glass apparatus. Buffer solutions and solutions of sodium hydroxide are freed from traces of heavy metals by treatment with 0.05–0.1 volume of Dowex A-1 chelating resin (50–100 mesh, Na^+ form). Other reagents are purified or prepared as indicated in the appropriate sections.

Adjustments of pH are controlled by a glass microelectrode assembly inserted into the magnetically stirred solution.

The Aminco-Bowman spectrophotofluorometer (American Instrument Company, Silver Spring, Md.) is used for the fluorimetric measurements. Slits are adjusted to a width of 4 mm. Other fluorimeters, whether monochromator or filter instruments, may of course be used instead.

To avoid losses of catechol compounds through autoxidation, the pH of urine should be kept below 2 during the collection. We use wide-mouthed 1-gallon jars for the collection. The jars are roughly graduated on the outside with pencil marks showing volumes of approximately 400, 800, 1200, 1600 ml, etc. At the beginning of the collection, 4 ml of $10N$ H_2SO_4 is added to the empty jar, and further additions of 4 ml are made by the attendant in charge of the collection every time the urine volume has reached a graduation mark.

III. HYDROLYSIS OF URINE

The conjugates of catecholamines and 3-*O*-methyl catecholamines in human urine are mainly ethereal sulfates which are easily hydrolyzed by heating with acid. Only insignificant amounts appear to be excreted as glucuronides (5,6). The hydrolysis is preferably carried out under nitrogen. We use 100-ml, round-bottom flasks connected to condensers by spherical ground joints and provided with a 2½-in. long side tubulature through which a glass tube tapered to a fine point is introduced. To prevent foaming, the stream of nitrogen is passed over the surface of the liquid. The nitrogen is slowly passed through a wash bottle equipped with a sintered-glass gas dispersion disk of medium porosity; it contains 100 ml of Fieser's anthraquinone β-sulfonate reagent (7).

Recoveries of catecholamines were not appreciably affected by hydrolysis (2). About 50% of the total catecholamines, on the average, were found to be excreted as conjugates (2).

Fruehan and Lee (8) recommend extraction of the free catecholamines by adsorption on alumina and subsequent hydrolysis of the filtrate for the estimation of the conjugated fraction. When free and conjugated fractions are to be estimated separately this procedure appears to be logical.

Procedure. Anthraquinone–sulfonate reagent: dissolve 13.3 g of NaOH in 30 ml of water. Shake 16 g of sodium dithionite and 2 g of anthraquinone β-sulfonate with 60 ml of water in a glass-stoppered, 100-ml measuring cylinder until as much as possible has been dissolved. Add the sodium hydroxide solution, mix, and make the volume up to 100 ml. The solution must be freshly prepared.

Measure out 25 ml of urine and adjust its pH to 1.5 with 10*N* H_2SO_4. Reflux gently for 20 min on a hot plate in a stream of nitrogen purified by passage through anthraquinone β-sulfonate solution. Cool under nitrogen.

IV. ADSORPTION ON ALUMINA

1. Column versus Batch Adsorption

Catecholamines may be adsorbed on alumina either by passing the urine through a column of the adsorbent or by mixing it intimately with a batch of suspended alumina. In the batch procedure, which is preferred by several authors (e.g., refs. 9–12), an equilibrium between adsorbed and free catecholamines is reached. The amount of catecholamines remaining unadsorbed depends, among other factors, on the quantity of adsorbent added. Whereas we are dealing with a closed system in the batch procedure, the column procedure is an open system which can be regarded, as a first approximation, as a consecutive series of closed systems. As the solution, descending the column, passes from one closed system to another, new equilibria are established leaving fewer and fewer molecules unadsorbed, until adsorption is virtually complete. To achieve the same result by batch adsorption would require repeated treatments of the supernatant with fresh batches of adsorbent, and similar considerations apply also to the operations of washing and eluting. There is no doubt therefore that, theoretically, the column procedure is more efficient. From a practical standpoint, too, the column procedure has advantages. One person can easily handle 12–16 columns simultaneously, probably in less time than would be required for the tedious series of centrifugations imposed by the batch procedure. Moreover, the granules of alumina are very brittle and are liable to be reduced to a fine dust, difficult to centrifuge, by the constant agitation and stirring necessary in the batch procedure. This criticism applies particularly to the procedure proposed by Crout (1) as a compromise between the column and batch procedures. In this method the urine sample is added to the requisite amount of alumina and the pH adjusted to 8.4. The mixture is stirred during the pH adjustment and for 7 min afterward on a magnetic stirrer. It is then poured into an adsorption tube to form a column which is washed and eluted in the usual way. The purpose of neutralizing the urine in the presence of the adsorbent is to protect the catecholamines from oxidation since they are adsorbed as soon as the pH is raised. Although Crout expressly warns against a too severe grinding of the alumina, we have been unable to avoid excessive grinding even at the minimum speed required to keep the alumina in suspension. Fragmentation of the adsorbent led to clogging and to seepage of finely suspended alumina from the column. For this reason, Crout's modification is not recommended.

2. Pretreatment of Alumina and Preparation of Columns

Heating of alumina in dilute HCl has been recommended (13) to remove alkaline, fluorescent, and heavy metal impurities. Another important function of the preliminary treatment is the removal of dust and fine particles which are likely to impede the flow qualities of the column. This is achieved during the washing operations by repeatedly stirring up the powder in water and decanting the turbid supernatant after a brief interval. The washed powder is dried at 300° in a muffle oven. This temperature was chosen since it is known to be optimal for the preparation of highly adsorbent alumina hydrates from aluminum hydroxide (14). Others have recommended drying at 100° (1) or 200° (12). Figure 1 shows that the adsorbing capacity of a product dried at 300° is definitely superior to that of a sample dried at 100° and also, though less markedly, to that of a sample dried at 200°.

The alumina thus prepared is acid. It is neutralized before use by an amount of alkali sufficient to establish a pH of about 8 in the suspension.

The adsorption tubes consist of a 50-ml spherical bulb and a stem of 5–6 mm bore with a constriction about 15 cm below the bulb. The top of the bulb is fitted with a spherical socket joint, size 18/9, for attachment to a pressure manifold previously described (15) and capable of delivering pressure of 1 psi to a series of 8 tubes in such a way that each tube can be operated independently without affecting the pressure in any of the other tubes. The adsorption tubes and the glass components of the pressure manifold are obtainable from Kontes Glass Co., Vineland, N.J.

Procedure. Mix 200 g of alumina ("for chromatographic adsorption analysis," British Drug Houses, U.S. distributor Gallard-Schlesinger Chemical Mfg. Corp.,

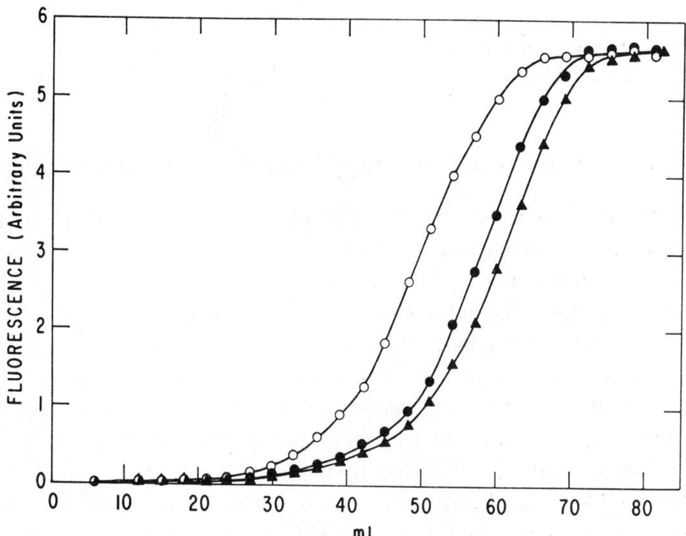

Fig. 1. Retention of norepinephrine on alumina dried at 100, 200, and 300°C. Solutions of 100 ml containing 0.1M ammonium acetate, 5mM EDTA, and 10 mg of norepinephrine hydrochloride and adjusted to pH 8.4 were passed through columns of 0.5 g of alumina. The filtrate, in fractions of 3 ml, was collected in tubes containing 0.05 ml 10N H_2SO_4. Fluorescence was measured at 320 mμ with activation at 275 mμ. (○) Alumina heated at 100° for 2 hr. (●) Alumina heated at 200° for 2 hr. (▲) Alumina heated at 300° for 2 hr.

Carle Place, Long Island, N.Y.) with 1 liter of 2N HCl and heat to boiling for 20 min, under vigorous stirring (**caution:** inefficient stirring may cause violent bumping). Filter on a sintered-glass funnel and wash with 1 liter of hot 2N HCl followed by ample water. Transfer the powder to a large beaker, stir it up repeatedly in about 1 liter of water, and decant the supernatant when the heavier particles have settled to the bottom. Filter again by suction and dry in a muffle oven at 300° for 2 hr. Cool in desiccator and store in closed bottle.

Suspend 0.7 g in 10 ml of 0.1M ammonium acetate solution adjusted to pH 8.0 with 1M ammonia and add 0.2–0.3 ml of 1N NaOH. Gently invert the mixture a few times. The exact amount of NaOH required to bring the pH of the mixture up to 8.3 $\pm$ 0.1 is determined at the glass electrode once for each new batch of alumina and the sample used is discarded. Subsequent neutralizations are not controlled by the glass electrode since the stirring damages the alumina granules.

Place a small wad of glass wool over the constriction of the adsorption tube, close the lower end of the tube with a finger, and fill the stem with water. Pour in the alumina suspension through a funnel and at the same time allow the liquid to drain from the tube. Wash remnants of alumina from the test tube, funnel, and bulb into the stem with a jet of water. The column must not be allowed to run dry and additions should be made when the meniscus reaches the top of the adsorbent.

3. Adsorption and Elution of Urinary Catecholamines

Adsorption is carried out at a pH of 8.4 and $0.2M$ acetic acid is used for elution according to Lund (16). It is customary to add EDTA or ascorbic acid or both to the urine before adjusting the pH in order to protect the catechols from autoxidation. Actually, urine usually contains sufficient chelating and antioxidant material to prevent losses of catechols for a limited time even at pH 8.4. This was noted during the development of a method for the estimation of DHMA (17). EDTA as well as ascorbic acid caused interference, and had to be omitted. Neither their omission nor their replacement by chelating resin had any effect on the recoveries of DHMA added to urine, despite the fact that DHMA is extremely susceptible to autoxidation.

Procedure. To the hydrolyzed urine (25 ml) add 0.5 ml of 10% EDTA. Adjust the pH to 8.4 with $5N$ NaOH, and finally with $0.5N$ NaOH. Centrifuge at 17,000 g for 10 min in a "Servall-Angle" centrifuge (about 11,000 rpm). Pass the supernatant through the column, wash with 10 ml of water and elute with 5 ml of $0.2M$ acetic acid, followed by 5 ml of water. Save the urine effluent and the first 3 ml of the washings for later estimation of 3-*O*-methyl catecholamines. Run a water blank with each batch of estimations ("column blank").

Filtration should proceed at the rate of 0.5–1 ml/min. As a rule, filtration by gravity is fast enough, and application of pressure is rarely required.

V. ADSORPTION ON CATION EXCHANGE RESIN

In Table I a comparison is shown between catecholamine estimations obtained from alumina eluates and those from the same eluates after further purification through a column of cation exchange resin. The figures obtained by the first method are higher than those obtained by the second method, particularly for epinephrine where the difference is almost threefold. Yet the recoveries are not significantly different for any of the amines. The conclusion is therefore justified that the differences in apparent concentration are due to unspecific fluorescence. The estimation of catecholamines in the alumina eluates is also complicated by the facts that the "faded blanks" are much higher than those observed in eluates after the two-step purification, and that the internal standards often read much less in the former than in the latter. The faded blank and the internal standards obtained after the two-step purification are usually of the same order as the corresponding values obtained with the "column blank." It is concluded, therefore, that alumina eluates are not suitable for the fluorimetric estimation of the catecholamines, at any rate when acid-hydrolyzed urine is analyzed.

TABLE I

Estimation of Catecholamines in Extracts of Hydrolyzed Urine Prepared by (*a*) Chromatography on Alumina Only and (*b*) Successive Chromatography on Alumina and Cation Exchange Resin

	Alumina			Alumina + Amberlite CG 50		
	Epinephrine	Norepinephrine	Dopamine	Epinephrine	Norepinephrine	Dopamine
Mean excretion, μg/24 hr	27.0	67.3	226	9.6	29.1	132.3
± SEM, 24 experiments	±5.45	±23.3	±45.9	±1.13	±4.28	±21.6
P[a]				0.006	0.1	0.1
Mean recoveries, %	105.2	75.6	69.9	85.8	81.4	54.5
± SEM, 12 experiments	±15.5	±10.8	±18.2	±6.13	±9.34	±7.44
P[a]				0.3	0.7	0.5

[a] Significance of difference from corresponding value obtained after alumina purification.

Before use, the resin is recycled repeatedly through the H^+ and Na^+ forms and fine particles are removed during the washing process by decantation. Eluate and resin are both brought to pH 6. It should be borne in mind that the equilibrium of the exchange reaction is approached slowly. The rate of flow should therefore be reduced so as not to exceed 0.5 ml/min, at least during the phases of adsorption and elution; faster rates are permissible during washing. When the column is eluted, the acid solution causes the resin to shrink. To avoid an unduly rapid rate of flow at this stage, the column is closed at its lower end and left to soak for ½ hr after the eluant has filled the resin bed.

A series of urine samples can comfortably be carried through both purification procedures in one day. Whereas catecholamines are fairly stable in the resin eluate, they are less stable in the eluate from the alumina column and undue delay at this stage should be avoided. If delay is unavoidable, the addition of EDTA, described below, should be made before the eluates are stored.

Procedure. Suspend 100 g of the carboxylic cation exchange resin, Amberlite CG 50, Type 2 (200–400 mesh), in 1 liter of $2N$ HCl and leave overnight. Wash several times with water by decantation; then add 1 liter of $2N$ NaOH and stir slowly for about 30 min. Wash again with water and suspend for 30 min in 1 liter of $2N$ HCl. Repeat the process twice more. Finally, suspend the resin, in the Na^+ form, in 1 liter of M sodium acetate buffer pH 6.0 and adjust the pH to 6.0 with $1M$ acetic acid until a stable equilibrium is reached. Store the resin in $1M$ acetate buffer pH 6.0 in the refrigerator. Before use, filter an appropriate quantity on a sintered-glass funnel, wash thoroughly with water, and suck free of adhering moisture.

Prepare columns in adsorption tubes of the same type as those used for the alumina columns. Place 0.5 g of the moist resin into the bulb, fill the stem with water, and wash the resin into the stem with a jet of water, draining the stem at the same time.

Add 1 ml of 1% EDTA to the eluate from the alumina column and titrate to pH 6.0 $\pm$ 0.1 with $1N$ NaOH. Pass through the resin column, wash with 20 ml of water, and elute with 10 ml of $1M$ acetic acid. When about 4 ml of eluate have been collected, close the column and leave it for ½ hr before continuing the elution.

VI. FLUORIMETRIC ESTIMATION OF CATECHOLAMINES

1. The Trihydroxyindole Method

The fluorimetric estimation of urinary epinephrine and norepinephrine by the trihydroxyindole (THI) reaction is now practically universal and hardly needs an introduction. Briefly, epinephrine is oxidized to adrenochrome and the latter is rearranged to fluorescent *N*-methyl-3,5,6-trihydroxyindole (adrenolutin) in alkaline solution in which the compound is very unstable unless protected from oxidation by a suitable reducing agent. An analogous product, noradrenolutin, is presumably formed from norepinephrine. The history of the THI reaction and its chemistry have been competently reviewed (18–20).

The new and unpublished modification of the THI method which will be presented here represents, I believe, a significant advance in several directions, despite the very large number of modifications already on record. It is superior to other forms of the method in intensity of fluorescence, low level, and stability of the blank and the discrimination between epinephrine and norepinephrine.

In the classical form of the THI method, ascorbic acid is used to stabilize the lutins in alkaline solution. It is well recognized, however, that ascorbic acid may itself form fluorescent oxidation products, especially when ferricyanide is used as the oxidizing agent. This results in high and slowly rising blanks, depending on the concentration of ascorbic acid present. Various remedies have been suggested; a sure way of obtaining a low and stable blank is to omit the oxidizing agent (12) but such a blank can hardly be considered to be a valid correction of extraneous fluorescence. Other workers have added a further reducing agent, such as ethylene diamine (21) or sodium borohydride (22), in order to prevent the oxidation of ascorbic acid.

Häggendal (23) replaced ascorbic acid by 2,3-dimercaptopropanol. Blanks in this procedure are low and stable. We have further studied this modification (24) and found that, when norepinephrine was oxidized at pH 5.8–6.0 with ferricyanide and converted to the lutin with sodium hydroxide, the fluorescence was unstable and decayed with a half-life time of about 30 min. It could be stabilized by readjusting the pH to about 5, 4 min after alkalinization when the rearrangement of the quinone to the indole configuration was completed. Such a readjustment of the pH after rearrangement is familiar to those who use the Carlsson-Waldeck technique of dopamine estimation (25). It has also been used by Häggendal (26) in the estimation of metanephrine and normetanephrine. It is less well known that Price and Price (27) recommended reacidification to pH 5 in their version of the THI method, with ferricyanide as the oxidizing agent and ascorbic acid as the reducing agent. They found the fluorescence intensity to be increased and the blank reduced, but the fluorescence was unstable. Contrary to the findings of Price and Price, the fluorescence formed from norephinephrine under the conditions here described is relatively stable at pH 5–6. It was found to have decayed by about 10% after 30 min and by about 20% after 2 hr.

Whereas high readings are obtained when norepinephrine is oxi-

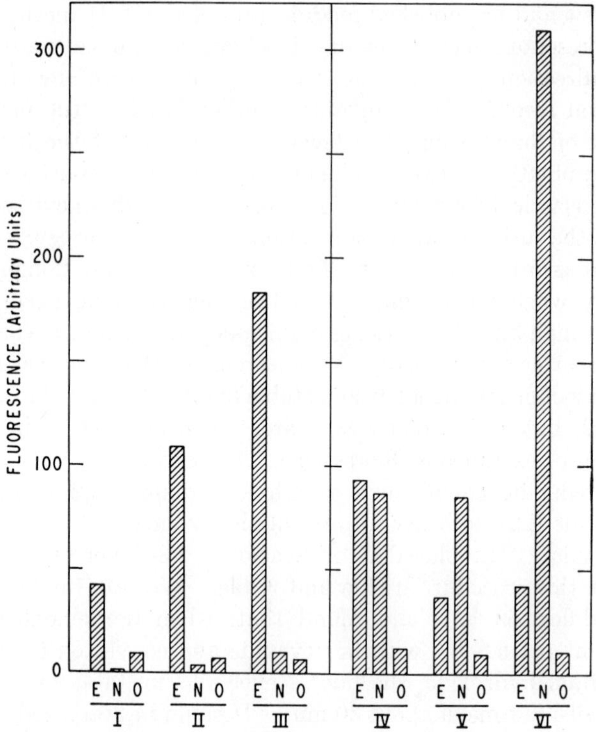

Fig. 2. Fluorescence intensities obtained by various modifications of the trihydroxyindole method. (I) Ferricyanide-ascorbic acid, pH 3: 2 ml of eluate from resin column (column blank) + 0.25 ml of M formic acid + 0.1 ml of 0.25% $K_3Fe(CN)_6$ + (after 5 min) 0.5 ml of $5N$ NaOH-2% ascorbic acid, 9:1. Read at 415/515 mμ. (II) Iodine-ascorbic acid, pH 3: 2 ml of eluate from resin column (column blank) + 0.25 ml of M formic acid + 0.1 ml of $0.1N$ I_2 + (after 5 min) 0.25 ml of $0.1N$ $Na_2S_2O_3$ + 0.5 ml of $5N$ NaOH-2% ascorbic acid, 9:1. Irradiated for 3 min. Read at 415/515 mμ. (III) Ferricyanide-Cu^{2+}-thiol, pH 3: 1 ml of eluate from resin column (column blank) + 0.3 ml M formic acid + 0.05 ml $0.01M$ Cu^{2+} + 0.05 ml of 0.25% $K_3Fe(CN)_6$ + (after 5 min) 0.3 ml of $10N$ NaOH–1% mercaptoethanol in 20% Na_2SO_3, 1:1, + (after 4 min) 0.3 ml of $10M$ acetic acid, + 0.9 ml of water. Centrifuged and read at 415/500 mμ. (IV) Ferricyanide-ascorbic acid, pH 6: 2 ml of $0.33M$ acetate buffer pH 5.9 + 0.1 ml of 0.25% $K_3Fe(CN)_6$ + (after 5 min) 0.5 ml $5N$ NaOH–2% ascorbic acid, 9: 1. Read at 400/505 mμ. (V) Iodine-ascorbic acid, pH 6: 2 ml of $0.33M$ acetate buffer pH 5.9 + 0.1 ml of $0.1N$ I_2 + (after 5 min) 0.25 ml of $0.1N_2Na_2S\ O_3$ + 0.5 ml $5N$ NaOH-2% ascorbic acid, 9: 1. Irradiated for 4 min. Read at 400/505 mμ. (VI) Ferricyanide-thiol, pH 6: 2 ml of $0.33M$ acetate buffer pH 5.9

dized, those obtained from epinephrine at pH 6 are low, ranging from 10 to 30% of those given by an equivalent amount of norephinephrine (Fig. 2). This is due to a much more rapid fading of the epinephrine fluorescence in the interval between alkalinization and acidification. The greater instability of epinephrine fluorescence than of norepinephrine fluorescence in the presence of a thiol compound as reducing agent has been utilized in the automated procedure of Merrills (28), where the lutins are formed in an alkaline thioglycollate solution. In our hands, thioglycollate did not result in fluorescence readings for norepinephrine which were as high or as stable as those obtained with dimercaptopropanol or mercaptoethanol; others have also noticed a rapid decline of norepinephrine fluorescence in the alkaline thioglycollate solution (29).

Owing to the low fluorescence readings given by epinephrine after oxidation at pH 6, the contribution of epinephrine to the fluorescence obtained by the technique described is almost negligible in most urine extracts. Interference from an equal quantity of dopamine amounts to about 0.3% of norepinephrine fluorescence and is neglected.

To estimate epinephrine in the presence of norepinephrine, use is made of the fact that epinephrine is rapidly oxidized at a pH as low as 3 where the oxidation of norepinephrine is very slow (16,30). Nevertheless, the fluorescence intensity obtained from epinephrine after oxidation at pH 3 by ferricyanide was unsatisfactory, whether ascorbic acid or dimercaptopropanol was used as reducing agent. The addition of zinc ions as a catalyst (31) brought little improvement. A systematic search revealed that cupric ions alone out of 18 different heavy metal ions examined had a strong catalytic effect and increased the fluorescence about tenfold. Häggendal (23) had already studied the effect of Cu^{2+} on the reaction, but only at the oxidizing pH of 6. Under these conditions Häggendal found, in agreement with our own results, that Cu^{2+} has no effect on the fluorescence intensity but stabilizes the fluorescence in the alkaline solution.

+ 0.1 ml 0.25% $K_3Fe(CN)_6$ + (after 5 min) 0.5 ml of 5*N* NaOH-1% mercaptoethanol in 20% Na_2SO_3, 1:1, + (after 4 min) 0.3 ml of 10*M* acetic acid. Read at 395/475 mμ. (I–VI) E is 0.1 ml of epinephrine standard (1 μg/ml). N is 0.1 ml of norepinephrine standard (1 μg/ml). O is 0.1 ml of water. The fluorescence readings for E and N have been corrected by subtracting the reading for O.

Almost immediately two complications were noticed. Whereas a first sample of dimercaptopropanol gave satisfactory results, two subsequent samples turned yellow in the presence of Cu^{2+}, completely quenching all fluorescence. This difficulty was overcome by substituting mercaptoethanol for dimercaptopropanol. Mercaptoethanol used in a concentration equivalent with respect to thiol groups proved to be a perfect substitute for dimercaptopropanol also in the reaction carried out at pH 6. Since it is less expensive, more soluble in water, and apparently purer than dimercaptopropanol, we prefer it for both reactions.

The second complication arises from the fact that the stability of the fluorescence obtained by epinephrine oxidation at pH 3, like that formed from norepinephrine at pH 6, is greatly increased if the solution is acidified to pH 5 after the alkaline rearrangement. In the presence of a copper salt a white flocculent precipitate appears at this stage which has to be removed by a brief centrifugation.

If Cu^{2+} is omitted during the oxidation step but added along with the reducing agent, no catalytic effect is observed, indicating that Cu^{2+} catalyzes the oxidation and not the rearrangement.

When the oxidation of epinephrine was studied in a solution of $0.67M$ acetic acid adjusted to various values of pH, a curve with two pH optima, at pH 3.0 and 4.0, separated by a minimum at pH 3.4 was obtained (Fig. 3). However, when epinephrine was added to an eluate from the resin column, whether containing urinary constituents (urine extract plus internal standard) or not (column blank plus internal standard), a plateau was observed at pH values above 3.0; the small trough shown by the curve of the urine extract at pH 3.4 is of doubtful significance.

The oxidation of norepinephrine at the acid pH is also catalyzed by Cu^{2+}. Whereas norepinephrine in the absence of Cu^{2+} is not oxidized until the pH rises to about 4, its oxidation in the presence of Cu^{2+} begins at a pH of about 2.8 and rises steeply between pH 3 and 4. A pH of 2.8–2.9 is therefore chosen for the differential oxidation of epinephrine. The oxidation of norepinephrine seems to depend on other factors which are not entirely clear at present. If norepinephrine is added to an eluate from the resin column, particularly an eluate containing urine constituents, its oxidation is negligible below pH 3.4 (Fig. 3). In numerous analyses of urine eluates which were controlled by internal standards of epinephrine and norepineph-

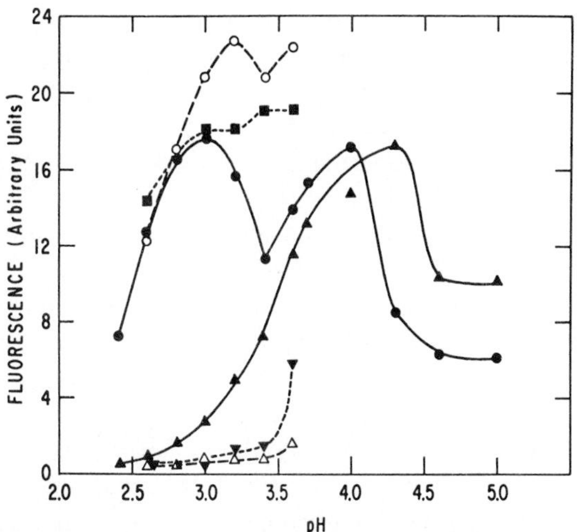

Fig. 3. pH–Activity curve of ferricyanide-Cu^{2+}-thiol method. (●) Epinephrine, 0.1 μg, added to 0.67*M* acetate buffer pH 3.6. Acetate buffers pH 3.7–5.0 were used for measurements above pH 3.6. (▲) Norepinephrine, 0.1 μg, added to 0.67*M* acetate buffer pH 3.6, or pH 3.7–5.0. (■) Epinephrine, 0.1 μg, added to eluate from resin column (column blank). (▼) Norepinephrine, p.1 μg, added to eluate from resin column (column blank). (○) Epinephrine, 0.1 μg, added to eluate from resin column (urine sample). (△) Norepinephrine, 0.1μ g ,added to eluate from resin column (urine sample). All readings were corrected by subtracting readings obtained in the absence of added epinephrine or norepinephrine. The pH was adjusted to the desired level below 3.6 by the addition of formic acid (10 or 1*M* solutions).

rine, we have come across only exceptional samples in which the fluorescence formed from norepinephrine at pH 2.9 exceeded 5% of the fluorescence formed from a corresponding amount of epinephrine. The usual fluorescence from norepinephrine is 0–3% of that from epinephrine. Although it is not required for the calculation, we always include an internal norepinephrine standard in the pH 3 reaction as a control.

To enable the reader to cope with the contingency of an undesirably high norepinephrine oxidation, an alternative method will be described in which iodine is used as oxidizing agent and ascorbic acid as reducing agent, essentially according to Crout (1) and incorporating the irradiation procedure recommended by Kahane and Vestergaard (32). This method seems to have a larger margin of safety as regards the oxidation of norepinephrine, but its fluorescence yield does not reach that of the ferricyanide-Cu^{2+}-thiol method (Fig. 2). It also has a larger blank, even though it is stable (Fig. 4). Easily the worst method for the oxidation of epinephrine at pH 3 is that combining ferricyanide and ascorbic acid: it has a low fluorescence yield and a high and unstable blank (Figs. 2 and 4).

The pH of the resin eluate is 3.5–3.6. It is reduced to pH 2.8–2.9 by the addition of 1*M* formic acid to eluate 3:10 (v/v). Kahane and Vestergaard (32) reported that a buffer of sodium metaphosphate and phosphoric acid had a specific inhibitory effect on the oxidation of norepinephrine in the region of pH 3. Unfortunately, metaphosphate even at less than 1/20 of the concentration used by Kahane and Vestergaard, inhibits not only the oxidation of norepinephrine but also that of epinephrine in the ferricyanide–copper method, probably owing to a chelation of copper. The system of Kahane and Vestergaard is therefore not applicable under our conditions.

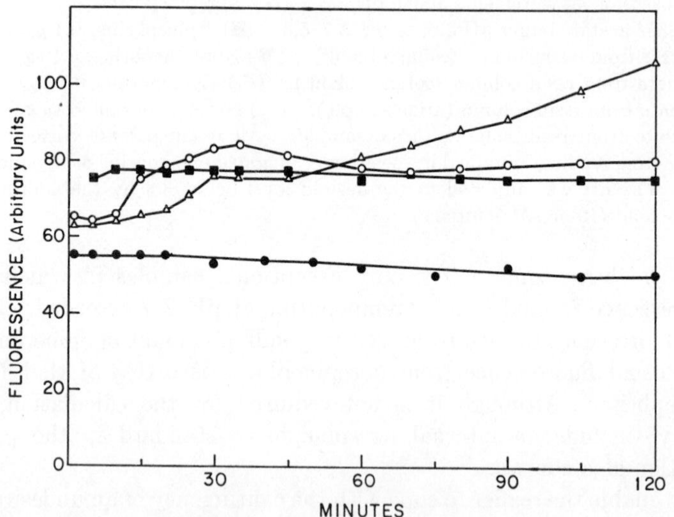

Fig. 4. Stability of blanks in modifications of the trihydroxyindole reaction. (●) Ferricyanide–thiol procedure. (■) Iodine–ascorbic acid procedure. (○) Ferricyanide–ascorbic acid procedure. (△) Ferricyanide–ascorbic acid procedure; 5*N* NaOH containing 2% (by volume) ethylenediamine (21). Experimental details as described in legend of Figure 2.

Most authors who use ascorbic acid as the reducing agent stipulate that it should be mixed with the sodium hydroxide solution before addition according to Lund (16), but several authors (27,33,34) specify that it should be added in a separate solution before the sodium hydroxide. I have been unable to find a discussion of the rationale for one or the other procedure in the literature. We have ourselves practiced addition of the two reagents in separate solutions, since separate solutions were required for the preparation of the faded blank in any case (2), but we have since become aware of the fact that the fluorescence produced by catecholamines is considerably higher when the two reagents are added as a mixture rather than separately. This effect is especially marked when the oxidation is performed at pH 3. It is probable that at this pH ascorbic acid exists largely in the ketohydroxy configuration. On the addition of alkali it will be isomerized to the much more strongly reducing enediol configuration. Presumably this transformation is not rapid enough to prevent entirely the oxidative breakdown of the lutins. Anton and Sayre (12) report that under certain conditions a higher fluorescence is obtained when ascorbic acid is mixed with 10*N* than with 5*N* NaOH. This puzzling observation may perhaps be explained on similar lines. Although it is unlikely that the shift in the ketohydroxy-enediol equilibrium would be of practical significance when the NaOH concentration is raised from 5 to 10*N*, the ionization of the two hydroxyl groups will be greater in the more concentrated solution of NaOH and therefore the initial reducing capacity of the mixture will be increased.

These considerations do not apply when thiol compounds are used as reducing agents. Results are identical whether the solutions of thiol and base are added separately or in mixture. In the following procedures, addition of the two reagents as a mixture is described, but their separate addition is optional, provided the reducing agent is added first.

Many authors emphasize the need to adhere exactly to a rigid time schedule when adding the various reagents. As a matter of fact we did not find that timing was excessively critical in our procedures. Nevertheless, a reasonable degree of standardization should be observed.

Whereas the activation maxima of the lutins are not appreciably changed the fluorescence maxima are shifted to shorter wavelengths when the pH is changed from 13 to 5. Readings are taken at 415/500 mμ after oxidation at pH 3, for activation and emission wavelengths, respectively, corresponding to the maxima for adrenolutin, and at 395/475 mμ after oxidation at pH 6, corresponding to the maxima for noradrenolutin.

2. The Fluorimetric Estimation of Dopamine

Carlsson and Waldeck (25) introduced a method in which dopamine is oxidized by iodine at pH 6. The oxidation product is protected by sodium sulfite and isomerized in strongly alkaline solution to a dihydroxyindole. The fluorescent product is stabilized by acidification to about pH 5.

This method has been modified by a number of authors (9,35–40). For instance, manganese dioxide (39) or periodate (40) has been used as the oxidizing agent instead of iodine; another modification concerns the final acidification for which mineral acid rather than acetic acid has been proposed (9).

It has always been recognized that the fluorescence formed from dopamine by iodine oxidation continues to rise slowly for many hours in acid medium, and various remedies have been suggested. Some authors delay the readings for 45 min (9), 2 hr (38), or 20 hr (36). Carlsson and Waldeck (25) irradiated the samples with a mercury arc lamp (peak emission 254 mμ) in silica tubes for 10 min, a period later extended to 15 min (35). This produced a fivefold increase in the intensity of fluorescence. Irradiation in Pyrex tubes also increased the fluorescence to about two-thirds of the value obtained in silica tubes. After 17 hr the fluorescence was approximately the same whether the sample was irradiated in silica tubes, in Pyrex tubes, or not at all. According to Udenfriend (37), heating the samples at 45° for 30 min is as effective as ultraviolet irradiation in silica tubes and produces a fluorescence which is stable for 24 hr.

We have compared the fluorescence obtained in irradiated, heated, and untreated samples of urine eluates. Irradiation was carried out in borosilicate glass (Corning disposable tubes) with light from two General Electric fluorescent tubes (F20 T12 Black Light, peak emission 350 mμ) arranged as described by Kahane and Vestergaard (32) (Fig. 5). Both heating and irradiation produced a similar increase in fluorescence of about 50% when the samples were read immediately after the incubation. However, the fluorescence continued to increase slowly for the next 6 hr in all samples, whether irradiated, heated, or untreated, although the gain produced by irradiation or heating was reduced to about 10% after that time. We have no data on the effect of irradiating in the region of 254 mμ, but we are unable to confirm that heating at 45° for 30 min completely stabilizes the fluorescence. On the other hand, when the results were calculated with the aid of internal standards treated in the same way and read at the same time, fairly constant values were obtained at different times; in other words, the changes in fluorescence do not seriously impair the method, provided the conditions and the timing are standardized.

It has been suggested (37,39) that the gradual increase in the fluorescence formed from dopamine is due to a continuing rearrangement from the quinoid to the dihydroxyindole configuration. This is

Fig. 5. Arrangement for irradiation of solutions. The cover standing behind the lamps is placed over the lamps during irradiations.

doubtful since the rearrangement requires a strongly alkaline reaction. The level of fluorescence and its further increase are the same when acidification is delayed for more than 5 min. Moreover, fluorescence increases not only in the sample containing dihydroxyindole, but also in the reagent blank and the faded blank, suggesting the occurrence of secondary reactions.

Anton and Sayre (40) reported that the oxidation of dopamine with periodate yields a more intense fluorescence than oxidation with iodine; they further claim that this fluorescence is stable from the outset. Their procedure provides for the simultaneous differential estimation of dopamine and 3,4-dihydroxyphenylalanine (dopa). Since passage through the cation exchange resin would in any case have removed dopa from our eluates, the different pH adjustments prescribed by Anton and Sayre would be pointless. We have simplified their method by omitting the addition of ethanol and citrate buffer. We have retained the use of phosphoric acid for acidification since we found that readings and results reached stability sooner than when hydrochloric or sulfuric acids were used. By increasing the periodate concentration we have increased the reproducibility of results with our urine eluates. We can confirm the superiority of periodate as regards fluorescence yield, but find the same slow increase of fluorescence under our modified conditions and the same susceptibility to irradiation or heating as after iodine oxidation. However, we do not regard this as a critical flaw, for the reasons already discussed.

The wavelength maxima for excitation and emission are 330 and 380 mμ, in accordance with published data (35,38). At these wavelengths and with 4-mm slits, norepinephrine fluorescence is less than 1% of the dopamine fluorescence and is neglected.

Internal standards and faded blanks are set up for each measurement in the estimation of epinephrine and norepinephrine as well as in that of dopamine. For the preparation of the faded blank, the addition of the reducing agent is omitted and only sodium hydroxide solution is added. After an interval of ten minutes or more the mixture is acidified and the reducing solution is added last.

3. Methods

Reagents. Potassium ferricyanide, 0.25%. Keep refrigerated; renew after 1 month.

Sodium metaperiodate, 2%, freshly prepared.

Iodine, 0.1N: 1.27 g of I_2 and 25 g of KI in 100 ml.

Sodium thiosulfate, 0.1N: 2.48 g of $Na_2S_2O_3 \cdot 5\ H_2O$ in 100 ml.

Sodium sulfite, anhydrous, 20%, freshly prepared.

Sodium sulfite, anhydrous, 25%, freshly prepared.

Ascorbic acid, 2%, freshly prepared.

Mercaptoethanol, 1% (v/v) in 20% sodium sulfite, freshly prepared.

10N NaOH–mercaptoethanol reagent: mix equal volumes of 1% mercaptoethanol in 20% Na_2SO_3 and 10N NaOH immediately before use.

5N NaOH–mercaptoethanol reagent: mix equal volumes of 1% mercaptoethanol in 20% Na_2SO_3 and 5N NaOH immediately before use.

Alkaline sulfite reagent: 8 vol of 5N NaOH mixed with 2 vol of 25% Na_2SO_3 immediately before use.

Alkaline ascorbate reagent: 9 vol of 5N NaOH mixed with 1 vol of 2% ascorbic acid immediately before use.

Phosphate buffer, pH 7.2, 1M: dissolve 13.8 g of $NaH_2PO_4 \cdot H_2O$ in about 80 ml of warm water, adjust the pH to 7.2 with 5N NaOH, and make the volume up to 100 ml. The reagent tends to deposit crystals on standing which have to be dissolved by gentle heat before use.

Cupric acetate, $Cu(C_2H_3O_2)_2 \cdot H_2O$, 0.2% (0.01$M$).

Formic acid, 1M: 43.5 ml of 88% acid per liter.

Acetic acid, 10M: 575 ml of glacial acetic acid per liter.

Phosphoric acid, approximately 7.2M: 500 ml of 85% H_3PO_4 per liter.

a. Standard Solutions. Stock solutions, containing 1 mg of base per milliliter, were made up in 0.1N HCl as follows: L-epinephrine bitartrate, 181.9 mg/100 ml, DL-norepinephrine HCl, 121.6 mg/100 ml; and dopamine HCl, 123.8 mg/100 ml. The substances were purchased from Calbiochem, Los Angeles, California. Dilute standards containing 10 or 1 μg/ml in 0.01N HCl were prepared from the stock solutions and stored at 3° for up to two weeks.

A. ESTIMATION OF EPINEPHRINE

a. Ferricyanide-thiol Method. Dilute the eluate from the resin column to 15 ml. Place portions of 1 ml each into 4 tubes, a, b, c, and d. Add 0.1 ml of epinephrine standard (1 μg/ml) to tube b, 0.1 ml of norepinephrine standard (1 μg/ml) to tube c, and 0.1 ml of water to each of tubes a and d. To each tube add 0.3 ml M formic acid, 0.05 ml 0.01M cupric acetate, and 0.05 ml 0.25% ferricyanide. After 5 min add 0.3 ml of the 10N NaOH-mercaptoethanol reagent to tubes a, b, and c. Leave for 4 min, then add 0.3 ml 10M acetic acid. To tube d (faded blank) add 0.15 ml of 10N NaOH and after 10 min add 0.3 ml of 10M acetic acid followed by 0.15 ml mercaptoethanolsulfite reagent. Centrifuge the tubes for about 5 min at 600g. Read the fluorescence in the Aminco-Bowman spectrophotofluorometer at an excitation wavelength of 415 mμ and an emission wavelength of 500 mμ.

b. Iodine Method (Alternative Procedure). Transfer 2-ml portions of the eluate (diluted to 15 ml as above) to tubes a, b, c, and d and add water or internal standards as above. Add 0.5 ml 1M formic acid and 0.1 ml 0.1N iodine. After

5 min add 0.25 ml 0.1N sodium thiosulfate, followed, in tubes a, b, and c, by 0.5 ml of alkaline ascorbate reagent. To tube d add 0.45 ml 5N NaOH. Irradiate all tubes for 3 min in the light of two General Electric fluorescent tubes (F20 T12 BL) arranged as illustrated in Figure 5. After 10 min add 0.05 ml of 2% ascorbic acid to tube d. Read at 415 and 515 mμ.

B. ESTIMATION OF NOREPINEPHRINE

Withdraw a 6-ml portion from the remainder of the diluted eluate and adjust the pH to 5.9 $\pm$ 0.1, first with 5N, and later with 1N NaOH. Make the volume up to 12 ml. Place four portions of 2 ml (=1 ml of diluted eluate) into tubes a–d. As in Section VI-3-B-a, add 0.1 ml of epinephrine and norepinephrine standards (1 μg/ml) to tubes b and c, and add 0.1 ml water to tubes a and d. Add 0.1 ml 0.25% ferricyanide. After 5 min add 0.5 ml of the 5N NaOH-mercaptoethanol reagent to tubes a–c, and 0.25 ml 5N NaOH to tube d. Add 0.3 ml of 10M acetic acid to tubes a–c after an interval of 4 min and to tube d after an interval of 10 min. Finally add 0.25 ml of mercaptoethanol–sulfite reagent to tube d. Read at 395/475 mμ for excitation and emission wavelengths, respectively.

C. ESTIMATION OF DOPAMINE

After the withdrawal of 4 $\times$ 2 ml for the estimation of norepinephrine, 4 ml remains out of the neutralized part of the eluate. Three milliliters of it is used for the estimation of dopamine. Place 1 ml each into tubes a–c. Add 0.03 ml of dopamine standard (10μg/ml) to tube b. Now add to all tubes 0.3 ml M phosphate buffer pH 7.2 and 0.1 ml 2% sodium metaperiodate. After 1 min stop the oxidation by adding 0.3 ml 5N NaOH–25% Na_2SO_3 reagent to tubes a and b and 0.24 ml of 5N NaOH to tube c. Allow 4 min for rearrangement in tubes a and b and 10 min for fading in tube c before adding 0.3 ml 7.2M H_3PO_4. Finally add 0.06 ml 25% Na_2SO_3 solution to tube c. Irradiate for 5 min in the light of the two G.E. tubes. After 1 hr read at 330/380 mμ.

4. Calculations

A. EPINEPHRINE

No correction for the presence of norepinephrine is required. If the internal standard indicates a norepinephrine oxidation exceeding 5% of the epinephrine fluorescence it is better to repeat the estimation using the alternative method. Since only 10 ml of the eluate have been used for the estimations of epinephrine, norepinephrine, and dopamine, the remaining 5 ml are sufficient for the alternative method if either the volumes specified in the description of the procedure are halved or if the remaining eluate is diluted to 10 ml.

The epinephrine content of the eluate is obtained by subtracting the reading for the faded blank (tube d) from that of tube a and expressing the difference in terms of micrograms of epinephrine by comparison with the internal standard, given by the difference of readings of tubes b and a.

$$\mu g \text{ E per 25 ml of urine} = 0.1 \times 15\ (a - d)/(b - a)$$

where a, b, and d stand for the readings given by the respective tubes.

B. NOREPINEPHRINE

The readings are corrected for the amount of epinephrine present. If E μg of epinephrine per milliliter of eluate was found, and if a, b, c, and d represent the fluorescence readings of the sample, internal epinephrine standard, internal norepinephrine standard, and faded blank, respectively, then

$$\mu g \text{ NE per 25 ml urine} = 0.1 \times 15\ [(a - d) - (b - a)\ \text{E}/0.1]/(c - a)$$

Usually the correction $(b - a)\text{E}/0.1$ is small compared with $(a - d)$.

C. DOPAMINE

Since 0.5 ml of eluate and 0.3 μg of internal standard were used

$$\mu g \text{ D per 25 ml urine} = 0.3 \times 30\ (a - c)/(b - a)$$

where a, b, and c represent the readings of the sample, the internal standard and the faded blank, respectively.

VII. THE FLUORIMETRIC ESTIMATION OF METANEPHRINE AND NORMETANEPHRINE

The method described is essentially that of Smith and Weil-Malherbe (6) with some recent modifications designed to correct certain shortcomings in the original method which occasionally led to losses (41). In this method, as in similar methods described by other authors (26,42–44), metanephrine and normetanephrine are oxidized to fluorescent products under conditions very similar to those used in the estimation of epinephrine and norepinephrine. The spectra of fluorescence activation and emission of the reaction products formed from metanephrine and normetanephrine are indistinguishable from those of adrenolutin and noradrenolutin. This may indicate an oxidative demethylation, but the exact mechanism of the reaction has not been established and the formation of *O*-methylated congeners of adrenolutin and noradrenolutin cannot be excluded.

According to Armstrong and his associates (43,45), 10–20% of metanephrine and normetanephrine is excreted in the free form, the rest is conjugated, predominantly in the human species, in the form of the ethereal sulfates (6,46). The free and conjugated fractions may be separated by passing the unhydrolyzed urine either through cation exchange resin where only the free bases are retained (45) or through anion exchange resin where only the conjugated amines are retained (46). Usually the two fractions are determined together after acid hydrolysis at 100°.

Whether or not the simultaneous estimation of epinephrine and norepinephrine is desired, these amines have to be eliminated before metanephrine and normetanephrine can be determined, and the easiest and mildest way of accomplishing this is by adsorption on an alumina column. The initial steps of the procedure, up to and including the filtration through the alumina column, are therefore identical with those described for the estimation of the catecholamines.

After passing through the alumina column, the 3-*O*-methylated amines are isolated from the filtrate by adsorption on a column of a cation exchange resin. Resins of the carboxylic or the sulfonic acid type have been used for the purpose. The sulfonic acid resins, such as Dowex 50 or Amberlite CG 120, have the advantage of greater capacity, but they are also more difficult to elute. On the other hand, elution from carboxylic acid resins, such as Amberlite CG 50, is usually easy, but the retention of amines is liable to be incomplete unless the salt concentration is low. Eluates from sulfonic acid resins were reported to be unsuitable for fluorimetry because they were frequently colored and had high blanks (47). Eluates from carboxylic resins, on the other hand, contain fewer impurities and give satisfactory results on fluorimetric analysis. The required degree of desalination may be achieved by dilution (42), by a prior adsorption–elution cycle on the sulfonic acid resin Dowex 50 (43) or by electrodialysis (6). Dilution, to be effective, has to be drastic and thus entails a considerable reduction in sensitivity. Adsorption on Dowex 50 is time consuming, particularly as it is necessary to evaporate the eluate to dryness. Electrodialysis, in our experience, is quick, simple, and efficient. Our first electrodialyzer (6) has later been modified by the incorporation of a heat exchanger in the center compartment so as to prevent an excessive rise of temperature during dialysis (2,48). The center compartment, which has a capacity of about 50 ml, is separated from the two electrode compartments by a cation exchange membrane opposite the cathode and an anion exchange membrane opposite the anode (Nepton membranes, Ionics, Inc., Cambridge, Mass.). The electrodes are platinum disks of 30 mm diameter with a platinum wire attached to the center and led out through a rubber gasket. Power is supplied from a selenium rectifier. For illustrations of the electrodialyzer, the reader is referred to our previous publications (2,6,48). Both the old and new models were made for us by the Instrument Fabrication Section of the National Institutes of Health; they are not available commercially.

Electrodialysis is started with a current of 0.85–0.90 A and continued until the current has dropped to 0.4–0.5 A, a process which is usually completed within 20–30 min. In our laboratory two electrodialyzers are operated simultaneously in parallel from the same rectifier.

To prevent losses of the amines, the pH of the solution has to be maintained close to their isoelectric point, i.e., at about 10.5. Instead of adding drops of ammonia from time to time as originally described (6), the urine is now buffered with glycine. Even after electrodialysis cations may remain in the urine in concentrations sufficient to impair the retention of the amines on the cation exchange resin. Some dilution at this stage was therefore found advantageous although it can be kept within manageable proportions.

Anton and Sayre (44) have described a procedure for the isolation of urinary metanephrine and normetanephrine by solvent extraction. It is, however, not very efficient. The authors report recoveries of 50%. In our hands, the method gave recoveries of 20–30%.

Ferricyanide, iodine, and periodate have been used for the oxidation of the 3-*O*-methylated amines. Of these, ferricyanide which requires the presence of zinc ions in relatively high concentration gives the highest fluorescence intensity. In the method here described metanephrine alone is oxidized by ferricyanide at pH 3, whereas metanephrine and normetanephrine are oxidized together at a pH near neutrality. For this step iodine is used since it produces fluorescent compounds which have greater stability and better additivity than those obtained with ferricyanide.

In keeping with the general practice in our laboratory, internal standards are prepared with every estimation. The reading of the unknown sample is corrected by deducting the reading obtained with a reagent blank. It was found unnecessary to include a faded or unoxidized sample blank since it did not differ significantly from the reagent blank (6). The modifications described have resulted in figures for the 24-hr excretion more in keeping with those found by others. In 125 recent estimations the average 24-hr excretion of metanephrine was 54.2 μg, that of normetanephrine 161 μg; these figures were not corrected for average recoveries of 65%.

1. Methods

Reagents. Citrate buffer pH 3.0, 0.1M: dissolve 2 g of citric acid in about 80 ml of water; bring to pH 3.0 by the addition of 5N NaOH, and make up to 100 ml. Renew after 1 week.

Tris buffer pH 8.0, 0.5M: dissolve 60.5 g of Tris (hydroxymethyl) aminomethane in about 500 ml of water; add 5N HCl to pH 8.0 and make the volume up to 1 liter.

Zinc sulfate, 0.5%: 0.5 g $ZnSO_4 \cdot 7\ H_2O$ per 100 ml.

Iodine, 0.005N.

Ascorbic acid, 2%, freshly prepared.

Alkaline ascorbate: 9 parts of 5N NaOH plus 1 part of 2% ascorbic acid mixed immediately before use.

Formic acid, 1M, potassium ferricyanide, 0.25%, and Amberlite CG 50 resin, as previously described (Section V-1 and Section VI-3-A).

H_2SO_4, 0.2N; NaOH, 0.2 N; acetic acid, 1M.

Standards: stock solutions containing 1 mg of base per milliliter are prepared by dissolving 118.5 mg of DL-metanephrine-HCl and 119.9 mg of DL-normetanephrine–HCl (Calbiochem, Los Angeles, California) in 100 ml of 0.1N HCl. Dilute standards containing 1 μg/ml of 0.01N HCl are prepared from the stock solutions. They may be stored at 3° for 2 weeks.

A. PREPARATION OF URINE EXTRACT

Add 0.6 g of glycine to the hydrolyzed urine (25 ml) which has been passed through a column of alumina, as described (Section IV-2 and Section IV-3). Adjust the pH to 10.5 with concentrated ammonia. Transfer the sample to the center compartment of the electrodialyzer. Fill the electrode compartments with 0.2N H_2SO_4 on the cathode side and with 0.2N NaOH on the anode side. Start electrodialysis with a current of 0.85–0.90 A and continue until the current has dropped to 0.4–0.5 A. Run out the sample through the stopcock at the bottom of the center compartment, and dilute it with 4 volumes of water. Adjust the pH to 6.0 with M acetic acid and pass through a column of 0.5 g CG 50 resin which has been thoroughly washed with water. When the sample has passed through, wash with 20 ml of water and elute with 5 ml of M formic acid, followed by 2 ml of water. Close the column at the bottom for ½ hr when about 2 ml of eluate has been collected.

B. ESTIMATION OF METANEPHRINE

Place 1-ml portions of the eluate and 1 ml of $0.1M$ citrate buffer pH 3.0 into each of two tubes, a and b. Add 0.25 ml of metanephrine standard (1 μg/ml) to tube b, and 0.25 ml of water to tube a. To both add 0.5 ml of 0.5% zinc sulfate and 0.2 ml of 0.25% ferricyanide. After 10 min add 0.5 ml of alkaline ascorbate solution. Read within 20 min at an activating wavelength of 415 and an emission wavelength of 515 mμ.

C. ESTIMATION OF METANEPHRINE PLUS NORMETANEPHRINE

Titrate 4.5 ml of the eluate to pH 8.0 ± 0.1 and divide into three equal portions, a, b, and c. Dilute to 3 ml with $0.5M$ Tris buffer pH 8.0. Add 0.25 ml of metanephrine standard (1 μg/ml) to sample b, and 0.5 ml of normetanephrine standard (1 μg/ml) to sample c. Add 0.5 ml of water to tube a and 0.25 ml water to tube b. To each tube add 0.15 ml of $0.005N$ iodine solution and, after 1 min, 0.5 ml of alkaline ascorbate solution. Leave for 1–1½ hr before reading at 400/505 mμ.

2. Calculation

Since 1 ml was used for the estimation of metanephrine and 1.5 ml for that of normetanephrine, out of a total of 7 ml we have

$$1\ \mu\text{g metanephrine per 25 ml urine} = 0.25 \times 7\ (a - d)/(b - a)$$

and

$$1\ \mu\text{g normetanephrine per 25 ml of urine} = 0.5 \times 7\ [(a - d) - 1.5\ (b - a)\ M/0.25]/1.5\ (c - a)$$

where a is the reading obtained with the sample, b is the reading obtained with the sample plus added metanephrine standard, c is the reading obtained with the sample plus added normetanephrine standard, d is the reading obtained with the reagent blank, and $M = \mu$g of metanephrine per milliliter of eluate.

VIII. THE ESTIMATION OF 3-METHOXY-4-HYDROXYMANDELIC ACID (VANILLYLMANDELIC ACID, VMA)

The most convenient methods for the routine estimation of VMA are based on its oxidation to vanillin, introduced by Sandler and

Ruthven (49,50). Ferricyanide (51), periodate (52), alkaline solutions of cupric ions (53), and catalytic oxidation under pressure (49,50) have been used to bring about this conversion. Homovanillic acid turned out to be a source of interference in the copper method (53). Periodate oxidation is also subject to interference from unknown constituents of urinary extracts. Whereas the oxidation of VMA by periodate proceeds smoothly at room temperature in pure solutions, the temperature had to be increased to 50° for the oxidation of VMA in urine extracts (52). This procedure seems to work well in the majority of urine samples, but we have come across occasional samples where the oxidation was strongly inhibited even at 50°, as indicated by internal standards. The method of Sunderman et al. (51), in which ferricyanide is used as the oxidizing agent, has proved entirely satisfactory in our experience. We have introduced some slight modifications, such as absorptiometry at 360 mμ, in preference to the color reaction with indole, and the inclusion of two internal standards with each sample. For the sake of simplicity the internal standards are only added after the extraction of urine with ethyl acetate, but control experiments have shown that results are not significantly different whether the standards are added at this point or at the beginning. An unoxidized blank is prepared for every sample to allow for the presence in urine of vanillin from dietary sources. In contrast to the trihydroxyindole method where, in some modifications, interfering fluorescence arises from the interaction of ascorbic acid and ferricyanide, an unoxidized blank is an appropriate correction here. Although the absorption maximum of vanillin is at 348 mμ in alkaline medium, readings are taken at 360 mμ to minimize interference from *p*-hydroxybenzaldehyde (λ_{max} 330 mμ), as recommended by Pisano et al. (52).

1. Methods

Reagents. Potassium carbonate, 1*M*: dissolve 138.2 g of K_2CO_3 anhydride to 1 liter of H_2O.

Potassium ferricyanide, 0.6%. Keep refrigerated; renew after 1 month.

Potassium phosphate, dibasic, 4*M*: dissolve 139.3 g of K_2HPO_4 to 200 ml.

Zinc sulfate, $ZnSO_4 \cdot 7\ H_2O$, 1.2%.

Ammonia, 0.5*N* solution containing 6% sodium chloride.

Sulfuric acid, 10*N*; hydrochloric acid, concentrated, sodium hydroxide, 5*N*.

Ethyl acetate (reagent grade); toluene (reagent grade).

Florisil (Floridin Co., Pittsburgh, Pa.).

Standard, 100 μg/ml: dissolve 10 mg VMA in 100 ml of 0.1*N* HCl. Keep refrigerated; renew after 3 months.

Procedure. Shake 50 ml of urine, 5 ml of 10*N* H_2SO_4, and 4 g of Florisil for 10 min, centrifuge and filter the supernatant. Saturate 44 ml of the filtrate (corresponding to 40 ml of urine) with solid NaCl and extract first with 88, and then with 44 ml of ethyl acetate by mechanical shaking for 10 min. Reextract the combined extracts with 10 ml of 1*M* potassium carbonate by shaking for 5 min. Separate the aqueous layer as completely as possible and wash the ethyl acetate extract with 2 ml of water. Measure the volume of the aqueous solution, including the washing (*p* ml). Transfer to a small beaker and add slowly, with stirring, 5 ml of concentrated HCl at 0°. When evolution of CO_2 has ceased, measure out four samples of 3 ml (a, b, c, and d). Add 0.1 ml of standard (=10 μg VMA) to b, 0.2 ml of standard (=20 μg VMA) to c, and 0.6 ml of zinc sulfate solution to all four samples. To a–c add 0.6 ml of ferricyanide solution; to d add 0.6 ml of water (unoxidized blank). Incubate for 2 hr in a water bath at 37° protected from bright light. If available, tubes of low-actinic (red) glass may be used. Cool, add 0.5 ml of 4*M* phosphate solution, and adjust the pH to 7.0 with 5*N* NaOH. Extract with 5 volumes of toluene by shaking for 10 min. Reextract the vanillin formed into 3 ml of 0.5*N* ammonia–6% NaCl reagent (5-min shaking). Read samples a–c in an ultraviolet spectrophotometer at 360 mμ against the unoxidized blank (sample d).

2. Calculation

The corrected reading of the 20-μg internal standard is usually within ±15% of twice the corrected reading of the 10-μg standard. A result is calculated based on the 10-μg standard and another one based on the 20-μg standard, and the mean of the two is used as the concentration of the unknown sample.

Since four portions of 3 ml were taken out of a total of $p + 5$ ml, the result is multiplied by the factor $(p + 5)/12$ to obtain the concentration of VMA in 10 ml of urine.

IX. THE ESTIMATION OF 3,4-DIHYDROXYMANDELIC ACID (DHMA)

DHMA is of interest as an important intermediate of catecholamine metabolism. Its excretion in increased amounts might be expected not only as a result of increased catecholamine output, but also in situations when the activity of monoamine oxidase is increased relative to that of catechol-*O*-methyltransferase.

The fact that DHMA is a very unstable product, and that it normally occurs in urine in low concentration, makes its estimation a difficult problem. DHMA may be converted to protocatechuic aldehyde in a reaction analogous to the oxidation of VMA to vanillin. The oxidation of DHMA to protocatechuic aldehyde has been utilized in two previous methods (54,55) but in neither of them has it been possible to eliminate interference from 3,4-dihydroxyphenylacetic acid (dopac). A method has recently been developed in this laboratory which is sufficiently specific and simple for routine use (17). It is based on the observation of Crowell and Varsel (56) that aromatic aldehydes in methanol solution are nonfluorescent owing to the quenching effect of the carbonyl group, but become fluorescent after acidification when the aldehyde is converted to an acetal. The method consists of the following steps: (*1*) adsorption of DHMA on a column of alumina and elution with 1N H_2SO_4, (*2*) removal of inhibitors, presumably amino acids, by passage through a column of cation exchange resin, (*3*) oxidation of DHMA to protocatechuic aldehyde by shaking an ethyl acetate extract with ammonia solution, and (*4*) extraction of the aldehyde and fluorimetry before and after acidification.

As a catechol derivative DHMA is adsorbed on alumina from weakly alkaline solutions, but its elution requires stronger acid than that of the catecholamines.

When an ethyl solution of DHMA is shaken with dilute ammonia solution for 3 min, DHMA is quantitatively oxidized, as shown by DeQuattro et al. (55). Under these conditions dopac is oxidized to an aldehyde which differs from protocatechuic aldehyde in several respects. The two aldehydes can be separated by extraction from aqueous solution at pH 7; moreover, only protocatechuic aldehyde shows the fluorescence increase on acidification.

In the method of DeQuattro et al. (55) the concentration of protocatechuic aldehyde in the ammonia extract is determined absorptiometrically. Since the absorption spectrum of the reaction product formed from dopac overlaps that of protocatechuic aldehyde, DeQuattro et al. suggested a correction to eliminate the interference of dopac in the estimation of DHMA. We were, however, unable to confirm the validity of this correction (48). In view of the fact that dopac occurs in urine in concentrations at least ten times as high as DHMA, even a slight amount of interference seriously upsets the results.

After the passage of urine through alumina, the eluate from the alumina contains substances which are extracted by ethyl acetate from acid solution and which frequently inhibit the oxidation of DHMA. The inhibitors could be largely, though not completely, eliminated by passing the alumina eluate through a column of a strongly acidic cation exchange resin at pH 2, suggesting that they are acidic amino acids known to be adsorbed on alumina (57,58) and to occur in urine (59). Since urinary extracts thus treated still show some residual inhibition of variable strength, it is necessary to include an internal standard with each analysis. Addition of ascorbic acid or EDTA to the urine before the passage through the alumina column greatly increased the inhibition of DHMA oxidation.

Complete dryness of reagents and glassware is a prerequisite for the success of the fluorescence reaction. Cuvettes, therefore, should not be washed with water between estimations. To avoid contamination by traces of acid it is advisable to take the preacidification reading in one cuvette and then transfer the sample to a second cuvette in which the acidification is performed. The reading before acidification serves as the sample blank.

In agreement with the results of Wada (60,61) the mean daily excretion rate was found to be of the order of 100 μg. Excretion rates in eight cases of pheochromocytoma varied from approximately 200 to 2000 μg/24 hr. No evidence for the excretion of an acid-hydrolyzable conjugate of DHMA was found.

1. Method

Reagents. Aluminum oxide (see Section IV-2).

Cation exchange resin, AG 50 W, X 4, 100–200 mesh (Calbiochem, Los Angeles, Calif.) recycled repeatedly through the Na^+ and H^+ forms as described for resin CG 50 (Section V-A). Finally, the resin, in the H^+ form, is thoroughly washed with water and stored in water at 3°. Before use, an appropriate amount is sucked free of adhering water and portions of 0.5 g of moist resin per column are weighed out.

Ethyl acetate (Merck and Co., Inc., Rahway, N.J.) and absolute methanol "B & A" (Allied Chemical, General Chemical Division, Morristown, N.J.). These solvents may be used without purification. Other brands had to be treated with charcoal and distilled to reduce the fluorescence of the blank.

Approximately 5*N* H_3PO_4: dilute 11.5 ml of 85% phosphoric acid to 100 ml.

10% H_2SO_4 in methanol, freshly prepared: add slowly 1 ml of concentrated H_2SO_4 to 9 ml of absolute methanol at 0°.

Standard solution: 1 mg of DHMA (Calbiochem) per milliliter of 0.01*N* HCl, stored at 3°.

Ammonia, 5*N*; sodium hydroxide, 5*N*; sulfuric acid, 1*N*.

Procedure. Add 0.04 ml of DHMA standard, containing 40 μg of DHMA, to one of two 25-ml samples of urine. Adjust the pH to 5.5–5.6 with 5*N* ammonia solution and centrifuge at 1000*g* and 0° for 10 min. Continue the addition of ammonia until the pH reaches 8.4 and, if necessary, centrifuge again. Pass the samples over columns of 0.7 g of alumina, wash with 10 ml of water, and elute with 5 ml of 1*N* H_2SO_4 followed by 3 ml of water.

Adjust the pH of the eluate to 2.0 with 5*N* NaOH. Pass the solution through a column of 0.5 g of AG 50 W resin. Wash with 10 ml of water. Saturate the combined filtrate and washings with solid NaCl and extract twice with 3 volumes of ethyl acetate by shaking for 10 min.

Combine the extracts, allow suspended droplets of the aqueous phase to settle out, and remove them. Shake the ethyl acetate extract for 3 min with 5 ml of 5*N* ammonia. Run the aqueous phase into a beaker containing 5 ml of 5*N* H_3PO_4 at 0°. Shake the ethyl acetate phase with 3 ml of water and add the washings to the contents of the beaker.

Adjust the pH, if necessary, to fall between 6.0 and 7.0. Saturate with NaCl and extract twice with 3 vol of ethyl acetate. Dry the combined extracts over anhydrous Na_2SO_4 for at least 1 hr, preferably overnight. Evaporate the extract in a rotary evaporator at a bath temperature of 30–35° under reduced pressure. Treat the residue with three successive 1-ml portions of methanol and transfer to a clean, dry cuvette. Mix well. Read fluorescence at an activating wavelength of 280 mμ and an emission wavelength of 322 mμ. Add 0.01 ml of 10% H_2SO_4 in methanol, mix well, and repeat the reading.

2. Calculation

The calculation is based on the estimation of the internal standard. If Δ_1 is the difference of readings of the urine sample without addition before and after acidification, and Δ_2 is the corresponding difference of the sample with internal standard, then the urine sample contains $40 \times \Delta_1/(\Delta_2 - \Delta_1)$ μg of DHMA per 25 ml.

Acknowledgments

I am greatly indebted to my colleagues, Dr. Elizabeth R. B. Smith and Dr. Llewellyn B. Bigelow, for permission to use their unpublished results. I also wish to acknowledge the devoted technical assistance of Mrs. Grace Bowles and Mrs. Viola Epps.

References

1. J. R. Crout, in *Standard Methods of Clinical Chemistry,* Vol. 3, D. Seligson, Ed., Academic Press, New York, 1961, pp. 62–80.

2. H. Weil-Malherbe, *Z. Klin. Chem.*, *2*, 161 (1964).
3. H. Weil-Malherbe and A. D. Bone, *J. Clin. Pathol.*, *10*, 138 (1957).
4. R. I. Henkin and F. C. Bartter, *Federation Proc.*, *24*, 133 (1965).
5. McC. Goodall and L. Rosen, *J. Clin. Invest.*, *42*, 1578 (1963).
6. E. R. B. Smith and H. Weil-Malherbe, *J. Lab. Clin. Med.*, *60*, 212 (1962).
7. L. F. Fieser, *J. Am. Chem. Soc.*, *46*, 2639 (1924).
8. A. E. Fruehan and G. F. Lee, *Am. J. Clin. Pathol.*, *46*, 172 (1966).
9. B. D. Drujan, T. L. Sourkes, D. S. Layne, and G. F. Murphy, *Can. J. Biochem. Physiol.*, *37*, 1153 (1959).
10. A. Pekkarinen and M.-E. Pitkänen, *Scand. J. Clin. Lab. Invest.*, *7*, 1 (1955).
11. A. F. DeSchaepdryver, *Arch. Intern. Pharmacodyn.*, *115*, 233 (1958).
12. A. H. Anton and D. F. Sayre, *J. Pharmacol.*, *138*, 360 (1962).
13. H. Weil-Malherbe and A. D. Bone, *Biochem. J.*, *51*, 311 (1952).
14. A. S. Russell and C. N. Cochran, *Ind. Eng. Chem.*, *42*, 1336 (1950).
15. H. Weil-Malherbe, in *Methods in Medical Research*, Vol. 9, J. H. Quastel, Ed., Yearbook Medical Publishers, Chicago, 1961, pp. 130–146.
16. A. Lund, *Acta Pharmacol.*, *5*, 231 (1949).
17. H. Weil-Malherbe, *J. Lab. Clin. Med.*, in press.
18. U. S. von Euler, *Pharmacol. Rev.*, *11*, 262 (1959).
19. A. Vendsalu, *Acta Physiol. Scand. Suppl.*, *49*, 173 (1960).
20. J. Häggendal, *Pharmacol. Rev.*, *18*, 325 (1966).
21. U. S. von Euler and F. Lishajko, *Acta Physiol. Scand.*, *45*, 122 (1959); *51*, 348 (1961).
22. E. C. Gerst, O. S. Steinsland, and W. W. Walcott, *Clin. Chem.*, *12*, 659 (1966).
23. J. Häggendal, *Acta Physiol. Scand.*, *59*, 242 (1963).
24. H. Weil-Malherbe and L. B. Bigelow, to be published.
25. A. Carlsson and B. Waldeck, *Acta Physiol. Scand.*, *44*, 293 (1958).
26. J. Häggendal, *Acta Physiol. Scand.*, *56*, 258 (1962).
27. H. L. Price and M. L. Price, *J. Lab. Clin. Med.*, *50*, 769 (1957).
28. R. J. Merrills, *Anal. Biochem.*, *6*, 272 (1963).
29. R. F. Vochten and A. F. DeSchaepdryver, *Experientia*, *22*, 772 (1966).
30. U. S. von Euler and U. Hamberg, *Acta Physiol. Scand.*, *19*, 74 (1949).
31. U. S. von Euler and I. Floding, *Acta Physiol. Scand. Suppl.*, *33*, 118, 45 (1955).
32. Z. Kahane and P. Vestergaard, *J. Lab. Clin. Med.*, *65*, 848 (1965).
33. T. B. B. Crawford and W. N. Law, *J. Pharm. Pharmacol.*, *10*, 179 (1958).
34. D. F. Sharman, S. Vanov, and M. Vogt, *Brit. J. Pharmacol.*, *19*, 527 (1962).
35. A. Carlsson and M. Lindqvist, *Acta Physiol. Scand.*, *54*, 87 (1962).
36. F. Bischoff and A. Torres, *Clin. Chem.*, *8*, 370 (1962).
37. S. Udenfriend, *Fluorescence Assay in Biology and Medicine*, Academic Press, New York, 1962, p. 137.
38. E. G. McGeer and P. L. McGeer, *Can. J. Biochem. Physiol.*, *40*, 1141 (1962).
39. V. J. Uuspää, *Ann. Med. Exp. Biol. Fenniae* (*Helsinki*), *41*, 194 (1963).
40. A. H. Anton and D. F. Sayre, *J. Pharmacol.*, *145*, 326 (1964).
41. E. R. B. Smith, unpublished observations.
42. S. Brunjes, D. Wybenga, and V. R. Johns, Jr., *Clin. Chem.*, *10*, 1 (1964).

43. K. Taniguchi, Y. Kakimoto, and M. D. Armstrong, *J. Lab. Clin. Med.*, *60*, 212 (1962).
44. A. H. Anton and D. F. Sayre, *J. Pharmacol.*, *153*, 15 (1966).
45. Y. Kakimoto and M. D. Armstrong, *J. Biol. Chem.*, *237*, 208 (1962).
46. E. H. LaBrosse and J. D. Mann, *Nature*, *185*, 40 (1960).
47. A. Randrup, *Clin. Chim. Acta*, *6*, 584 (1961).
48. H. Weil-Malherbe and E. R. B. Smith, *Pharmacol. Rev.*, *18*, 331 (1966).
49. M. Sandler and C. R. J. Ruthven, *Lancet*, *1959-II*, 114, 1034.
50. M. Sandler and C. R. J. Ruthven, *Biochem. J.*, *80*, 78 (1961).
51. F. W. Sunderman, Jr., P. D. Cleveland, N. C. Law, and F. W. Sunderman, *Am. J. Clin. Pathol.*, *34*, 293 (1960).
52. J. J. Pisano, J. Crout, and D. Abraham, *Clin. Chim. Acta*, *7*, 285 (1962).
53. H. Weil-Malherbe, *Anal. Biochem.*, *7*, 485 (1964).
54. H. Miyake, H. Yoshida, and R. Imaizumi, *Jap. J. Pharmacol.*, *12*, 79 (1962).
55. V. DeQuattro, D. Wybenga, W. von Studnitz, and S. Brunjes, *J. Lab. Clin. Med.*, *63*, 864 (1964).
56. E. P. Crowell and C. J. Varsel, *Anal. Chem.*, *35*, 189 (1963).
57. T. Weiland, *Z. Physiol. Chem.*, *273*, 24 (1942).
58. F. Turba and M. Richter, *Ber. Deut. Chem. Ges.*, *75*, 340 (1942).
59. W. H. Stein, *J. Biol. Chem.*, *201*, 45 (1953).
60. Y. Wada, *Tohoku J. Exptl. Med.*, *79*, 389 (1963).
61. Y. Wada and N. Watanabe, *Tohoku J. Exptl. Med.*, *84*, 161 (1964).

The Automated Analysis of Absorbent and Fluorescent Substances Separated on Paper Strips

ALAN A. BOULTON* *Medical Research Council Unit for Research on the Chemical Pathology of Mental Disorders, The Medical School, Birmingham, United Kingdom*

* Presently with the Psychiatric Research Unit, University Hospital, Saskatoon, Saskatchewan, Canada.

I. INTRODUCTION

The exponential growth in chromatographic methodology, begun in 1944 (1) is now being paralleled by the increasing utilization of scanning techniques for direct quantitative assessments of thin-media chromatograms. There are at least 12 different commercial machines currently on the market, varying from relatively specific devices designed for the assessment of cellulose acetate electrophoresis strips up to devices on which 15-cm square thin-layer plates or paper strips of any length can be accommodated. In addition to these scanning devices, there are numerous attachments designed to fit most of the existing spectrophotometers and several spectrophotofluorimeters. In order to obtain meaningful data from these devices, it is important that certain precautions be taken during the preparative, separative, and chromogenic/fluorogenic localization stages. The precautions necessary for the production of adequate paper chromatograms have been recently reviewed (2,3), and the timely paper by Hamilton (4) illustrates most convincingly the need for great care during the handling of chromatograms, especially when they are damp. The numerous solute application devices (5–12) illustrate clearly the awareness of workers in the field for careful application of precise volumes of the sample into or onto limited areas; a careful selection of the solvent system and the nature of the support media are, however, equally important. For instance, by selecting a solvent system in which the substance of interest possesses a low R_f value and a fine-grained "slow" paper followed by a considerable "overrun," using the descending technique, often produces a well defined and cleanly separated zone when the chromatogram is finally developed. Because these conditions can usually be obtained overnight, the time factor is often unimportant. The final stages of a quantitative analysis are greatly facilitated if the separated zone adheres closely to theoretical predictions, since in these cases simple methods of analysis such as triangulation or even the measurement of peak height may be sufficient for an adequate quantitative result. Since, in the case of fluorescent substances, a linear relationship usually exists between concentration of the separated fluorophore and the amount of fluorescence, at least at the concentration levels frequently encountered on chromatograms, a direct photoelectric measurement of the amount of emitted fluorescent light is adequate regardless of the shape of the fluorescing zone, so long as the whole of the separated zone is evenly illuminated by the activating light. Unfortunately, matters are not so convenient in the case of colored zones, and in the absence of relatively sophisticated cross-scanning systems it is imperative that a uniform deposition of chromogen extend across the full width of the scanning slit (see ref. 3 for a detailed discussion, also refs. 13–15). Devices incorporating a cross-scanning mechanism have been described (15,16), but so far they have not been incorporated into existing commercial apparatus. In a recent paper Bush (17) mentions a device, now nearing completion, in which background compensation and a cross-scanning mechanism are included. With such a device it should be possible to detect and quantitatively estimate very small amounts of chromogen regardless of the zone geometry. Until such machines are routinely available, however, it is important that absorbent substances be applied to the origin line of the chromatograms, so that after development the colored zone extends uniformly across the width of the chromatographic strip.

With the increased utilization of scanning techniques the problem of suitable storage and handling of data becomes important. It is now quite routine in some laboratories for upwards of 100 different estimations to be made in a single working day. Clearly, in these circumstances manual methods of analysis become time consuming (e.g., Table I) and inadequate. Section II describes how, in our laboratory, paper chromatograms in strip form are analyzed by direct scanning techniques followed by computer evaluation of the scanning records.

II. PREPARATION, DEVELOPMENT, AND DRYING OF CHROMATOGRAPHIC STRIPS

1. Preparation

The amount of time and effort expended in the preparation of paper strips (usually 5 × 57 cm as supplied by Whatman) suitable for chromatographic separations can be substantially reduced by the use of simple pieces of apparatus designed to streamline the marking, folding, and holding of the strips. The initial marking is accomplished by gripping one end of the strip between "markers" and placing pencil dots to delineate the positions for subsequent folding and a pencil line to represent the origin. The positions of these pencil marks are obviously dictated by the shape and size of the chromatogram development tank to be used and the contours of the solvent dish within the tank. The "marker" itself is manufactured from two bars of stainless steel or aluminum riveted together, the upper bar possessing smaller dimensions than the lower bar. The positions for the pencil dots are cut or etched out of the upper, smaller bar. After marking, the strips are clipped in a dexion assembled apparatus so as to be fairly rigidly held, and a horizontal cross bar is positioned so as to tilt that part of the strip containing the origin line for ease of access during the solute application. The sample dissolved in some suitable solvent is manually applied directly to the origin line from micropipets with or without the aid of a syringe. We have found that a fairly even deposition of the solute is possible using this technique, although obviously the various mechanical devices for solute application could be employed instead. Alternatively, for very small amounts of sample dissolved in relatively large solvent volumes the running-up technique devised by Bush (2,13) could be utilized. Cold air as produced from a conventional hair dryer can be blown across the strips from the right. After drying, the strips are folded using "folders" (stainless steel or aluminum bars of equal dimensions riveted together) along the lines indicated by the pencil dots and then threaded onto stainless steel frames.

A large sheet of plate glass placed over the solute application area and retained on the dexion framework produces a clean working area for all of these operations.

2. Chromatographic Development

The paper strips threaded onto the stainless steel and glass frames are placed in the glass chromatography tanks. For solvent systems of low polarity or low boiling points, two layers of Whatman No. 4 quality paper are used to line the inside of the tank, thus ensuring a solvent-saturated atmosphere within the tank. The lining papers are prevented from falling inward by a retaining stainless steel framework. Two solvent dishes can be placed on recessed platforms and a maximum of 16 strips (4 per frame) accommodated at any one time. After equilibration when necessary, the solvent is added through holes in the lid, and after a suitable development time the strips are removed for drying. It is quite possible to mark the strips and, after threading them on frames, to place them in the chromatography tank and remove them again without touching them at any time.

3. Drying

After removal from the tanks, the strips are most conveniently dried by hanging the frames from hooks in parallel slats of wood in a fume cupboard and allowing air at room temperature to stream between them. For faster drying it would be equally feasible to hang the frames inside an adequately ventilated oven. A much more detailed description of this section has been given in an earlier publication (3).

In those cases where naturally colored or fluorescent substances have been separated, the strips may be scanned immediately after drying; in other cases some form of chemical modification is necessary before suitable chromogenic or fluorogenic zones are produced.

III. DIRECT SCANNING OF PAPER CHROMATOGRAPHIC STRIPS

1. Apparatus for the Production of Analog Records

Numerous scanning devices (2,3,13–15,18–39) have been described for the direct assessment of colored and/or fluorescent zones on thin-media chromatograms, and although some of these devices were designed to operate with reflected light (18–24), most are based on the principle of transmission (2,3,13–15,22–39) presumably because, in this latter case, the instruments are easier to design and construct. The levels of sophistication vary from straightforward filter instruments (2,3,15,28,29,31–34,36–39) to those employing one or two monochromators; sometimes the sample is propelled by belts or rollers (2,3,13,38,39), while on others the sample is fixed to a carriage. In nearly all cases the photomultiplier response is fed directly to an analog recorder, although some devices possess a mechanical means of linearizing the signal (21,55). The device described by Hamman and Martin (30) introduces monochromatic light to the sample via a glass fiber and similarly collects the emitted and transmitted light in a cluster of glass fibers. The recent paper by Salganicoff et al. (26) describes a most elegant transmission device in which, by dual wavelength scanning followed by electronic subtraction of one signal from the other, a much more stable background is obtained, allowing at

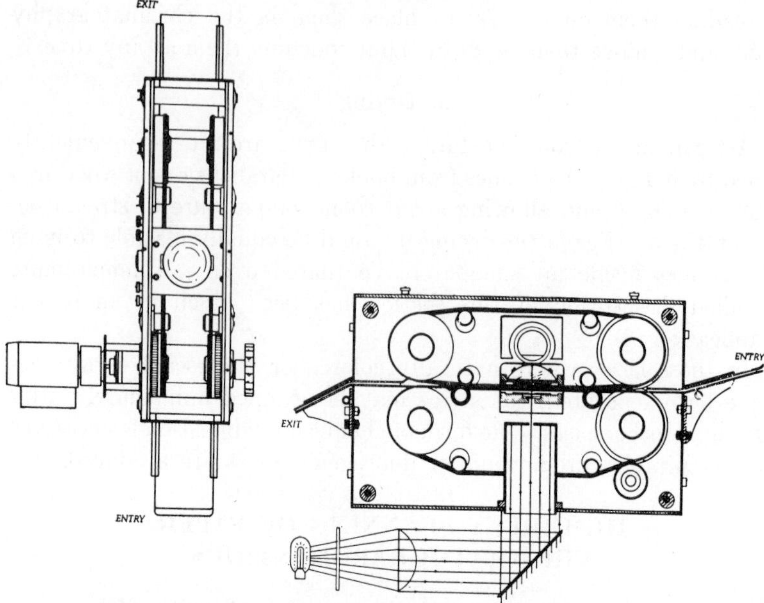

Fig. 1. Principles of the scanning device. See text and Figures 2–4 for further details.

least a twofold increase in sensitivity over earlier scanning devices. There is absolutely no doubt that the current massive advances in electronic and optical technology, when incorporated into conventional scientific instrumentation, will produce vast improvements in design, operation, and sensitivity.

In our laboratory we have also chosen a filter transmission device simply because it is cheaper and relatively simple to design and construct. Figure 1 illustrates the mechanical and optical principles adopted in our apparatus where we are required to scan paper strips 5 cm wide. It is clear that any other flexible medium, so long as it is 5 cm wide, could be accommodated. By adopting a different method of sample transport such as a rack and pinion or spring wheel type of drive on the sample holder, any medium possessing any dimensions, whether flexible or not, could be scanned.

The light source for fluorescence measurements is a mercury arc lamp (Wotan Hg/3) producing the usual intense lines in the ultraviolet. The stabilized voltage supply is the same as that used in the Locarte filter fluorimeter instrument.* In the case of absorptio-

* Locarte Co. Ltd., 24, Emperer's Gate, London, S.W. 7.

metric analyses, the light source is a quartz–iodine 50 W lamp (Atlas Lighting Ltd.) connected directly through a voltage stabilizer to the 50 cps main supply. Since the spectrum of the quartz–iodine lamp is continuous from 3000 Å and significant quantities of energy are available in the blue-violet region (40), it may be that this lamp would also be useful for some fluorescent measurements.

The heat absorbing filter (Chance HA1) is supplemented in the case of the absorptiometer by a fan-driven ventilation system fixed into the base of the scanning device (Fig. 3). The glass (quartz is only necessary at wavelengths below 3000 Å) plano-convex lens, focal length 9 cm, serves to partially collimate the light beam which, after reflection through 90° from a front-aluminized glass plate, is directed into the scanning head (Figs. 1 and 2). In the case of the absorptiometric device (Fig. 3) the quartz–iodine lamp is mounted

Fig. 2. Scanning apparatus arranged for fluorescence measurements. The stabilized voltage supply for the mercury arc lamp is seen on the left and the lamp itself is attached to the retort stand. This very crude arrangement allows rapid adjustment of the light path. The small box next to the retort stand houses the magnetic clutch apparatus. The scanning device shown in Figure 1 is mounted on a wooden box which encloses the heat absorbing filter, the collimating lens and the mirror. The assembly bolted to the front of the scanning device is the heater; hot air is delivered to this attachment via the flexible tube.

below the slits, thus obviating the need for a mirror. The primary filter position is either immediately below the lower slit (Figs. 1 and 2) or in the scanner base (Fig. 3). The reason for this is that the fluorimeter and absorptiometer were constructed separately, and in the case of fluorescence measurements some colored glass filters (5 × 0.5 cm) are usually adequate for the isolation of the peak activation and fluorescent wavelengths. Since these filters can be inexpensively cut to any size it seemed sensible to fit them close to the sample. For absorptiometric analyses however, it is frequently necessary to isolate a particular wavelength, and in the absence of a monochromator an interference filter is required; as these are usually expensive it is preferable to use the standard sizes (2-in. square or 2 in. in diameter in the U.K.) and position them in the base away from the confines of the slits, belts, etc. Immediately below the paper

Fig. 3. Scanning apparatus arranged for absorptiometric measurements. In this apparatus thes canner illustrated in Figure 1 is mounted directly over the base holding the quartz–I_2 lamp. A fan-driven ventilation system maintains a relatively constant lamp-operating temperature; the air intakes of this system are clearly visible. The photomultiplier mounting can be seen in the left-hand wall of the scanner. The slit and filter system shown in Figure 4 fits in the right-hand wall. The logging device is on top of the analog recorder.

strip a slit (5 cm long and 0.25 mm wide and variable widthwise in the range 0–3 mm by 0.25-mm steps) serves to illuminate a small segment of the chromatogram as it passes across the slit. Mounted immediately above the paper strip and in line with the lower slit, a second slit of equal dimensions and variability allows emitted and transmitted light through a secondary colored glass filter (it is sensible to inspect these filters in ultraviolet light before use, since some of them exhibit intense fluorescence) when necessary, through the plano-convex collector lens (focal length 6 cm) and onto the grid of the photomultiplier tube (RCA 931A). The small dimensions of the slits and their close proximity to the paper reduce the effects of light scatter which are quite appreciable in an unoiled, nonhomogenous material like paper (16,22,24,26,36,41,42). Servicing and changes to the slit and filter system are accomplished simply by unscrewing four nuts, after which the whole assembly (Fig. 4) can be withdrawn through the right-hand wall of the scanning head.

Fig. 4. Slit and filter assembly. This assembly may be withdrawn from the right-hand wall of the scanner by simply unscrewing four nuts. The width of the upper and lower slits may be altered after slackening the appropriate screws. The primary filter fits below the lower slit and the secondary filter in the rectangular slot above the upper slit. The collector lens fits in the circular recess above the secondary filter.

The paper strips are propelled across and between the slits at constant speed by the endless moving neoprene belts (Fig. 1). These belts are driven by the right-hand upper and lower pulleys which are connected through a gear train to a synchronous motor. The gear is engaged and disengaged through a magnetic clutch. An adequate grip on the edges of the chromatographic strips to ensure a constant speed is provided by adjustment of the concentrically mounted torsion pulleys. The speed of scanning is usually fixed at 1 cm/sec for large strips (5 × 57 cm), but it can be reduced by changing the gear ratios when smaller chromatograms need to be accommodated. Scanning records of small chromatograms or electrophoretograms are obtained by attaching such strips to a larger carrier strip in which a suitable segment may be removed. The slit length in these cases is reduced by masking the ends with black tape.

After slight linear amplification, a portion of the photomultiplier response is fed either directly to a conventional analog recorder in the case of fluorescence measurements or through a logging device in the case of absorptiometric measurements. The logarithmic conversion is based on the circuit described by Sweet (43) and produces signals which bear linear relationships with concentration up to at least an optical density value of 3.0 when colored solutions in a 1-cm cell are placed in the paper strip position. It is now possible to produce a solid-state device which is much more convenient than the existing logging device with respect to both size and function (44).

Direct transmission scanning of colored zones on paper does not usually produce linear calibration curves (Figs. 6*b*, 7*b*, and 8*b*, and refs. 3,24,26,36,45) and large deviations from Beers law can occur. By altering the refractive index of the support medium however (in the case of paper this can be by dipping in liquid paraffin followed by gentle blotting), a more nearly linear relationship between optical density and concentration is obtained (e.g., Fig. 6*b*). The procedures involved however, are time consuming and rather messy, so in the main it is preferable to scan dry strips and obtain quantitative values in the concentration ranges where a relatively linear relationship exists. The dotted lines in Figures 6*b*, 7*b*, and 8*b* are an attempt to show how accurate such approximations are likely to be in some specific cases.

By adopting a recorder speed of 1 cm/sec the scanned chromatogram can be superimposed on the analog trace, which facilitates a

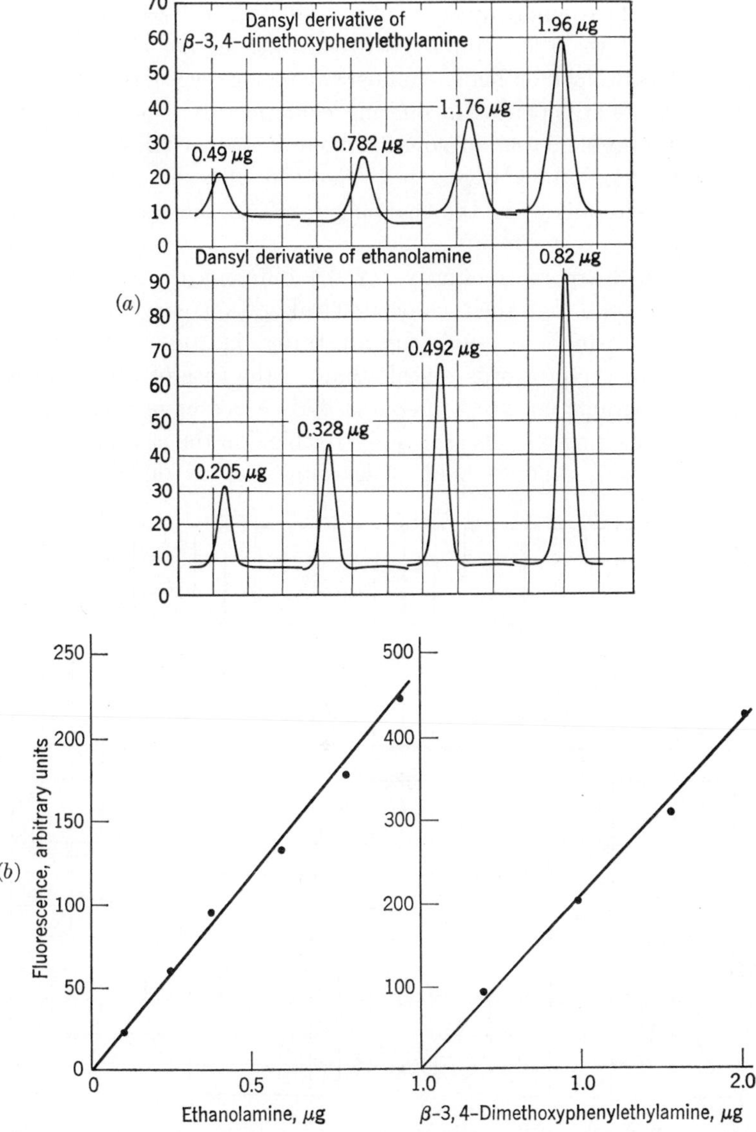

Fig. 5. (*a*) Typical analog records, produced by scanning chromatograms on which prepared fluorescent derivatives have been separated: in this case, the dansyl derivatives of β-3,4-dimethoxyphenylethylamine and ethanolamine. (*b*) Calibration curves obtained from scanning records similar to those shown in Figure 5*a*.

speedy identification of the peaks of interest. A full-scale deflection (fsd) time of 1 sec or less, and a sensitivity range of 0.5–50 mV for fsd can now be obtained on many commercial analog recorders, but most manufacturers still persist in offering chart speeds which are very slow. This problem was surmounted in our laboratory by connecting a synchronous motor and gear box, identical with the ones used to drive the belt pulleys in the scanning head, to the recorder chart drive roller.

Figures 5–8 are representative of the different types of analog scanning records and their associated calibration curves that can be obtained using the apparatus illustrated in Figures 2 and 3. Inspection of these records reveals that, in the case of the analysis of chromatograms on which prepared derivatives have been separated, very pleasing records are obtained which, in the case of fluorescence (Fig. 5*a*), produce bold peaks superimposed on almost flat

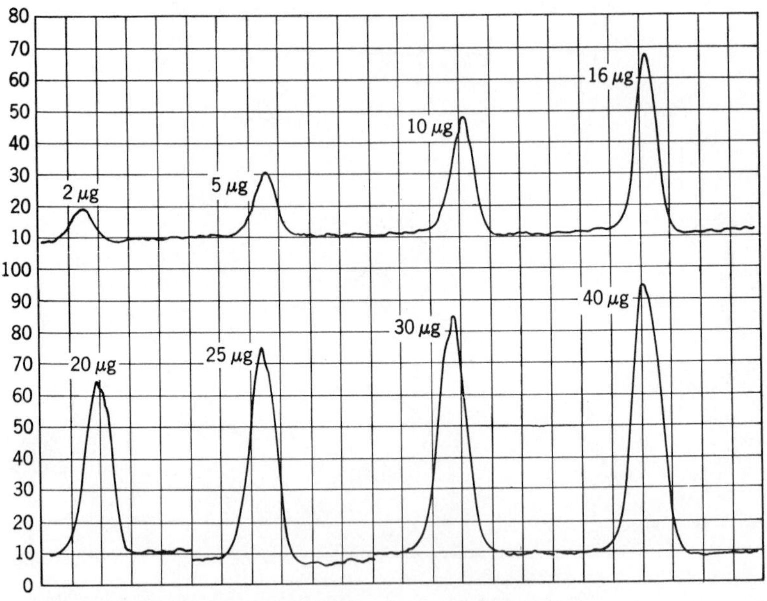

Fig. 6*a*. Typical analog records produced by scanning chromatograms, on which prepared colored derivatives have been separated. In this case the conventional diazo derivative of phenol.

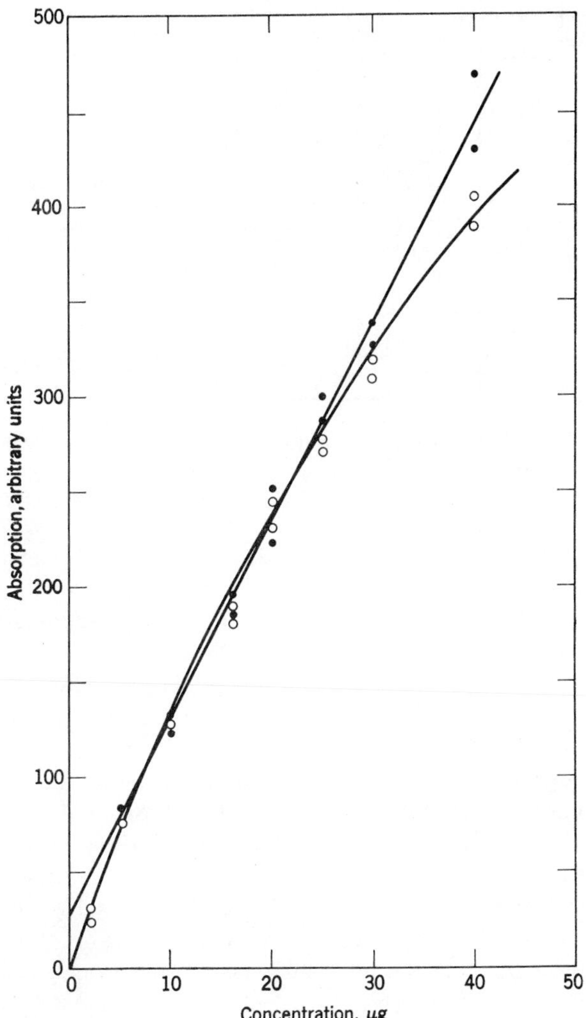

Fig. 6*b*. The calibration curve produced from the above and other scanning records. Note the linearizing effect (closed circles) produced when the paper is rendered translucent by oiling.

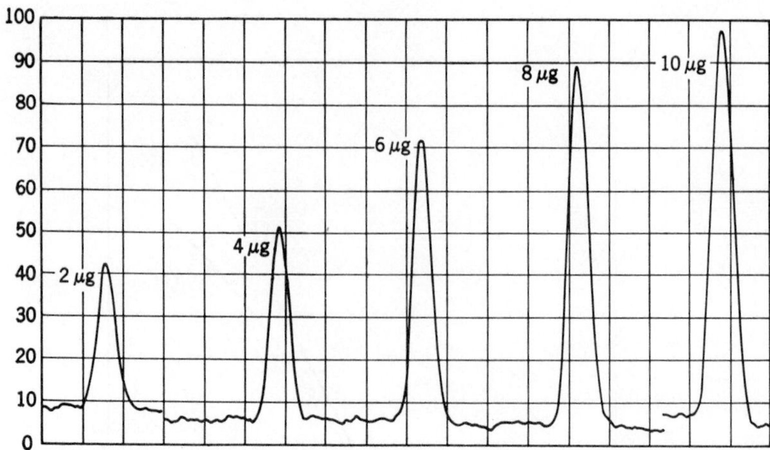

Fig. 7*a*. Typical analog records produced by scanning chromatograms after chemical modification. In this particular case the colorless amino acid alanine was located by reaction with the ninhydrin–cadmium acetate reagent.

and featureless base lines. In the case of colored derivatives (Fig. 6*a*) similar bold peaks are obtained on base lines which, although essentially flat, do contain the ups and downs representing the inhomogeneties of the paper. Figures 7*a* and 8*a* illustrate how the smoothness of the base line disintegrates as less carefully defined conditions of zone location are utilized so that by Figure 8*a* the base line is quite contaminated. It is important to note, however, that in all cases meaningful quantitative data can be obtained by peak area estimation, and only in the case of Figure 8*a* would small peaks possibly be confused with base-line irregularities. The use of mechanically controlled techniques of chemical treatment of the paper chromatograms as described by Bush (46) and/or dual wavelength scanning followed by electronic smoothing as described by Salganicoff et al. (26) would obviously result in much-improved sensitivity and analog traces.

2. Heating Apparatus

In the case of certain fluorescent substances it has been noticed that heating the chromatograms prior to scanning caused, in some

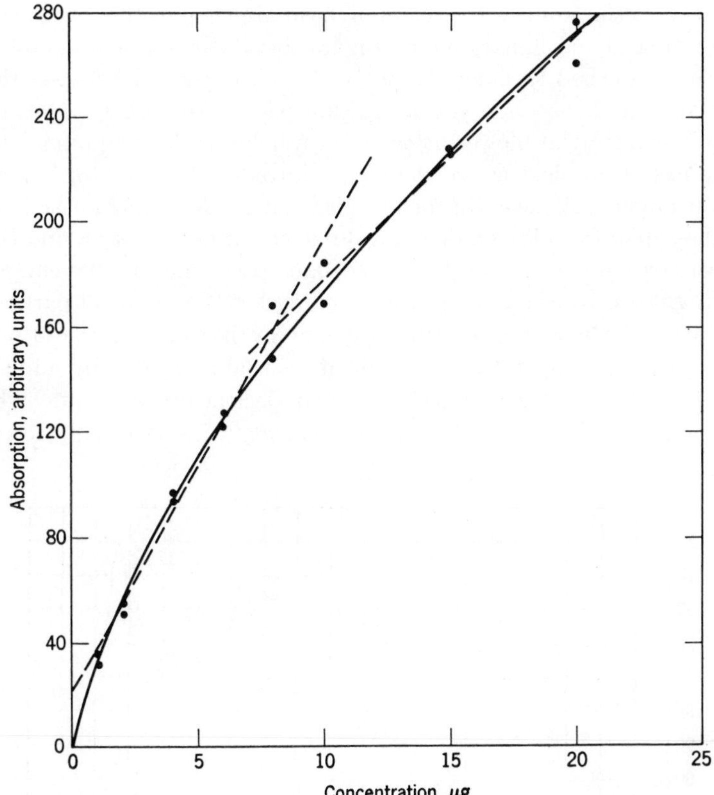

Fig. 7*b*. The calibration curve produced from the above and other scanning records. Note in this case that approximately linear relationships between optical density and concentration may perhaps be used in the ranges 0–8 and 10–20 μg.

cases, a quite considerable increase in the amount of fluorescence. In order to explore this phenomenon in greater detail a heating chamber was constructed in such a way as to be easily fitted to and detached from the front of the scanning head. The device is simply a single-belt drive system in which paper strips may be heated to any desired temperature in the range 30–150°C for a fixed time interval (30 sec) before entering the scanning chamber. The heater which bolts on to the front of the scanning head after removal of the entry plate (Fig. 1) is illustrated in Figure 2. Paper chromatographic strips are pushed through the entry slot until gripped between the belt system on one side and a pressure pad on the other. Hot air from a commercial blower attached to a hollow tube containing the windings of an electric fire element is introduced at the top of the heating device by means of the flexible tubing. A grating in the base of the apparatus allows a downward movement of hot air around the paper strip moving upward at a constant speed (initially 0.9 cm/sec until gripped by the belts in the scanning head) so that all parts of the paper are heated in the same way and for the same length of time. The required temperature is obtained by trial and error by adjustment of the rheostat connected to the electric fire element. The temperature of the second and subsequent strips is usually lower

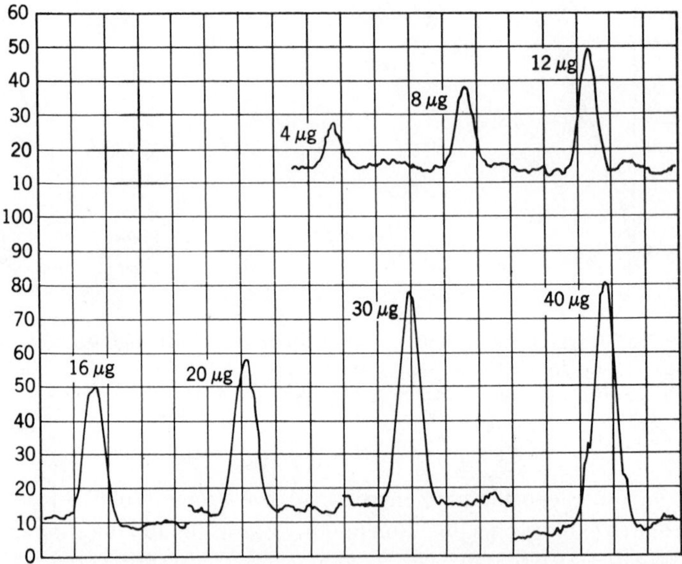

Fig. 8*a*. Typical analog records produced by scanning chromatograms after chemical treatment of the paper. In this case isobutyrohydroxamic acid has been located with ferric chloride solution. These particular records are the worst we have seen to date, and it is of interest to note how the base line tends to disintegrate as less precise methods of location are used.

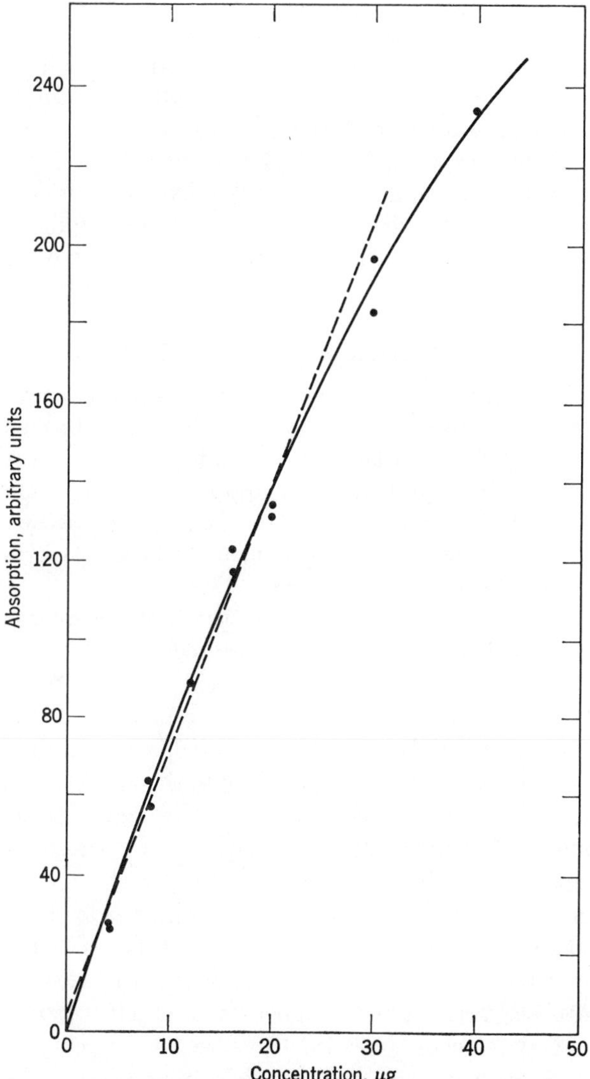

Fig. 8*b*. The calibration curve produced from the above and other scanning records. In this case a reasonably linear relationship between optical density and concentration exists in the range 0–25 μg.

than that of the first. A space of 1 cm is left between the heater and the scanning head to minimize heat conduction, although the paper strip passage is, of course, surrounded and light tight. In order to prevent any upset in the photomultiplier response as the whole apparatus warms, the PM tube is enclosed by a cooling jacket through which cold water is allowed to circulate. Using this apparatus paper strips are uniformly heated so that any particular zone on the strip is heated for the same length of time, acquires the same final temperature, and cools to the same extent before passing across the slits.

3. Apparatus for the Production of Digital Records

Although numerous commercial analog-to-digital converters are available, the one used in our apparatus, illustrated in Figure 9, was developed by Milligan Electronics Ltd. to fit our more specialized requirements. The prototype apparatus shown in Figure 9 also includes the PM voltage supply, the analog and digital gain and zeroing controls, the paper tape punch (Addo Ltd.) as well as the A → D converter itself.

A portion of the PM voltage is fed to the A → D converter where it is compared with an internal ramp-generated voltage; simultaneously with this, pulses from a 200 kc/sec oscillator pass to the binary counters. When the internal generated voltage equals the external applied voltage the pulse train is terminated, at which time the total number of pulses is directly proportional to the applied external voltage. An internal clock mechanism gives the signal "read" so that these pulses, or counts, after amplification, are transferred to the paper tape punch. The state of the pulse and ramp voltage generators are then reset to zero and the whole process is repeated. The rate of sampling of the input voltage is synchronized with the paper tape punch and the whole controlled by an internal clock mechanism which can be varied from an upper value of 110/sec to any lower value. We have chosen (47,48) 10 values/sec as being sufficient to produce an adequate profile of the chromatogram when the scanning speed is 1 cm/sec. For smaller samples, say cellulose acetate electrophoresis strips, a sampling speed of 10/sec is still a reasonable choice since the scanning speed is usually slowed to between 0.4 and 0.1 cm/sec. In order to achieve paper punching speeds in excess of

Fig. 9. Prototype apparatus for the production of digital records. In this photograph the photomultiplier voltage supply (left) and the Addo paper tape punch (right) are clearly visible. The analog and digital zeroing and gain controls are seen at the left of the meter board and the analog-to-digital conversion equipment in front of the meter board. The eight meters indicate the state of the count.

22 characters/sec, as might be the case if EEG records, for example, were being digitized, it would be necessary to use a more sophisticated paper tape punch than the Addo-X employed here.

An 8-bit character on eight-hole paper tape in binary values has been chosen because this allows economies in expensive electronic apparatus. Because of this, therefore, there is no spare tape channel on which to punch parity checks, nor are there any memory units or special circuits to generate "end of measurement" marks. These disadvantages are not serious, since errors on the digital record for each particular chromatogram are detected and corrected in the subsequent computer analysis procedure.

It has been arranged that a "full house" on the binary counters of 255 $\pm$ 1 is equivalent to 2.55 V. It is clear, therefore, that a signal of 0.4 V produces a count of 40 and one of 1.7 V a count of 170. Internal calibration voltages of 0.1 and 2.55 V are provided as a check on the accuracy of the counters.

As the response of the A $\rightarrow$ D converter is linear up to at least 11 V and the scanner amplifier up to about 9 V, it is possible to overshoot the maximum of 255 counts by up to a factor of 3.5. As this respective overshoot is additive in the computer evaluation (Section IV-3), this means that the digital gain controls may be set high so that very small and very large peaks are both analyzed in the computer from the same digital record and at the same time. This is clearly not possible on conventional analog systems where small peaks are reduced to insignificance among the background in attempts to accommodate large peaks on the recorder chart. Similarly large peaks considerably overshoot when the analog controls are set to maximum sensitivity to resolve small peaks. There are times when sequential scanning to adequately resolve large and small peaks on the same chromatogram are difficult and, in any case, such a tedious procedure is a nuisance.

Typical punched tapes are shown in Figures 10 and 11. In Figure 10 the overall appearance of a single chromatogram is illustrated, whereas in Figure 11 a section from a different tape indicates a peak going "over the top" and three other peaks. For convenience and to prevent the confusion which frequently accompanies large amounts of data, the following coding system for each single digital tape is used. In binary characters the number 129, which means start, is punched, then the date, in the order day/month/year, the number of the operator, and last the number of the tape for the day in question (Fig. 10). As it is often convenient to record about 50 chromatograms on a single tape, this last number will rarely exceed 10, representing, even in this case, 500 different chromatograms. A short blank space is usually allowed on the tape before the start of any particular chromatogram, and a series of erases (maximum number of punched holes, i.e., full house) in the range 7–50 inclusive signifies the end of a particular chromatogram (see Fig. 10). The chromatograms are themselves numbered automatically, each set of erases advancing the number by one. The end of the tape is indicated by a sequence of erases exceeding 50 in number. Adequate lengths (10 ft or so) of blank tape are left at the beginning and end of each sequence of records to facilitate loading and unloading operations.

Although it is possible to turn the paper tape back through the punch up to a maximum of 22 characters which can then be erased simply by activation of the appropriate button on the punch, it is simpler to use the digital display buttons on the production model A $\rightarrow$ D converter. On the production apparatus, 8 buttons represent the various binary counters and they can be activated simply by pushing them in. Activation of the paper tape punch by one character

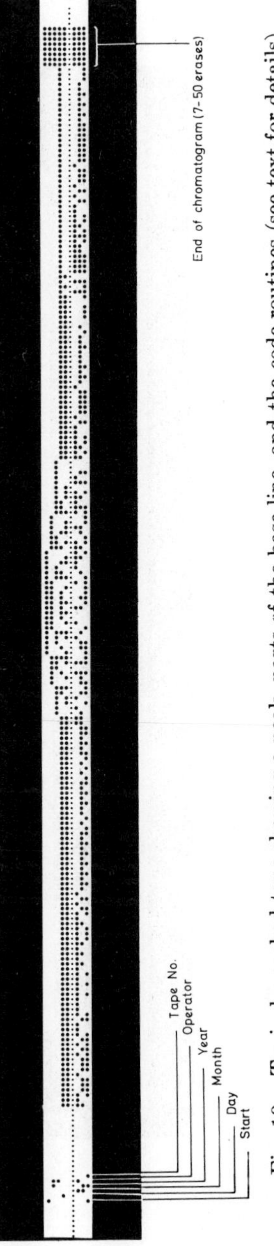

Fig. 10. Typical punched tape showing a peak, parts of the base line, and the code routines (see text for details).

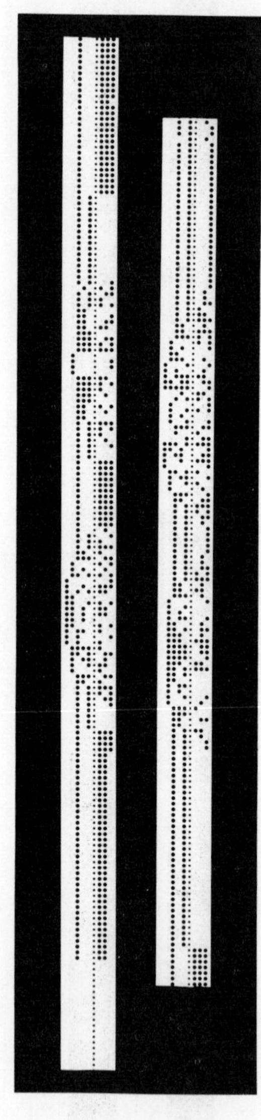

Fig. 11. Section of punched tape showing four peaks, one of which goes "over the top" (see text for details).

followed by release of the digital display button produces a record of the binary value, which was set up. In this way, synthetic data tapes may be produced as well as the computer code previously described.

A commercial scanning apparatus based on the techniques described in this section is now available from the Shandon Scientific Co., Ltd.* It incorporates a carriage system of sample propulsion rather than continuous belts. A production model of the A → D converter is also available from the Milligan Electronics Co., Ltd.†

IV. COMPUTATIONAL PROCEDURE

1. General Description

The simplest, most obvious, and most straightforward method of obtaining the areas under the peaks recorded on the digitized records is to add the digitized values over the relevant range of the record. The effects of a nonzero base line may be allowed for by subtracting the area of a suitable trapezium from the sum thus obtained. Methods utilizing this approach have been described (17,49–54,94); if large quantities of data are to be analyzed, however, it is necessary for such a procedure to have some sort of mechanism with which to decide over what range of the data the integration is to take place. Methods which rely on a change of ordinate value or even change of gradient to do this cannot be used with confidence in situations where there is a wide variation in peak sizes or an absence of relatively flat, stable base lines (e.g., Fig. 8*a*). In such cases it is necessary to make use of the shape characteristics of the peaks to obtain clear information about its behavior as it merges into the base line. This can be achieved by using a curve-fitting procedure to obtain peak parameters which can then be used for a determination of the area. Such procedures have their advantages even for the analysis of single, well-resolved peaks; in the cases of overlapping sequences of peaks they become essential. A computer program utilizing such methods has been written by Ross and Holder of the University of Birmingham; preliminary descriptions have been published (48,56). Since it is written in a language not widely available beyond the local computer (English Electric Leo Marconi KDF 9), it does not seem

* Shandon Scientific Co. Ltd., 65, Pound Lane, Willesden, London, N.W. 10.

† Milligan Electronics Co. Ltd., 36, Clapham Crescent, London, S.W. 4.

desirable to give a complete listing here. The following description of some of the methods used has been adapted from one provided by Ross and Holder.

In view of the requirement that the program should be able to deal with sequences of overlapping peaks, its central element is a nonlinear least squares fitting routine. This routine, in common with other methods of this sort (57–59,95), requires certain preliminary information about the number, locations, and sizes of the peaks contained on the digital record; consequently, the bulk of the program is devoted to techniques devised to produce this information. Since it was intended from the outset that the program be used with other types of data of a basically similar sort (for instance, ion-exchange chromatography), it was necessary to avoid implicitly building in the specifications for any one type of data, but to leave as much flexibility as possible. The way in which this has been achieved has been to read in the information required to specify the characteristics of the data from a particular source as a set of control parameters read from a "meta-data" tape. The flexibility is further improved by constructing the program from several more or less autonomous modules called chapters. In the current version of the program there are seven of these, and their functions are given below.

Chapter 0	Data input and control.
Chapter 1	Data fault detection and correction.
Chapter 2	Data smoothing.
Chapter 3	Location of significant maxima and minima and the fitting of a "by-eye" base line.
Chapter 4	Generation of the first estimates of the peak parameters.
Chapter 5	Division of the record into blocks of overlapping peaks.
Chapter 6	Nonlinear least squares fit to a block of overlapping peaks.

The normal path through the program is from Chapter 0 serially through to Chapter 5 with Chapter 6 as a subroutine of Chapter 5. Chapter 1 may be optionally omitted, and if the first estimates of the peak parameters are provided, Chapters 2–4 may also be left out. The course adopted is specified by control information supplied from the meta-data tape as is the amount of print-out provided. The amount of print-out is usually kept as brief as possible, commensurate with the provision of adequate information (Fig. 12), but in the event that additional diagnostic information is required, this can readily be obtained by alteration of the control words on the meta-data tape (Fig. 13, pp. 355–357). A virtue of this approach is that it has proved possible to retain the bulk of the program unchanged, except for Chapter 0, through all the different types of data met with so far. A more detailed description of the principal algorithms used is set out below.

2. Data Input and Control

Chapter 0 of the computer program is concerned with the data input and the administration of the rest of the program. It reads the meta-data tape and adjusts the internal parameter of the program accordingly. Because it contains the routines necessary to read-in and decode the data from the punched data tapes it must usually be written specifically for a given data type. This chapter therefore, departs from the principle of complete generality which is adhered to in all the other chapters and acts as a sort of standarizing interface between them and the external data sources. It contains options to print or graph the raw data and it can select from the total data those subsets which are to be analyzed. This latter feature is useful if one particular record in a series has been awkward to analyze, in which case it becomes necessary to subject it to a detailed diagnostic print-out or some of the modified types of treatment which may be specified by variation of the control parameters.

3. Detection and Correction of Faults in the Data

The detection and correction of errors in the data are dealt with in Chapter 1 of the computer program. It seemed necessary to include such a section since the 8-bit code used (Section III-3) did not permit error detection on a parity basis. The method tests for discontinuities by serially examining the second differences of all the 3-point sections on the data tape and comparing them with a parameter specified on the meta-data tape. This parameter is virtually constant for any particular data source in that it is a function of the noisiness of the data and the curvature of the sharpest peak. It can be selected by trial and error or by a careful scrutiny of the data (either analog or digital). When a fault is diagnosed the error

```
START 09.39   DATE  27/07/67


AB/MD/ROSS/52.1


PSA  1 358 530 888
PSA  2 363 519 882
PSA  3 484 410 894
PSA  4 471 424 895
PSA  5 730 165 895
PSA  6 67 825 892


   0  1000     2    100    200       0        0        0        0       0   201     201      0       0       0       0       0       0      0

 1.0000, +0   5.0000, +0   1.0000, +2  -4.0000, +0   7.0000, +0   5.0000, -1   1.0000, +2   3.5000, +1   1.0000, +1   1.0000, +1
 0.0000,-99   0.0000,-99   0.0000,-99   0.0000,-99   1.0100, +0   1.0000, +0   0.0000,-99   0.0000,-99   0.0000,-99   5.0000, -2
 1.0000, -2   1.0000, -1   1.5000, +0   1.5000, +0   4.0000, +0   4.0000, +0   0.0000,-99   0.0000,-99   0.0000,-99   0.0000,-99
TRANSFORMATION SET FOR  FLUORESCENT LIGHT


EXPERIMENT OF   16   /   6   /  67        TAPE     1



CHROMATOGRAM          1


   0   DATA FAULTS DETECTED       0   CORRECTED.     DISPLACEMENT        0       OT   0          LM     1      UM   155



SMOOTH BUFFER START   163     FINISH   200        MEAN RESIDUAL    0.0000



PEAK     1  POSITION    180.768  WIDTH    2.3359  HEIGHT   192.210  AREA    1125.4401
BASELINE HEIGHT AT ORDINATE    180  IS    14.5549  SLOPE IS   -0.30416  RESID M.S. =        4.189  NET DEVIATION =      0.000011



CHROMATOGRAM          2


   0   DATA FAULTS DETECTED       0   CORRECTED.     DISPLACEMENT        0       OT   0          LM     1      UM   154
```

```
SMOOTH BUFFER START   164     FINISH   200        MEAN RESIDUAL    0.0000

PEAK     1  POSITION    178.479  WIDTH     2.3267  HEIGHT   193.161  AREA   1126.5608
BASELINE HEIGHT AT ORDINATE   178  IS    19.6267  SLOPE IS   -0.26021  RESID M.S. =        1.876  NET DEVIATION =      0.000238

CHROMATOGRAM          3

    0   DATA FAULTS DETECTED       0   CORRECTED.     DISPLACEMENT       0      OT   0          LM      1       UM    162

SMOOTH BUFFER START   162     FINISH   200        MEAN RESIDUAL    0.0000

PEAK     1  POSITION    183.005  WIDTH     2.3298  HEIGHT   187.470  AREA   1094.8271
BASELINE HEIGHT AT ORDINATE   182  IS    20.8950  SLOPE IS   -0.31793  RESID M.S. =        2.166  NET DEVIATION =     -0.000019

CHROMATOGRAM          4

    0   DATA FAULTS DETECTED       0   CORRECTED.     DISPLACEMENT       0      OT   0          LM      1       UM    205

SMOOTH BUFFER START   151     FINISH   200        MEAN RESIDUAL    0.0000

PEAK     1  POSITION    175.791  WIDTH     2.3220  HEIGHT   188.868  AREA   1099.2925
BASELINE HEIGHT AT ORDINATE   175  IS    26.8602  SLOPE IS   -0.35728  RESID M.S. =        6.440  NET DEVIATION =      0.000005
```

Fig. 12. Brief computer output. This print-out is usually sufficient, since it records all the relevant and salient features of the analysis. On the occasions when the residual mean square or net deviation seems odd or questionable, a more detailed print-out as shown in Figure 13 is easily produced.

corrector routine is called in. This operates by attempting to restore continuity by one of two methods: (*a*) by moving, in a vertical direction, the single point about which the discontinuity occurs (i.e., mispunched point) or (*b*) moving vertically all the points to the right of the fault (i.e., a displacement fault). The choice between these two alternatives is determined by whichever gives the least mean square residual when a second-order polynominal is fitted through the faulty point and five adjacent ones. If the mean square residual for both alternatives is unacceptably large, as in the case, for example, when several faults occur close together, then the record is truncated at that point and all subsequent parts abandoned. This situation usually indicates that something has gone wrong with the production of the digital paper tape record. Our first paper tape punch was a "rogue" and produced numerous mispunched points per record; a more recent model behaves very much more predictably, so that it is now relatively rare to need to call on the mispunched-point correcting routine. Whatever the outcome of the error-correcting routines the program prints a record of what it has done; mispunched points are fault coded MP and displacement DP.

A particular type of displacement, as opposed to a sudden base-line shift, is the one in which the digital record goes "over the top" (Fig. 11, Section III-3). Special attention has been paid to this phenomenon. When the digitizing device runs past its limit of 255 counts it normally continues, lacking only the 256 digit (Fig. 11); the displacement corrector allows such records to be restored to their correct continuous form. Instead of being coded DP such records are coded OT (over the top) and when the record returns below 256 or 512 it is coded UB (under the bottom). At the end of the record there should be an equal number of OT's and UB's and at all times the excess of OT's over UB's must be in the range 0 to 2; if it is not the record is truncated (fault code TR).

4. Data Smoothing

Chapter 2 of the computer program smooths the data and locates on the record those regions where steep rises or falls occur. It is able to deal with a variety of data point densities, peaks, and amount of noise by containing a number of alternative procedures to deal with each of these situations. The appropriate procedure is called

```
AT   261   (   179   209   209    11    36    59   )   I.F.   135   OT    1   RESIDUAL   6.8370, -1   GIVES  -273   -245
AT   270   (    65    45    22   249   219   190   )   I.E.   132   UB    0   RESIDUAL   1.1501, +0   GIVES   278    249
AT   314   (    41    41    41    10    41    41   )   I.E.   121   MP   11   CORRECTION    31.0       GIVES    41     41

    5   DATA FAULTS DETECTED        5   CORRECTED.      DISPLACEMENT       0      OT   0            LM     1       VM   318



SMOOTH BUFFER START   123     FINISH   200         MEAN RESIDUAL    0.0000



MAXS  134.0    145.0    174.0    185.0

MINS  123.0    140.0    163.0    180.0

BY EYE BASELINE IS A =    42.6       R =  -0.024

FIRST ESTIMATES OF PARAMETERS

PEAK   MEAN     S.D.       HEIGHT        AREA

   1    134     1.5057     281.4045    1062.0534
   2    145     1.5080      84.0000     317.5170
BASELINE HEIGHT AT ORDINATE    139   IS    40.5000   SLOPE IS    -0.04167
LOWER END OF FITTED REGION    127       UPPER LIMIT    151

RESID. M.S. =       410.7936       NET DEVIATION =     37.812564

PEAK     1  POSITION    134.436   WIDTH      1.5463   HEIGHT    278.865   AREA    1080.8537
PEAK     2  POSITION    144.998   WIDTH      1.4611   HEIGHT     82.565   AREA     302.3971
BASELINE HEIGHT AT ORDINATE    139   IS     40.6381   SLOPE IS    -0.06967  RESID. M.S. =       15.9292       NET DEVIATION =       0.320309



PEAK     1  POSITION    134.456   WIDTH      1.4722   HEIGHT    292.001   AREA    1077.5664
PEAK     2  POSITION    144.997   WIDTH      1.4564   HEIGHT     82.521   AREA     301.2653
BASELINE HEIGHT AT ORDINATE    139   IS     40.7815   SLOPE IS    -0.07401  RESID. M.S. =        0.5542       NET DEVIATION =       1.155815

PEAK     1  POSITION    134.454   WIDTH      1.4736   HEIGHT    292.407   AREA    1080.1128
PEAK     2  POSITION    144.996   WIDTH      1.4573   HEIGHT     82.551   AREA     301.5509
BASELINE HEIGHT AT ORDINATE    139   IS     40.7145   SLOPE IS    -0.06922  RESID. M.S. =        0.5282       NET DEVIATION =      -0.000959
```

Fig. 13. See caption on page 357.

```
PEAK     1  POSITION   134.454  WIDTH    1.4737  HEIGHT   292.406  AREA  1080.1222
PEAK     2  POSITION   144.997  WIDTH    1.4573  HEIGHT    82.551  AREA   301.5524
BASELINE HEIGHT AT ORDINATE   139  IS   40.7140  SLOPE IS   -0.06919  RESID. M.S. =      0.5282      NET DEVIATION =      0.000000

ORDINATE     OBSERVED      FITTED   OBSERVED-FITTED

   127          41.0        41.55        -0.55
   128          41.0        41.50        -0.50
   129          41.6        41.72        -0.10
   130          45.4        44.37         1.01
   131          60.9        60.01         0.84
   132         114.6       114.26         0.36
   133         219.6       220.82        -1.25
   134         320.8       319.90         0.91
   135         314.1       314.02         0.08
   136         208.7       209.60        -0.89
   137         107.4       106.61         0.77
   138          57.2        56.96         0.28
   139          43.0        43.24        -0.24
   140          40.8        41.12        -0.31
   141          42.2        42.51        -0.32
   142          50.0        50.48        -0.52
   143          72.0        72.73        -0.78
   144         106.6       105.71         0.91
   145         123.1       122.85         0.29
   146         104.1       105.36        -1.26
   147          72.7        72.25         0.47
   148          51.6        49.96         1.61
   149          41.3        41.92        -0.63
   150          39.9        40.18        -0.28
   151          40.0        39.90         0.10


FIRST ESTIMATES OF PARAMETERS

PEAK  MEAN     S.D.       HEIGHT       AREA

   3   174    1.7382     25.7381    112.1426
   4   185    1.2878    167.3810    540.3315
BASELINE HEIGHT AT ORDINATE   178  IS   39.8033  SLOPE IS     0.01242
LOWER END OF FITTED REGION   167      UPPER LIMIT   190



RESID. M.S. =      207.0778       NET DEVIATION =     24.006113

PEAK     3  POSITION   174.394  WIDTH    1.6450  HEIGHT    24.221  AREA    99.8705
PEAK     4  POSITION   185.459  WIDTH    1.3046  HEIGHT   164.354  AREA   537.4524
BASELINE HEIGHT AT ORDINATE   178  IS   39.9526  SLOPE IS    0.07792  RESID. M.S. =     11.9148      NET DEVIATION =    -0.338090
```

```
PEAK      3  POSITION    174.436   WIDTH      1.6191  HEIGHT     25.492  AREA     103.4605
PEAK      4  POSITION    185.481   WIDTH      1.2509  HEIGHT    175.308  AREA     549.6778
BASELINE HEIGHT AT ORDINATE    178  IS    39.2612  SLOPE IS    0.02391  RESID. M.S. =     4.6207     NET DEVIATION =     1.079129

PEAK      3  POSITION    174.436   WIDTH      1.6266  HEIGHT     25.558  AREA     104.2109
PEAK      4  POSITION    185.476   WIDTH      1.2570  HEIGHT    175.369  AREA     552.5732
BASELINE HEIGHT AT ORDINATE    178  IS    39.1575  SLOPE IS    0.01689  RESID. M.S. =     4.5692     NET DEVIATION =     0.005581

PEAK      3  POSITION    174.436   WIDTH      1.6258  HEIGHT     25.553  AREA     104.1341
PEAK      4  POSITION    185.476   WIDTH      1.2564  HEIGHT    175.390  AREA     552.3582
BASELINE HEIGHT AT ORDINATE    178  IS    39.1695  SLOPE IS    0.01782  RESID. M.S. =     4.5688     NET DEVIATION =    -0.000064
```

ORDINATE	OBSERVED	FITTED	OBSERVED-FITTED
167	39.7	38.97	0.69
168	40.0	39.00	1.00
169	40.0	39.10	0.90
170	39.9	39.64	0.26
171	41.9	41.78	0.12
172	47.2	47.39	-0.19
173	56.3	56.38	-0.10
174	64.2	63.75	0.44
175	62.5	63.18	-0.65
176	55.7	55.22	0.50
177	46.6	46.52	0.10
178	41.1	41.48	-0.38
179	39.0	39.68	-0.64
180	39.0	39.29	-0.29
181	39.5	39.54	-0.06
182	39.6	43.06	-3.49
183	58.1	64.41	-6.32
184	129.0	127.24	1.76
185	205.6	202.54	3.03
186	196.2	200.09	-3.90
187	123.8	123.38	0.39
188	67.4	62.66	4.76
189	44.6	42.80	1.77
190	40.0	39.65	0.30

Fig. 13. Detailed computer output as obtained from a more complex chromatogram. Examination of this record shows that four peaks were located, but that initially poor fits were obtained. After several iterations a best fit for peaks *1* and *2* and peaks *3* and *4* was obtained and the values calculated during the various iterations listed. The observed and fitted points in the regions of the peaks are also listed. This computer analysis is of the section of punched tape illustrated in Figure 11. Please note, however, that the order of the peaks is reversed and the number of points has been reduced by a factor of 4 (see Section IV for detailed description).

by computing a formula from data on the meta-data tape. Smoothing is accomplished by fitting orthogonal polynomials in a completely orthodox least squares way. The mean or constant term of the fitted curve gives the value of the smoothed point. The coefficient of the linear term determines the slope. The amount of variability removed by the linear term is compared with the residual mean square (after possibly fitting a variable number of higher order terms if the particular procedure being used requires it). If the amount of variability removed is large compared with the residual mean square, the gradient at that point is labeled significantly steep. This approach allows, to some degree, the noisiness of the record to be allowed for in the assessment of steepness. Since toward the beginning of the record the residual mean square is based on few degrees of freedom and therefore is liable to be inaccurate, the program allows a supplementing estimate to be optionally included via the meta-data tape, along with an appropriate number of degrees of freedom with which the estimate is to be weighted. It has been found that this residual remains very constant from record to record within one data type so that after only a few runs with a new type of data enough information is available to give good starting estimates for the residual. The smoothed data points, the slopes with their significant assessments, and the residual around each point are all optionally printable, as is the distribution of residuals against ordinate.

5. Location of Significant Maxima and Minima and "By-Eye" Base-Line Fit

The procedures in Chapter 2 have produced a smoothed set of data with the graininess and undesirable short-range abnormalities removed so that only the salient features of the record remain. Significant maxima and minima can now be detected.

The first part of Chapter 3 is a logical algorithm which converts patterns of "steep-ups," "steep-downs," and "flats" into sequences of maxima and minima. It does this by identifying each of four following patterns: "up-flat-down" as a maximum, "down-flat-up" as a minimum, "down-flat-down" as a minimum followed by a maximum, and an "up-flat-up" as a maximum followed by a minimum. The latter two cases are necessary to resolve the following situations into two peaks:

The remainder of Chapter 3 fits a "by-eye" base line by drawing a line through two minima in such a way that no minimum lies below the line (it is not, of course, sufficient to draw it through the two lowest minima).

6. First Estimates of Peak Parameters

First estimates of the parameters of each of the peaks detected by previous chapters is obtained in Chapter 4 of the computer program. For each peak the program must find a height, a position, and a width at 0.6065 of the height. The position of the peak is taken as the position of the maximum as determined in Chapter 3. The height is the height of the ordinate at this position above the "by-eye" base line. Determination of the width of the peak is more complicated. If the peak has been well separated from any neighbors the program locates a point about halfway up each side and then finds the horizontal distance of these two points from the maximum. It also determines the heights of these points. From this information it is possible to compute left-side and right-side estimates of the peak width at a height of 0.6065 of the full height. The difference between, or the ratio of, these two widths can be used as a measure of the skewness of a peak.

In the case of overlapping peaks a similar procedure based on the ratio of the height at the mimimum between two peaks to the two adjacent maximum heights is used. To avoid the confusion that arises from not knowing how much of the height of the minimum is attributable to each of the two peaks, an additional assumption is necessary. This assumption is that the right half width of the first peak of an overlapping pair is equal to the left half width of the second component of the pair. This process may be repeated at each minimum to give first estimates of the widths along an overlapping sequence.

7. Subdivision of the Record

This short section, Chapter 5 of the computer program, is concerned with using the first estimates of the peak position and peak width so that the record may be divided into blocks of overlapping peaks. This is accomplished quite simply by considering each minimum in turn and comparing the sum of the widths of the peaks on each side with their separation distance. If the ratio of separation to width is greater than a parameter specified on the meta-data tape, the block of peaks is sent down to the least squares routine in Chapter 6 for further analysis.

8. Nonlinear Least Squares Fitting Procedure

This final chapter (number 6 of the program) is the largest, but since it is principally composed of a least squares fitting routine the description need not be too detailed. The peaks are assumed to be Gaussian so that the function to be fitted to a block of n overlapping peaks is:

$$y(x) = a + bx + \sum_{j=1}^{n} h_j \exp \left\{-(x - \mu_j)^2/2\sigma_j^2\right\}$$

The parameters to be fitted are a and b, the base-line ordinate and slope, the heights h_j, the positions μ_j and the widths σ_j. Since the function is nonlinear in some of the parameters it is convenient to fit linear correction terms to estimates of each of the parameters rather than the parameters themselves. This is done iteratively and the process repeated until the amount of decrease in the residual mean square on successive iterations is smaller than a specified limit. Since the iterations do not always converge to a stable limit, steps have to be taken to minimize the number of situations in which instability occurs. The instability is generally due to high correlation between some of the parameters; thus when the program encounters an increase in the residual mean square it examines the correlation matrix and decides which variables to temporarily ignore. A new attempt is then made to fit the remaining variables. Another method that is used to minimize the effects of correlation is to prevent the program from applying very large corrections at any iteration. The use of these two methods has considerably improved the ability of the program to home onto a set of parameters giving a good fit even in situations where the starting estimates provided by Chapter 4 have been quite poor.

A comprehensive range of optional print-outs is available including the sums of squares and products matrix, its inverse, and the values of the fitted parameters and the observed and fitted points at each iteration. The residual mean square and net deviation are always printed after a set of fitted parameters. The net deviation (which is the algebraic sum of deviations about all points) should of course be zero if the residual sum of squares has been minimized. However, since the process is an iterative one which stops after a finite number of cycles, there is a deviation from zero. This value is included on the print-out to help in deciding whether the Gaussian fit has been good enough.

V. ANALYTICAL RESULTS

Examination of the records illustrated in Figures 5–8 as well as the specific examples mentioned in Section VII show that a quantitative assessment of the peaks produced from most types of separated zones whether colored or fluorescent on paper strips may be achieved using a scanning device. Alteration of the dimensions of the sample carrying system obviously allows a quantitative assessment of most other types of separation on solid support media whether it be electrophoresis on cellulose acetate, paper, or gel, or chromatography on paper or the various thin-layer materials; it would also be equally possible to place a flow-through cell between the slits of a scanning device and obtain both analog and digital outputs from separations on columns by allowing the effluent after suitable chromogenic or fluorogenic reaction to flow through the cell.

Since a Gaussian assumption is valid in all of these applications and an estimate of the goodness of fit is printed when the peak areas are evaluated using the computer, it is likely that a considerable speedup in analysis with a consequent easing of manual effort is possible. In those cases where a Gaussian assumption is not valid some other natural or derived function could be used instead, for instance Lorentzian in the case of gas chromatography separations.

The validity of the Gaussian assumption for the paper chromatographic separations is fairly well established from theoretical predictions (1,59) and practical observation. This latter point is well

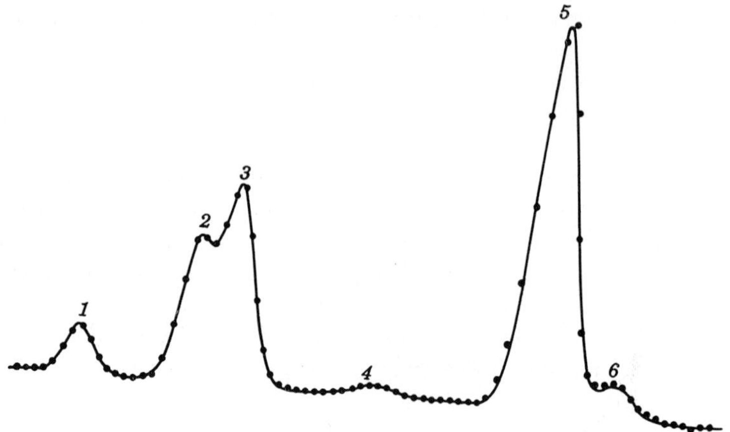

Fig. 14. Validity of the Gaussian assumption. The data shown in this figure was obtained from Harris and represents part of the output produced from an ion-exchange analysis of the free amino acids present in locust haemolymph. As can be seen the Gaussian assumption is valid and the fit between the observed unprocessed digital profile (· · ·) and the computer calculated profile (——) is good.

illustrated in Figures 5–8. Analysis of all the peaks in the sequence shown in Figure 14 took about 20 sec using the computer procedure, and as can be seen the fit between the calculated sequence and the raw data is very close.

The analysis of a single peak by the described computer process (English Electric Leo Marconi KDF 9, the program is currently written in Ferranti Mercury K Code) takes about 6 sec although this time is variable depending upon the state of the data, as can be seen from the following approximate timings of the various operations:

MP location and correction	1 sec
DP location and correction	1 sec
OT and UB location and correction	1 sec each
Data smoothing of a typical chromatogram containing about 500 points	1 sec
First estimates of peak parameters	<1 sec
Final estimates of peak parameters (per peak)	1 sec
Print-out	1 sec

The computer print-out is usually the one shown in Figure 12. It is kept brief to facilitate easy location and compilation of the various peak areas and the mean square residuals. More detailed print-outs (Fig. 13) are optional and can be selected by suitable instruction from the meta-data tape.

A comparison of various techniques for peak area estimation as they were available in this laboratory at the time is illustrated in Table I. It is clear that the computer procedure is not only at least 10 times as fast as any of the manual procedures we tried, but much more accurate. In more complex situations (Fig. 14) the computer evaluation would probably be faster by a factor of up to 100 than any manual technique. If a numerical integrating device were used to assess peak areas it is possible that for single peaks on uncontaminated base lines it would approach the accuracy of the computer procedure. This technique would, however, give quite erroneous results in overlapping situations similar to those in Figure 14 and the "down-flat-down" and "up-flat-up" situations described in Section IV-5.

A possible explanation of the fact that the computer evaluation method is more accurate than the adopted manual procedures in addition to the personal error factor is perhaps because the peak profile as produced on the analog recorder is obtained by an electro-mechanical technique and as such does not respond immediately to the photomultiplier signal, especially at a scanning speed of 1 cm/sec. The production of the digital record, however, is electrical (the paper tape punch although itself electromechanical has plenty of time to record the values held on the counters when the signal "read" is produced from the internal timer) and its response is essentially instantaneous. This probably explains why we frequently obtain computer results from the digital tape which are much nearer to the true Gaussian shape than the analog profile which was produced at the same time. If this explanation is correct, then it is quite feasible to scan at much higher speeds, 10 cm/sec if a fast 110 character/sec paper tape punch is used and a similar point density is required or higher still, if the data is stored magnetically. Although it is unlikely that any paper or sample carriage system could be adequately propelled at those higher speeds there is no reason why the sample should not remain stationary and be scanned by a moving beam of light such as in the flying spot microscope (60–64) or a television

TABLE I

Peak Area Evaluation of the Analog and Digital Records Obtained from the Direct Scanning of Paper Chromatographic Strips

	Planimetry of the analog record		Excision and weighing of a tracing of the analog record		Excision and weighing of the analog record itself		Computer analysis of the digital record	
Time taken to estimate a single peak in seconds	58	58	102	102	77	77	~6	~6
	S.E. %	95% F.L.	S.E. %	95% F.L.	S.E. %	95% F.L.	S.E. %	95% F.L.
Accuracy of the evaluation of a single peak	0.52	±1.1%	1.50	±3.2%	—	—	—	—
Evaluation by repeated scanning of a single chromatogram containing a small amount of separated fluorophore[b]	3.39	±7.3%	4.51	±9.7%	4.12	±8.9%	2.83	±6.1%
Evaluation by repeated scanning of a single chromatogram containing a medium amount of separated fluorophore[a]	2.59	±5.7%	2.54	±5.6%	2.33	±5.1%	1.33	±2.9%

[a] Equivalent to 2.0 μg of a highly fluorescent DNS derivative.
[b] Equivalent to 0.4 μg of a highly fluorescent DNS derivative.

tube (65–68). Although such systems would be technically relatively difficult to construct, they could operate very efficiently.

From Table I it is apparent that the accuracy of the scanning process as assessed by scanning a single strip containing a reasonable amount of fluorophore (10 μg of *p*-tyramine after conversion to a fluorescent derivative as described in Section VII-4; this amount of fluorescence is approximately equal to 2.0 μg of an amino acid as its highly fluorescent dansyl derivative) several times results in an overall error of $\pm 2.9\%$ (95% fiducial limits). The accuracy is reduced as the intensity of fluorescence decreases so that it is $\pm 6.1\%$ (95% fiducial limits) in the case of 2.0 μg of *p*-tyramine fluorogen (equivalent to about 0.4 μg of an amino acid as its dansyl derivative. The overall accuracy of the whole chromatographic and scanning process including strip preparation, separation, drying and, where necessary, zone visualization varies quite widely being $\pm 6.0\%$ (95% fiducial limits)in the case of reasonable quantities of separated prepared derivatives down to $\pm 15\%$ (95% fiducial limits) in the case of colorless urine extracts rendered colored or fluorescent by chemical modification. While these levels of accuracy are less than those obtained using conventional spectrophotometers, spectrophotofluorimeters or ion-exchange types of analyses (4,69) it must be remembered that at this time the process of chromatogram scanning has not acquired the sophistications built into the more conventional apparatus. It does seem likely that a scanning device incorporating dual wavelength cross-scanning techniques and electronic smoothing will begin to approach the accuracy of the more developed systems. The lack of accuracy is offset to some extent by the fact that in most cases answers correct to within ± 5–15% are meaningful when they are obtained from a relatively simple and cheap instrument and in the quantities usually produced in chromatography. In addition, a fairly specific zone localization technique coupled with the specificity offered by a suitable chromatographic separation means that one can be reasonably happy that a single substance is being analyzed rather than a mixture of related substances or contaminants, as is often the case in the spectrophotometric or spectrophotofluorimetric analysis of an extract.

It is now quite feasible to omit the log conversion device in the case of absorptiometric analyses as described in Section III-1 by simply including a log function in the computer program. In this way the

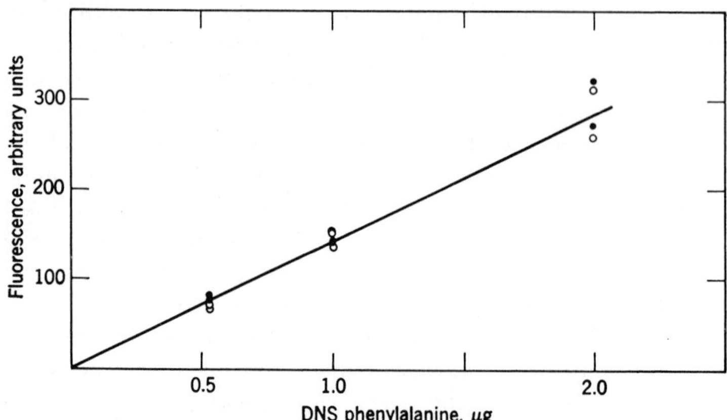

Fig. 15. Effect of scanning both sides of a chromatogram. In order to obtain this calibration curve, samples of DNS phenylalanine were separated overnight on Whatman No. 2 quality paper in the solvent system light petroleum/acetic acid/water, 10:9:1 (v/v): (○) the amount of fluorescence recorded when the strips are scanned with the pencil marks uppermost and (●) the fluorescent values when these same strips are scanned with the pencil marks down.

data are automatically logged and correct peak areas produced. While we have not yet incorporated this into analyses from our laboratory the technique works well in the case of ion-exchange data as produced from the usual type of Moore and Stein amino-acid analysis (70).

It is worth mentioning that R_f or R_x (x being a reference compound) data could be produced from the digital scanning record. Also, it is feasible to search back through digital records of previous chromatograms for concentrations, or other information, of a particular substance or set of substances as obtained on different occasions. Most of us would like to have the opportunity of recalling past experimental data for further perusal.

Although by an arbitrary decision we always scan our paper strips in the same way (pencil markings up, solvent front end to enter first), the results illustrated in Figure 15 would seem to suggest that in the case of separated fluorophores, at any rate, this is not necessary.

In our early scanning days we sometimes obtained duplicate values similar to those illustrated in Figure 16. The dotted line represents the best fit and was, of course, the one used, until it was noticed that

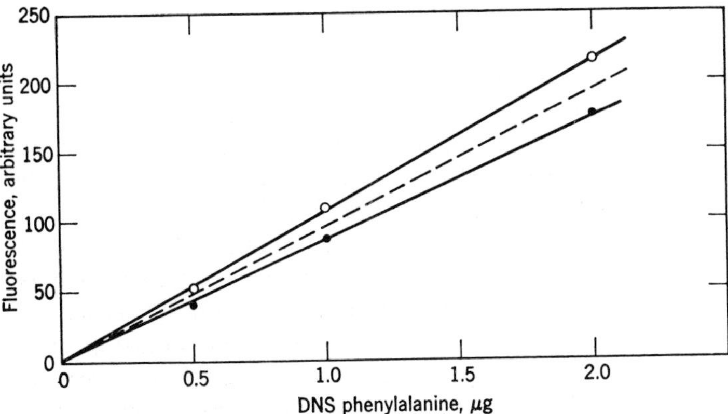

Fig. 16. Effect of paper thickness on the amount of recorded fluorescence. The curves shown above were obtained from the same quality of chromatographic paper. The dotted line was accepted as the true result until it was noticed that the papers, although possessing virtually identical chromatographic properties, could be separated into two distinct thicknesses. The upper line represents the thin variety and the lower line the thick.

the texture of the chromatograms seemed different. It was, in fact, possible to separate the strips into two distinct groups, thick and thin, simply by feeling them between finger and thumb. Measurements with a micrometer or the amount of light transmitted revealed that to a large extent the two groups were quite distinct from a thickness point of view; their chromatographic properties, however, were identical. The chief offender among the various grades of paper examined was the fast-running Whatman No. 4 quality. Complaints to the manufacturer quickly brought about an investigation and the thickness of these strips is now quite uniform.

VI. HEATING PHENOMENON

During investigations on the separation and analysis of amino acids as their so-called dansyl derivatives (Section VII-1, for details) it was noticed (72) that the amount of fluorescence obtained from the acidic amino acids was considerably less than might have been expected when these derivatives were directly compared with an equimolar amount of DNS phenylalanine, for instance. Attempts to increase the level of fluorescence by neutralizing the acid groups with alkali failed to effect an improvement, but on one occasion the treated chromatograms along with control strips were dried in an oven as opposed to a period in the hood at room temperature. After drying at the elevated temperature the strips were scanned (before complete cooling had taken place) to record the amount of fluorescence and it seemed as though the level of fluorescence on the control strip was higher than it had been earlier in the day. In further investigatory experiments strips containing amounts of separated dansyl derivatives were scanned after drying at room temperature for three hours or so and then after heating for different periods in an oven. It was seen that a scanning record as obtained as quickly as was feasible after removal of the strips from the oven resulted in a much increased peak size and that this increased amount of fluorescence decayed with time so that it was approximately the unheated value after about half an hour. In the case of dansyl derivatives of amines and amino acids the optimum temperature and time of heating as obtained by trial and error seemed to be above 70°C for more than 15 sec. Much lower values, however, such as 2 sec at 140°C, or 30 sec at 30°C, also caused appreciable increases in the amount of fluorescence. Two seconds at 140°C simply means putting a chromatogram into the oven and withdrawing it immediately.

Qualitative tests on this phenomenon clearly seemed to implicate quenching of the fluorescence by water, because dessication of the strips *in vacuo* or over P_2O_5 caused similar increases in the amount of fluorescence. By alternately heating the strips in an oven followed by damping in the steam jet produced from the spout of a boiling kettle with measurement of the amount of fluorescence after each operation showed that the process was quite reversible apparently for an indefinite time.

As any method of increasing sensitivity, especially for the very highly fluorescent DNS derivatives, is useful, some effort was invested in trying to understand the phenomenon more fully; consequently the heating attachment (73) described in Section III-2 was constructed and the photomultiplier housing of the scanning device was surrounded by a cooling jacket. In all subsequent scanning operations involving heated strips tap water was circulated through the cooling jacket to keep the PM tube cool.

In a carefully controlled experiment 2.0 μg of the DNS derivative of phenylalanine, after separation overnight in the solvent system light petroleum/acetic acid/water, 10:9:1 (v/v) on Whatman No. 4 quality paper was, after recording the amount of fluorescence after drying at room temperature for 5 hr, heated for 1 min at 135°C. Scanning records were then obtained at intervals over the next 30 min. The precise time of each "scan," after removal of the strip from the oven, was recorded from a stop watch as the peak maximum moved across the slits of the scanning device. The result of this experiment is illustrated by the points about the upper curve in Figure 17. A section of the same paper strip (cut from the chromatogram just before heating) was subjected to a similar sort of procedure in which the section was weighed, heated for 1 min at

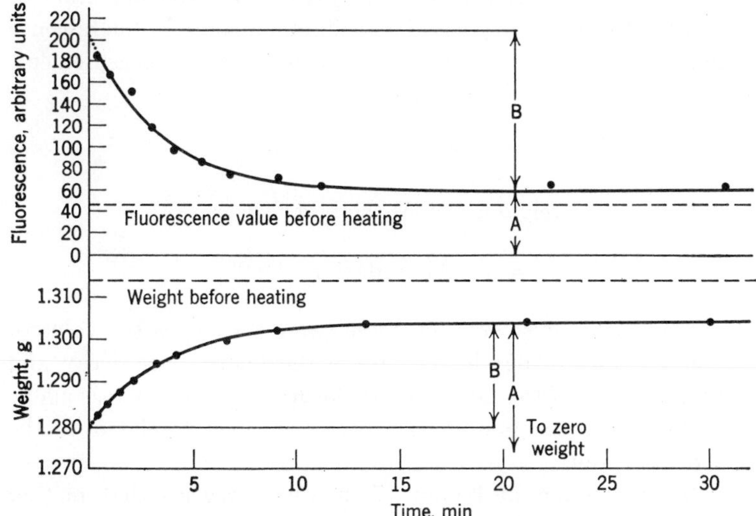

Fig. 17. Decay of fluorescence and reabsorption of water after a short period of heating. DNS phenylalanine (2 μg) was separated in the solvent system light petroleum/acetic acid/water, 10:9:1 (v/v) for 14 hr on Whatman No. 4 paper. After drying at room temperature for 5 hr, a small (5 × 5 cm) blank section was cut from the lower end of the chromatogram. The upper section, after scanning to record the unheated fluorescence value, was heated in an oven for 1 min at 135°C and then scanned at intervals. The points about the upper curve illustrate the values obtained. The other small section of the air-dried chromatogram was weighed and then, after heating for 1 min at 140°C, was weighed again at intervals during the next half hour. The points about the lower curve illustrate the values obtained. The continuous lines through the experimental points represent a computer-calculated best fit. Further details are given in the text.

140°C, and then reweighed at intervals over the next 30 min. The results of this experiment are shown by the points about the lower curve in Figure 17. Since in numerous experiments the shape of the curve obtained when the amount of fluorescence was plotted against time was approximately exponential, it seemed reasonable in the particular experiment illustrated in Figure 17 to fit a curve of the type:

$$y = A \pm Be^{-ct}$$

The best fit to the experimental data was very kindly obtained for me by Mr. R. L. Holder, using a computer, and is represented by the upper and lower continuous lines in Figure 17. Extrapolation of these curves to zero time gives the maximum amount of fluorescence obtainable on heating. In the case of the fluorescence measurement, the best fit is represented by the equation

$$y = 60.6 + 149e^{-0.0051t}$$

In the case of the weighing values the best fit was

$$y = 1.3041 - 0.0247e^{-0.0049t}$$

The point of interest between these two equations is the value of the time constant 0.0051 in the case of the fluorescence decay and 0.0049 in the case of the moisture reabsorption. It seems reasonable to conclude that the fluorescence decay is directly related to the water reabsorption.

Further inspection of Figure 17, however, reveals that neither curve returns to its unheated level; this is attributed to the fact that the two types of water (loose and tightly bound) (74) contained within the cellulose fibers of the paper chromatograms are lost during the period of heating and only the tightly bound is quickly reabsorbed. It may be that other volatile constituents or quenching substances are playing a role in this phenomenon, but further, more-detailed experiments would be required before this could be established.

The increase in fluorescence over the unheated value is given by

$$(A + B)/\text{unheated value} = 209.6/47 = 445\%$$

This represents a substantial increase in sensitivity, and to make use of this effect the heating attachment described in Section III-2 was constructed. Matters are arranged so that chromatographic strips on which susceptible fluorophores have been separated are heated usually to about 100°C during their passage (30 sec) through the attachment. The rate of cooling, moisture reabsorption, and thus decay of fluorescence is minimized and constant for all parts of the same strip and for different strips. Using this arrangement it is possible to obtain calibration curves similar to the one shown in Figure 18, in which nanogram quantities of various DNS derivatives are recorded.

Liberties were taken with the cell housing of an Aminco Bowman spectrophotofluorometer by blowing cold and hot air, in turn, across the fluorescent zone (DNS benzylamine) on a segment of a paper chromatogram, and plotting the absorption and fluorescence spectra; no appreciable spectral shift occurred when the paper was heated. The slight fluorescence shift observed (511 mμ → 497 mμ, maxima, uncorrected) certainly will not explain the large increase in the fluorescence yield.

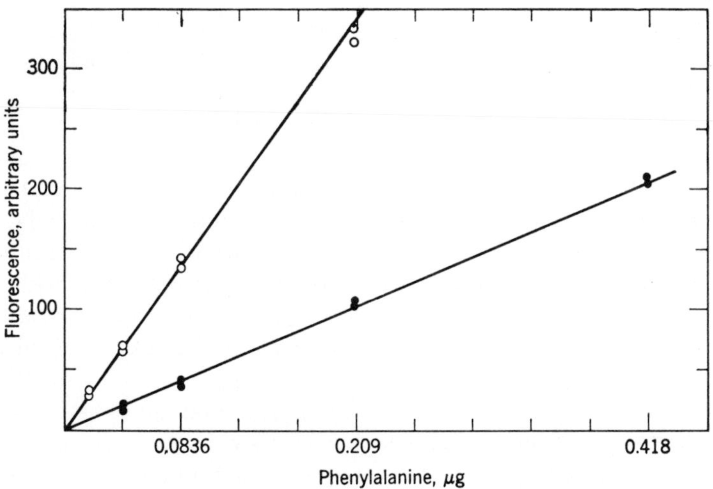

Fig. 18. Heated and unheated calibration curves. Samples of DNS phenylalanine separated overnight were scanned after air drying for 3 hr (●). These same strips were then rescanned after their passage through the heating device (○). In this particular experiment an increase of 320% was recorded.

Further investigation of the heating phenomenon produced some results which might be expected and others which were not. Table II shows the cold and hot fluorescence values and the mean percentage increase on heating, obtained when duplicate samples of a fluorescent derivative (2.56 μg DNS serine) was separated in benzene/diisopropyl ether/acetic acid/water, 1:1:1:1, (v/v), for the times shown on different types of Whatman chromedia (75).

The chromatograms were prepared and dried during 5 hr at room temperature as described in Section II and the scanning apparatus sensitivity controls were kept constant throughout the experiment except in the case of glass fiber strips where it was necessary to substantially increase the gain in order to obtain a meaningful scanning record. The various types of papers used are merely to illustrate the sort of sensitivities and heating increases achieved from the different media, and the list shown in Table II is not meant to be taken as an indication that these materials are necessarily suitable for the separation of DNS derivatives. In fact, most of them are not. An adequate separation of all the naturally occurring DNS amino acids, however, can be achieved on the conventional type of paper (76).

Inspection of the unheated fluorescence values listed in Table II shows that the type of support medium appreciably affects the amount of fluorescence recorded from the fluorescing zone. The percentage increase on heating shows how these vary, being quite high in the case of the usual nonimpregnated papers and decreasing quite considerably in the case of most of the loaded strips. Further points of interest are the remarkably high increases obtained for the cellulose phosphate-loaded P81 strips and the slight decrease observed in the case of the noncellulose glass fiber strips. This is further evidence of the quenching of the fluorescence by water, since one would expect water to have an additional effect in the case of cellulose-loaded strips, but no effect on glass.

A more unexpected result was obtained and investigated: the increase obtained on heating the same fluorescent substance separated, in duplicate, on the same quality paper, but in different solvent systems. DNS taurine (1.0 μg) was separated on Whatman No. 2 quality paper in the solvent systems shown in Table III, and then dried during 5 hr at room temperature. All the results listed were obtained at a constant sensitivity level in the scanning apparatus.

TABLE II

Heated and Unheated Fluorescence Values as Obtained on Different Types of Paper Media

Type of Whatman paper	Time of separation, hr	Intensity of fluorescence, arbitrary values		Mean increase, %
		Unheated	Heated	
No. 1	14	16.3	101.0	600
		18.0	105.3	
No. 2	14	17.0	92.9	550
		17.0	94.0	
No. 3	6	16.1	53.8	356
		14.4	55.0	
No. 3 MM	6	15.5	63.0	390
		17.5	66.0	
No. 4	6	23.9	102.0	400
		24.0	90.5	
No. 54, acid hardened	6	24.0	100.6	422
		25.0	106.4	
No. 541, acid hardened ashless	6	31.0	105.2	378
		25.0	106.4	
No. ST.82, silicone loaded	14	58.4	94.4	155
		59.1	87.6	
No. DE.81, ion-exchange diethylaminoethyl cellulose paper	14	65.9	119.5	164
		80.2	122.0	
No. AH.81, aluminum hydroxide loaded	14	17.1	—	176
		16.4	29.5	
No. AE.81, ion-exchange aminoethyl cellulose paper	14	18.5	42.9	202
		25.4	45.9	
No. SG.81, silica-gel loaded	14	4.4	—	577
		5.3	28.3	
No. ET.81, ion-exchange ecteola cellulose paper	6	54.5	108.7	193
		54.9	101.5	
No. P.81, ion-exchange cellulose phosphate paper	14	17.7	139.0	866
		16.0	152.0	
No. GF.81, glass fiber[a]	4	94.0	86.0	91.3
		101.0	92.0	

[a] In order to obtain these values, a shorter separation time of 4 hr followed by scanning at a higher sensitivity was necessary.

In the case of the heated fluorescence values, as recorded from the chromatograms obtained from the conventional type of solvent system containing an aqueous and nonaqueous part, it can be seen that the values are all about the same. It may perhaps be concluded from this that in absolute terms the amount of fluorescence obtained at the elevated temperature is constant, as would be expected from the design of the experiment. The variations seen at room temperature are probably caused by variations in the total amount of water and perhaps by volatile quenching substances deposited in the paper during the solvent development. The much higher fluorescence values at room temperature and the lower values at the elevated temperature in the case of the monophasic aqueous buffer system are not understood at this time. Although the solvent systems shown in Table III always produce this effect, sometimes with even wider discrepancies at the unheated level, several further experiments in many different solvent systems would be necessary before the above observation could be considered as a general rule.

TABLE III

Heated and Unheated Fluorescence Values as Obtained from Different Solvent Systems

Nature of solvent system	Intensity of fluorescence, arbitrary units		Mean increase, %
	Unheated	Heated	
Neutral	53.0	201	366
benzene/methanol/water	57.5	204	
2:1:1			
Alkaline	56.3	217	392
isopropanol/ammonia/water,	55.7	222	
20:1:2			
Aqueous salt, 0.75M sodium phosphate	121	172	135
buffer solution (71)	125	160	
Acid			
1. Butanol/acetic acid/water,	69.5	199	295
240:27:133	64.7	197	
2. Diisopropylether/methylethyl	42.5	195	390
ketone/acetic acid/water,	60.8	208	
8:2:5:5			

Finally, a few naturally fluorescent substances (1–5 μg) available off the laboratory shelf were separated, in duplicate, in the appropriate solvent system on Whatman No. 2 quality paper, and the amount of fluorescence recorded after room-temperature drying, and after heating. The results are listed in Table IV. In all cases the fluorescent zones on the paper chromatograms were activated at 365 mμ and the fluorescence passing a Chance 0Y8 colored glass filter (cut off below 440 mμ) measured.

It was a little disappointing, though perhaps not surprising, to find that only a few of the fluorophores increased their fluorescent yield on heating. In some cases heating is obviously disadvantageous since the sensitivity is markedly reduced. For any particular fluorophore it will be necessary to establish by trial and error the effects produced on heating. In the case of all dansyl derivatives,

TABLE IV

Heated and Unheated Fluorescence Values of Some Different Fluorophores

Fluorophore	Solvent system	Intensity of fluorescence, arbitrary units		Mean increase, %
		Unheated	Heated	
Xanthurenic acid	Butanol/acetic acid/water, 240:27:133	36 33	67 68	196
Kynurenine	Butanol/acetic acid/water, 240:27:133	310 322	280 287	90
Anthranilic acid	Isopropanol/ammonia/ water, 20:1:2	74 66	72 65	98
Quinine sulfate	Light petroleum/toluene/ acetic acid/water, 133:66:170:30	87 89	98 105	115
Rhodamine	Benzene/acetic acid/water, 2:1:1	38 34	51 52	143
Fluorescein sodium	Benzene/acetic acid/water, 2:1:1	32 31	24 24	76
Acriflavine	Butanol/acetic acid/water, 240:27:133	76 83	87 85	108
Acridine orange	Toluene/acetic acid/water, 10:7:3	85 78	34 35	42

however, very considerable improvements in sensitivity are achieved and quantitative assessments in the nanogram region become feasible. As it is possible to prepare DNS derivatives for a great majority of the interesting biogenic substances (76–93) the possibilities of analysis in these extreme regions is obvious.

Seiler (89) has recently shown that a similar increase on heating or drying is achieved for DNS derivatives separated on thin layers of silica gel after spraying the layers with triethanolamine/isopropranol, 20:80 (v/v).

Although it will not always be sensible or useful to include a heating or drying stage in the quantitative assessment of chromatographically separated fluorescent substances, especially when using different types of scanning apparatus, this section has illustrated some of the pitfalls that can arise during the direct scanning assessments of chromatograms. In susceptible cases, adequate drying of the paper chromatograms is essential; in England, at any rate, it is quite possible to record wide changes in relative humidity within a few hours, and these would obviously affect the levels of fluorescence recorded. In the analysis of different chromatograms the apparent errors that might arise from assessing strips of different qualities or those obtained from different solvent systems must also be borne in mind. As a rule it is well to consider direct scanning techniques as a relative method of analysis, and to require that strips containing standard amounts of absorbing or fluorescent material be obtained at the same time and in the same manner as the strips containing the substances under investigation.

VII. SOME SPECIFIC EXAMPLES

Very many applications of direct scanning techniques to many diverse problems have been reported recently (3,13,18,19,96–115). These specific applications have involved the analysis of colored and fluorescent substances as separated on all types and dimensions of supporting media. In this laboratory we have concentrated principally on the conventional type of paper strips rather than the smaller and faster procedures simply because time has not been a crucial factor and because the use of those strips has allowed an investigation of other interesting phenomena (see Section VI). Although a scanning apparatus arranged for the analysis of colored zones on chromatograms is available, we have preferred to opt for a fluorescent analytical procedure if this is at all possible. By doing this, much improved scanning records are obtained (compare, e.g., Figs. 5 and 8) and, of course, much greater sensitivity of detection is achieved.

Scanning records and their associated calibration curves as obtained simply as examples of direct scanning of paper chromatograms are illustrated in Figures 5–8. A few examples of the use of the apparatus in certain applications from our current research program are discussed below.

1. DNS Derivatives

Although considerable interest has been shown in these derivatives because of their intense fluorescence and extreme sensitivity, most of the publications to date (76–81,83,85,87–92,116) have been descriptions of separative procedures for different groups of substances or the resolution of overlapping pairs or groups. Although linear calibration curves even at the submicrogram level may be obtained (3,76,77,82,89) when prepared derivatives are separated, there have been no publications to date about the quantitative use of these derivatives in applications to biological extracts or fluids. The elegant studies of the protein chemists (117–119) are mainly concerned with a qualitative identification of the constituents of protein hydrolysates, and the analysis of the amino acids present in the superfusate of the cat cerebral cortex by Crowshaw et al. (86) was only semiquantitative. One of the difficulties (120), as far as amino acids are concerned, seems to be that during the dansylation reaction a further molecule of the dansyl reagent combines with the acid group of the amino acid, and the compound thus formed then hydrolyzes in the alkaline conditions of the reaction medium to form 1-dimethylaminonaphthalene-5-sulfonamide, 1-dimethylaminonaphthalene-5-sulfonic acid, and the aldehyde (or ketone) with one carbon atom less than the parent amino acid. In certain cases the formation of the aldehyde or ketone and other artefacts can be quite appreciable. The situation is even more complex when more than one amino acid is present. In the case of amines, however, this problem does not occur and it is possible to obtain nearly stoichiometric conversions to the DNS derivatives (Table V).

TABLE V

The Per Cent Conversion of Some Amines to Their Respective DNS Derivatives

Amine	Intensity of fluorescence, equimolar amount of crystalline standard = 100
β-3,4-Dimethoxyphenylethylamine	101.1 ± 6.4
Ethylamine	101.5 ± 4.3
Diethylamine	99.4 ± 7.4
Benzylamine	100.2 ± 3.3
Ammonia	96.1 ± 1.6
Piperidine	96.6 ± 3.7
Pyrrolidine	90.8 ± 7.3
2-Phenylethylamine	100.6 ± 7.1
Aniline	99.0 ± 1.9

Solutions suitable for quantitative analysis can be conveniently prepared by mixing, in ignition tubes, 25, 50, and 100-μl aliquots of an amine dissolved in 0.1*M* sodium bicarbonate (approximately 1 mg/ml) followed by the addition of 0.1*M* $NaHCO_3$ to make the final volume 100 μl. To these solutions 100 μl of the DNS reagent (1 mg/ml in acetone) is added. After shaking and standing the stoppered tubes at room temperature for 1 hr (overnight if preferred) excess acetone (800 μl) is added to precipitate the sodium bicarbonate. After gentle centrifugation in a bench centrifuge, aliquots (50 μl) of the supernatant may be applied directly to the origin line of paper chromatographic strips; alternatively, the supernatant may be transferred to clean tubes, in which case the precipitate is washed with a further 500 μl of acetone, and the final solution made up to 1.5 or 2.0 ml with acetone. After chromatographic separation in suitable solvent systems (76,77,121) and adequate drying, the chromatograms are scanned (excitation 3650 Å primary filter, Chance OXI; fluorescence 5100 Å secondary filter, Chance OY8) either at room temperature or after heating. Typical scanning records and calibration curves are shown in Figures 5 and 18. The conversion to the DNS derivative for several different amines, as assessed by comparing equimolar amounts of the derivative prepared in solution with a crystalline standard dissolved in ethanol, are shown in Table V.

It can be seen that, considering the errors involved in handling such small quantities of material and those involved during chromatography and scanning, the conversion to the DNS derivatives is virtually complete in all cases but one. Unfortunately when the technique was applied to amine extracts obtained from urine either by solvent extraction (122) or ion-exchange (123) chromatography, the scanning records obtained were nearly always poor. Inspection of the paper chromatograms in ultraviolet light revealed considerable fluorescent streaking along the length of the strips. Furthermore, the recovery of standard quantities of amines added to the urine was extremely variable.

Since the recovery of synthetic amine mixtures through the ion-exchange procedure was reasonably good, usually in the range 85–100%, it is concluded that the use of the DNS derivatives for quantitative assessments of the various urinary amino constituents is not really feasible. It is likely that acceptable results will be obtained in the cases of simple mixtures or relatively specific extraction procedures and in these instances the specificity and supreme sensitivity of the method will be very valuable.

2. Absorptiometry of a Urine Extract

This example is included because it involves an absorptiometric analysis of a urine extract and also because such a good result was obtained from what must be considered a very crude manner of chromogen formation. It is likely, therefore, that useful quantitative data may be obtainable from many qualitative procedures which would normally be considered too crude.

In 1962 Friedhoff and van Winkle (122) reported that alkaline chloroform extracts of urine from schizophrenic patients contained a substance, later identified as β-3,4-dimethoxphenylethylamine (DMPE) which exhibited a pink coloration, on paper chromatograms, after treatment with ninhydrin and Ehrlich's reagents. We repeated this work with a view to quantitating the daily excretion of this so-called pink spot [identified in our patients as *p*-tyramine (124)].

Extracts suitable for chromatography may be obtained in two ways; the first involves solvent extraction (122) and the second, separation on an ion-exchange column (125). In the former, urine containing 300 mg of creatinine (usually about 300 ml) is adjusted to pH 2.0 (narrow range indicator paper or the glass electrode) with concentrated HCl and then extracted three times with chloroform (3 × 150 ml). This $CHCl_3$ is discarded, although it may be used for an investigation of urinary acid metabolites. The extracted urine is then adjusted to pH 10.0 with 10*N* caustic soda and reextracted with $CHCl_3$ (3 × 150 ml). The $CHCl_3$ extract, after drying over sodium sulphate (20 g) is rotary evaporated to dryness at 50–60°C and the dried extract dissolved in 300 μl of $CHCl_3$ (it is usually sensible to transfer the reduced $CHCl_3$ volume to a 50-ml flask toward the end of the evaporation). This extract is applied as bands (Section II-1) to three Whatman No. 2 quality paper strips in the proportions 200:50:50 (v/v); 10 μg of an authentic amine solution (*p*-tyramine or DMPE) is applied over one of the 50-μl strips. This over-spotted strip acts as a convenient standard in the final visual assessment as to whether the pink spot is present or not. The chromatograms are developed overnight in the solvent system butanol/acetic acid/water, 12:3:5 (v/v), and after removal from the chromatography tank dried during 3 hr at room temperature.

In the second extraction procedure urine (100 mg creatinine) is percolated through an ion-exchange resin column (5 × 1.2 cm) in the hydrogen form described by Kakimoto and Armstrong (125). After washing the column with water (10 ml), 0.1*M* sodium acetate (20 ml), and then water again (10 ml), the absorbed phenolicamine fraction is eluted with 20 ml of aqueous ethanol/ammonia (1*N* NH_4OH in 65% ethanol) and rotary evaporated to dryness at 50–60°C. The dried residue is dissolved in 500 μl of 70% ethanol and aliquots (200, 100, and 100 μl, respectively) applied as bands to Whatman No. 2 quality paper strips. Ten micrograms of *p*-tyramine or DMPE is over-spotted on one of the 100-μl strips to aid in assessment. After development overnight in butanol/acetic acid/water, 12:3:5 (v/v), the chromatograms are dried during 3 hr at room temperature.

Both the above extraction procedures yield only basic substances existing in the urine in the unbound form; conjugated amino compounds may be obtained by repeating the above extractions after hydrolysis (30 min at pH 1.0 on a steam bath followed by neutralization) of either the neat urine or the initial column effluent.

The dry chromatograms are then dipped through ninhydrin-pyridine reagent (126) [0.2% in acetone/pyridine, 9:1 (v/v)] and after allowing evaporation of the excess acetone, heated at 120°C for 90 sec. At this stage numerous multicolored spots are visible on the chromatograms and a meaningful scanning record could not be obtained. By dipping the ninhydrin-developed strips through Ehrlich's reagent [2% *p*-dimethylaminobenzaldehyde in acetone/conc. HCl, 9:1 (v/v)] most of the ninhydrin-positive spots are bleached out and others change color. The very predominant yellow urea spot appears quickly at R_f 0.40. In positive urines a pink color develops at about R_f 0.60. This pinkness is at a maximum after 45 min.

Numerous substances can produce a pink coloration at this position on BuOH-HAc chromatograms, but if the color were solely due to DMPE or *p*-tyramine a quantitative value could be obtained by comparing the appropriate peak areas from scanning records similar to those shown in Figure 19, with a calibration curve (Fig. 20) obtained at the same time. The analog record sections shown in Figure 19 were obtained by scanning paper chromatograms at 5300 Å (interference filter peak transmission 5300 Å, band width

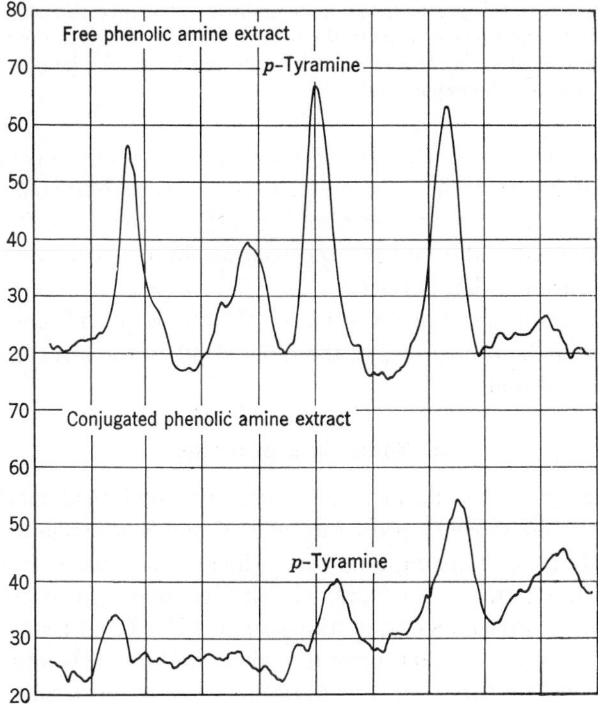

Fig. 19. Section of an absorptiometric scanning record obtained from a urine extract (see text for details).

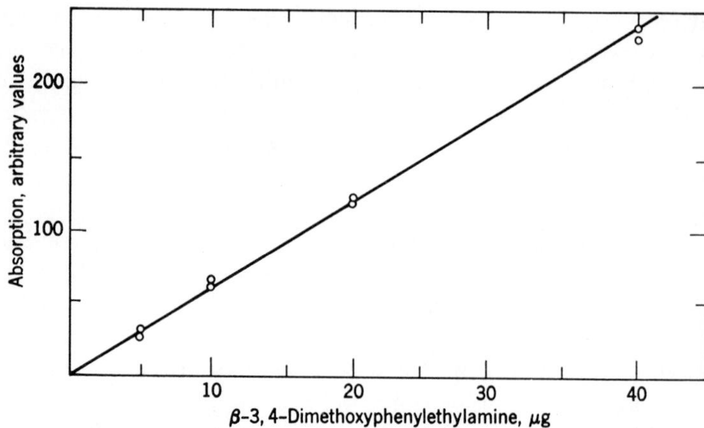

Fig. 20. Calibration curve for β-3,4-dimethoxyphenylethylamine as obtained using the ninhydrin–Ehrlichs reagents. Contrary to most other absorptiometric calibration curves, this particular example produces a linear relationship between optical density and concentration in the range 5–40 μg. Although the reason for this is not understood at this time, it may be related to the large amount of reagent present on the paper.

180 Å), it can be seen that the record obtained from the conjugated phenolicamine fraction is worse than the record obtained for the unbound phenolicamine (see also Fig. 24) fraction; both these fractions were produced from the same urine as obtained from a pink spot positive Parkinsonian patient.

The minimum level of detection of DMPE or *p*-tyramine by this technique is about 5 μg and in the range 5–40 μg the error is ±6.4% (95% fiducial limits).

3. Some Catecholamines

In attempting to obtain a more sensitive analytical method for DMPE the fluorescence procedure of Bell and Somerville (127) was investigated. Chromatograms on which susceptible substances have been separated are sprayed with acidified glycine solution (5% aqueous glycine solution adjusted to pH 3.0 with conc. HCl), partially dried in an oven for about 5 min at 65°C, and then suspended in formaldehyde vapor for 3 hr at 65°C. This is most conveniently achieved by hanging the chromatograms from glass rods in an old glass chromatography tank containing about 25 g of paraformaldehyde scattered on the bottom. Before treatment 10 ml of water is pipetted onto the paraformaldehyde and, after fitting a loose glass lid lined on its underside with a double thickness of filter paper, the whole is lifted into a large oven maintained at about 65°C. After 3 hr the tanks are very carefully removed from the oven and placed under a hood. The lid may be removed immediately, but preferably not until after cooling. If the strips are to be viewed immediately, it is necessary to wear a gas mask; alternatively, drying under the hood for 1 hr or so allows more comfortable viewing in ultraviolet light at 3650 Å. Susceptible catecholamines, phenolicamines, or amines exhibit characteristic fluorescent colors and in most cases can be seen at the submicrogram level.

It is possible to plot the spectrum and measure the amount of fluorescence of these fluorogens after elution from the chromatograms in 0.1*N* HCl. It is much more convenient and certainly a lot faster, however, to quantitate them directly in the scanning apparatus. A Chance 0Y8 secondary filter allows all the fluorogens listed by Bell and Somerville to be assessed. Typical calibration curves for DMPE, dopamine, and nordrenaline are shown in Figure 21, the

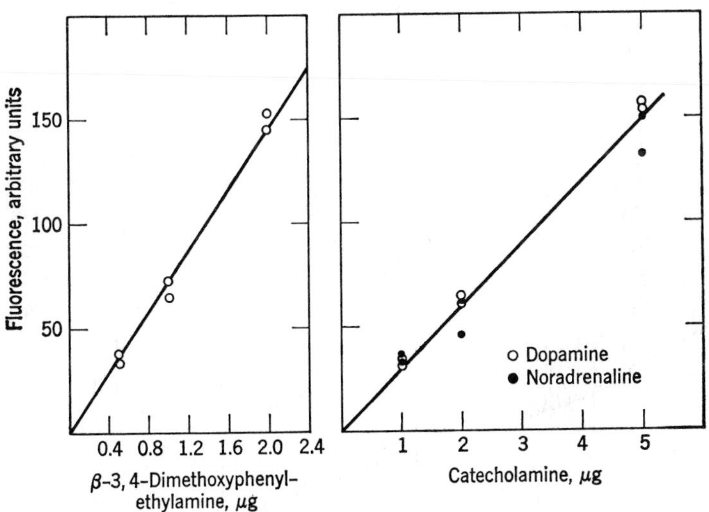

Fig. 21. Calibration curves of some catecholamines (see text for complete experimental details).

errors (95% fiducial limits) are ±10.4%, ±5.9%, and ±19.4%, respectively.

The fluorescence procedure is about 10 times more sensitive in the detection of DMPE than the absorptiometric technique (Fig. 20), but so far we have not yet detected any of this substance in the urine extracts from our schizophrenic or Parkinson patients.

The point to be emphasized in this application is that by employing a relatively specific fluorescent zone localization technique and with the attendant specificity of this zone after separation in a suitable solvent system, a somewhat more certain identification and analysis is possible than in the case of cruder, more general absorptiometric techniques. If a continuous interference filter or a monochromator were interposed between the paper and the photomultiplier tube in the scanning apparatus it would be possible to obtain, in addition to a quantitative evaluation of the separated fluorophore, a fairly diagnostic fluorescent spectrum of that fluorophore. Obviously a similar system on the primary side would allow an activation spectrum to be plotted (provided a lamp with a continuous ultraviolet emission were used) or an absorption spectrum in the case of colored zones. The obvious advantage of such a technique in the analysis of, say, a catecholamine or phenolicamine extract of a biological fluid is that one could chromatographically separate several constituents of the extract and then obtain individual quantitative values and in the crucial 0–5 μg concentration range.

4. Some *Para*-Substituted Substances

In attempts to develop a more sensitive technique for the estimation of *p*-tyramine in urinary extracts, the methods described by Smith (128) for the chromogenic localization of tyrosine on paper chromatograms and Udenfriend (129) for the analysis of *p*-tyramine in solution were modified as follows (130). Air-dried chromatograms are dipped through 1-nitroso-2-naphthol reagent (0.1 g of 1-nitroso-2-naphthol dissolved in 90 ml of ethanol, 0.5 ml of aqueous 2.5% sodium nitrite added, and the mixture diluted to 100 ml with concentrated nitric acid) and after draining for 5 min are heated for 15 min at 125°C in an oven. After removal from the oven the strips (now somewhat brittle) are allowed to stand at room temperature for 30 min. Inspection of the chromatograms under ultraviolet light at 3650 Å reveals a pleasing yellow fluorescence which is maximal at 30 min. Scanning records are obtained using a primary filter (Chance OXI 3650 Å) and a secondary filter Chance OY8 (cut off below 4400 Å). Typical calibration curves for *p*-tyramine and *p*-hydroxyphenylacetic acid are shown in Figures 22 and 23. The error involved in this quantitative analysis is ±5.2% (95% fiducial limits) for *p*-tyramine and ±5.3% (95% fiducial limits) for *p*-hydroxyphenylacetic acid in the convenient working range 0.5–5 μg. Although a comprehensive survey of this reaction with different substrates has not yet been undertaken, the following produce pleasant fluorophores which can be detected at the 1-μg level or less: tyrosine, *p*-tyramine, octopamine, *p*-sympatol, *p*-hydroxyphenylacetic acid, *p*-hydroxyphenylpropionic acid, *p*-hydroxyphenylpyruvic acid, and *p*-hydroxyphenylactic acid. In the main, the reaction seems reasonably specific for *para*-substituted substances but yellow fluorophores can be obtained with certain orthosubstituted compounds such as *o*-hydroxyphenylpropionic acid. One of the difficulties of a comprehensive survey with this particular range of substances is the difficulty in obtaining suitable samples of some of them.

In the application of this method to an analysis of *p*-tyramine in urine, the chromatograms obtained as described in Section VII-2 are treated with the 1-nitroso-2-naphthol reagent although, because of the greater sensitivity as compared with the ninhydrin–Ehrlich reagents, it is necessary to reduce the volume of extract applied to

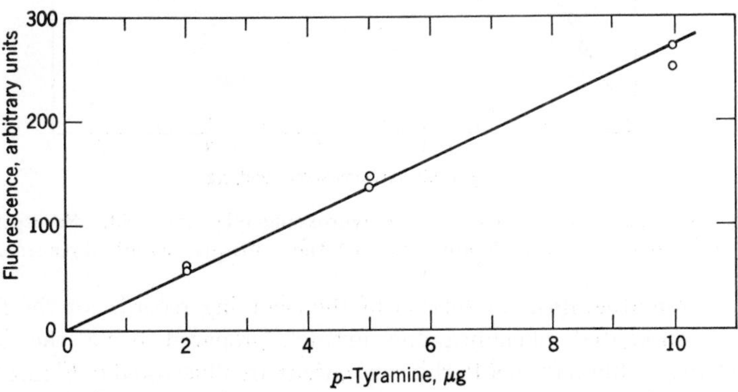

Fig. 22. Calibration curve for *p*-tyramine. Susceptible amines are converted to fluorophores using the 1-nitroso-2-naphthol reagent.

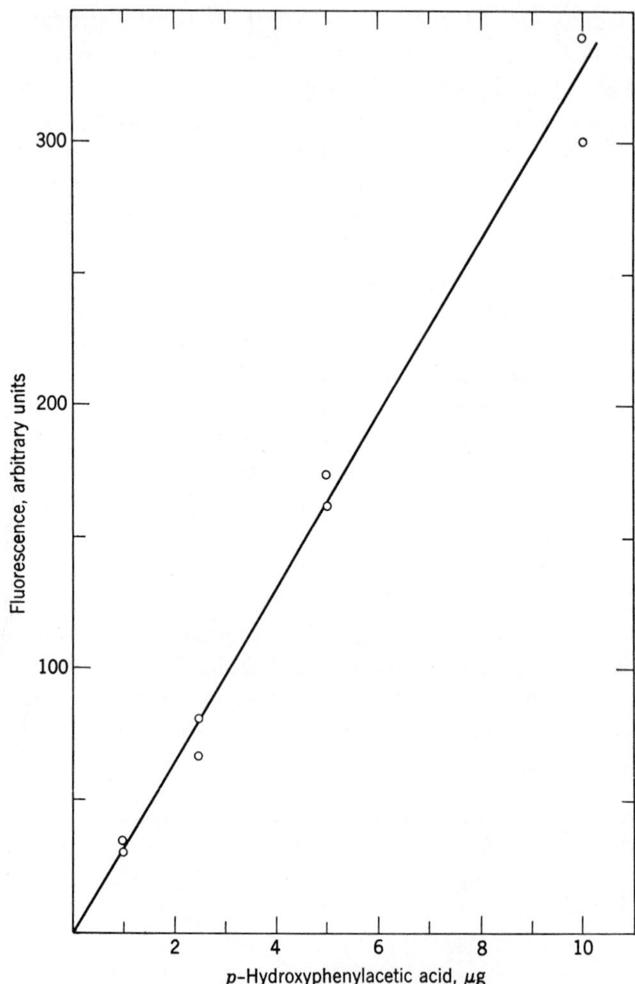

Fig. 23. Calibration curve for *p*-hydroxyphenylacetic acid. Susceptible acids are converted to fluorophores with the 1-nitroso-2-naphthol reagent.

the chromatograms. A section of the scanning records for the free and conjugated phenolicamine extract prepared using the ion-exchange column from a Parkinson patient are illustrated in Figure 24.

It is obvious that the quality of the records is much improved over those illustrated in Figure 19 which were obtained in a completely analogous fashion except in the final localization stage. It will be seen, however, that the record as produced from conjugated extracts is as usual more contaminated than the "free" form.

The peak next to the *p*-tyramine, in both the bound and unbound Parkinson extracts, exhibits a blue fluorescence after treatment with the 1-nitroso-2-naphthol reagent and is sometimes seen in extracts from these patients. It has not been seen in any extracts from healthy individuals to date. Its identity is unknown.

The acid metabolite of *p*-tyramine, *p*-hydroxphenylacetic acid may also be analyzed by the above procedure, after overnight (14 hr) separation on No. 2 paper in the solvent system benzene/acetic acid/water, 2:1:1 (v/v). Extracts suitable for chromatography are conveniently prepared by extracting an acidified urine (pH 2–0) sample (usually 5 ml) with ethyl acetate (3 × 5 ml). This urine sample can be neat urine or the initial column effluent as mentioned

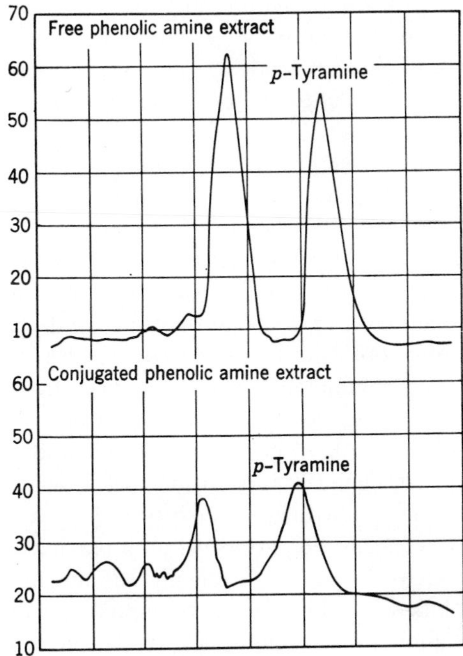

Fig. 24. Section of a fluorescent scanning record of a urine extract (see text for details).

in Section VII-2. In each extraction the ethyl acetate layer may be removed by a Pasteur pipet after low-speed centrifugation. After rotary evaporation the dried residue is dissolved in 500 μl of 70% ethanol and aliquots (usually 25 μl) applied to paper chromatographic strips. By running for 14 hr the *p*-hydroxyphenylacetic spot is found about 10 cm from the origin line. Several other zones, some exhibiting green or blue fluorescence, are frequently observed in these urinary acid extracts. The identities of these substances are not known, and, in fact, we have so far only encountered yellow fluorophores in our survey of synthetic materials.

VIII. CONCLUSION

It is clear that automated processes in the field of chromatography can considerably reduce the manual effort involved in the compilation, analysis, and to some extent the interpretation of data. One of the great advantages of chromatographic methods is the ease with which large amounts of data can be produced which, combined with an efficient method of analysis and the inherent specificity of chromatographic separations, offers a considerable speedup in our work.

It does seem that at the present time there is a need for two types of scanning apparatus, the first of which is a sturdy and cheap device for routine use. This device should possess a relatively fast scanning speed, permit the assessment of most of the conventional types of support media, operate for both fluorescent and absorptive substances, provide both analog and digital outputs, and be easy to operate. A simple optical system based on filters would suffice. Although numerous commercial scanning devices are available, most of them possess irritating faults, and they are either very slow and/or are restricted to a certain type of support medium. Because the above machine is designed for a routine role it ought to possess a fairly selective and sophisticated analytical process so that rapid quantitative analyses can be performed on a particular substance or series of substances. It is not difficult to imagine how specific parameters included in the meta-data of a convenient computer program would allow such an analysis. In the event of quantitative data being urgently required, as might be the case, for example, in the analysis of body fluids taken from a patient found in a drug-induced coma, then immediate results could be supplied by having the scanning device "on line" to the computer. In most cases, however, the chromatographic record on paper or magnetic tape is the most suitable. Later, routine scanning devices could incorporate a cross scanning and/or dual wavelength process with electronic smoothing of the signal.

In all quantitative applications attention must be given to the likely sources of error, such as variations during the drying of the chromatograms, the temperature during or preceding the scanning process, selection of the most appropriate quality of paper and the most appropriate solvent system, and the provision of proper quantitative standards for comparison.

The second type of scanning apparatus could be of a more advanced type, useful in research. This device can of course be more expensive and ought to provide for measurements by both transmission and reflection. Monochromators on both the primary and secondary sides should be included, as well as dual wavelength cross scanning with electronic subtraction and smoothing, continuous emission sources in the ultraviolet and the visible, variable scanning speeds, means to accommodate all or most types of support media, and several photomultiplier tubes, so that the whole spectral range may be covered. Such a device will allow the absorption, activation, and fluorescence spectra of absorptive and fluorescent substances to to be obtained. By recording the photomultiplier output in digital values during the measurement of a spectrum, it might be possible to correct it using a suitable computer program. An apparatus incorporating the features described above will be about twice as sensitive and accurate as most of the existing machines, and will also be useful in a qualitative sense by permitting the measurement of a spectrum.

For the future it is possible that the use of laser beams or polarized light and fiber optics will radically alter and improve the scanning process. The increased sensitivity that would accompany a reduction in the dimensions of the support media especially in the case of fluorescence measurements should also be seriously considered. If the levels of detection could be improved by a factor of 10 or more and if the substances separated on these modified supports (for instance cellulose threads or coated fibers) were scannable, the possibilities especially in the field of neurochemistry are immense.

Acknowledgments

I would like to thank Miss Carol Felton for her expert technical assistance; Mr. N. Chard, Mr. L. Grant, and Mr. E. Wakelam for their help in the design of the apparatus and also for the final constructions from these designs; Dr. I. Harris for the provision of the ion-exchange data; and Dr. Ivor Smith and Heinemann Medical Books Limited for permission to reproduce Figures 1, 5–8, and 18. I am particularly indebted to Mr. H. F. Ross and Mr. R. L. Holder for the interest they have taken in various aspects of this work and especially for the provision of the analytical procedure involving the computer.

References

1. A. J. P. Martin and R. L. M. Synge, *Biochem. J.*, *35*, 1358 (1944).
2. I. E. Bush, *The Chromatography of Steroids*, Pergamon Press, London, 1961.
3. A. A. Boulton, in *Chromatographic and Electrophoretic Techniques*, Vol. 1, I. Smith, Ed., William Heinman Medical Books Ltd., 1968.
4. P. B. Hamilton, *Nature*, *205*, 284 (1965).
5. E. G. C. Clarke and S. A. Sowter, *Nature*, *202*, 795 (1964).
6. M. F. Bacon, *J. Chromatog.*, *16*, 552 (1964).
7. W. M. Philip and H. G. Sammons, *Clin. Chim. Acta*, *11*, 285 (1965).
8. H. J. Monteiro, *J. Chromatog.*, *18*, 594 (1965).
9. R. E. Bailey, H. P. Peters, and J. H. Beek, *Nature*, *204*, 485 (1964).
10. C. R. Turner, *J. Chromatog.*, *22*, 471 (1966).
11. T. Darocha, C. H. Gray, and R. V. Quincey, *J. Chromatog.*, *27*, 497 (1967).
12. F. A. Vandenheuvel, *J. Chromatog.*, *25*, 102 (1966).
13. I. E. Bush, in *Methods of Biochemical Analysis*, Vol. 11, D. Glick, Ed., Interscience, New York, 1963.
14. R. Klaus, *J. Chromatog.*, *16*, 311 (1964).
15. R. J. Wieme, *J. Chromatog.*, *1*, 166 (1958).
16. R. R. Goodall and J. Goldman, personal communication.
17. I. E. Bush, *Science*, *154*, 77 (1966).
18. M. M. Frodyma, R. W. Frei, and D. J. Williams, *J. Chromatog.*, *13*, 61 (1964).
19. M. M. Frodyma, V. T. Lieu, and R. W. Frei, *J. Chromatog.*, *18*, 520 (1965).
20. F. Korte and H. Weitkamp, *Angew. Chem.*, *70*, 434 (1958).
21. A. L. Latner, L. Molyneux, and J. D. Rose, *J. Lab. Clin. Med.*, *43*, 157 (1954).
22. J. W. H. Lugg, *J. Chromatog.*, *10*, 272 (1963).
23. E. H. Winslow and H. A. Liebhafsky, *Anal. Chem.*, *21*, 1338 (1949).
24. R. B. Ingle and E. Minshall, *J. Chromatog.*, *8*, 369 (1962).
25. R. J. Block, E. L. Durrum, and G. Zweig, *A Manual of Paper Chromatography and Paper Electrophoresis*, 2nd ed., Academic Press, New York, 1958.
26. L. Salganicoff, M. Kraybill, D. Mayer, and V. Legallais, *J. Chromatog.*, *26*, 434 (1967).
27. J. A. Brown and M. M. Marsh, *Anal. Chem.*, *25*, 1865 (1953).

28. D. J. R. Laurence, *J. Sci. Instr.*, *31*, 137 (1954).
29. D. J. R. Laurence, *Biochem. J.*, *56*, xPv (1954).
30. B. L. Hamman and M. M. Martin, *Anal. Biochem.*, *15*, 305 (1966).
31. H. T. Gordon, *J. Chromatog.*, *22*, 60 (1966).
32. P. J. Ayres, S. A. Simpson, and J. F. Tait, *Biochem. J.*, *65*, 647 (1957).
33. W. Grassmann and K. Hannig, *Z. Physiol. Chem.*, *290*, 1 (1952).
34. W. Grassman, K. Hannig, and M. Knedel, *Deut. Med. Woch.*, *76*, 333 (1951).
35. D. M. Abelson, *Nature*, *208*, 784 (1965).
36. E. M. Crook, H. Harris, F. Hassam, and F. L. Warren, *Biochem. J.*, *56*, 434 (1954).
37. K. Semm and R. Fried, *Naturwiss.*, *14*, 326 (1952).
38. A. A. Boulton, N. Chard, and L. Grant, *Biochem. J.*, *96*, 82P (1965).
39. A. A. Boulton, N. Chard, and L. Grant, *Biochem. J.*, *96*, 69P (1965).
40. M. R. Yourag and J. A. Armstrong, *Nature*, *213*, 649 (1967).
41. P. Kubelka and F. Munk, *Z. Tech. Physik.*, *12*, 593 (1931).
42. F. A. Steele, *Paper Trade J.*, *100*, 37 (1935).
43. M. H. Sweet, *Electronics*, 105 (1946).
44. R. Spencer, personal communication.
45. V. H. Rees and A. J. R. Lawrence, *Clin. Sci.*, *1*, 329 (1955).
46. I. E. Bush, *J. Chromatog.*, *23*, 94 (1966).
47. A. A. Boulton, R. L. Holder, and H. F. Ross, *Biochem. J.*, *96*, 83P (1965).
48. A. A. Boulton, R. L. Holder, and H. F. Ross, *Biochem. J.*, *96* 70P (1965).
49. G. N. Graham and B. Sheldrick, *Biochem. J.*, *96*, 517 (1965).
50. F. H. Pollard, G. Nickless, D. E. Rogers, and D. L. Crone, *Proc. Soc. Anal. Chem.*, *1965*, 481.
51. A. Yonda, D. L. Filmer, H. Pate, N. Alonzo, and C. H. W. Hirs, *Anal. Biochem.*, *10*, 53 (1965).
52. R. D. B. Fraser, A. S. Inglis, and A. Miller, *Anal. Biochem.*, *1*, 247 (1964).
53. M. I. Krichevsky, J. Schwartz, and M. Mage, *Anal. Biochem.*, *12*, 94 (1965).
54. P. Vestergaard and S. Vedso, *J. Chromatog.*, *19*, 512 (1965).
55. F. Albert-Recht and J. A. Owen, *J. Chromatog.*, *10*, 577 (1964).
56. A. A. Boulton, R. L. Holder, and H. F. Ross, *Intern. Congr. Clin. Chem.*, *6th, Munich, July 1966.*
57. J. T. Bell and R. E. Biggers, *J. Mol. Spectry.*, *18*, 247 (1965).
58. J. A. Blackburn, *Anal. Chem.*, *39*, 100 (1967).
59. H. Svenson, *J. Chromatog.*, *25*, 266 (1966).
60. G. Z. Williams and R. G. Neuhauser, *Ann. N.Y. Acad. Sci.*, *97*, 346 (1962).
61. J. G. Hoffman, *Ann. N.Y. Acad. Sci.*, *97*, 380 (1962).
62. J. J. Freed and J. L. Engle, *Ann. N.Y. Acad. Sci.*, *97*, 412 (1962).
63. T. O. Caspersson and G. M. Lomakka, *Ann. N.Y. Acad. Sci.*, *97*, 449 (1962).
64. L. L. Hundley, *Ann. N.Y. Acad. Sci.*, *97*, 514 (1962).
65. A. Ward and G. W. McMaster, *Nature*, *208*, 428 (1965).
66. W. A. Bonner, *Ann. N.Y. Acad. Sci.*, *97*, 408 (1962).
67. V. E. Cosslett, *Ann. N.Y. Acad. Sci.*, *97*, 464 (1962).
68. P. O'B. Montgomery, *Ann. N.Y. Acad. Sci.*, *97*, 491 (1962).
69. D. H. Spackman, W. H. Stein, and S. Moore, *Anal. Chem.*, *30*, 1191 (1958).

70. I. Harris and H. F. Ross, personal communication.
71. G. L. Mills, in *Chromatographic and Electrophoretic Techniques*, Vol. 1, 2nd ed., I. Smith, Ed., Heineman Medical Books, 1960, p. 143.
72. A. A. Boulton, *Fed. Eur. Biochem. Soc., 2nd Meeting, Vienna, 1965*, A265.
73. A. A. Boulton, N. Chard, and L. Grant, *Biochem. J.*, *96* (1965).
74. R. Consden, in *Stationary Phase in Paper and Thin Layer Chromatography*, (Proceedings of 2nd Symposium, Liblice Czechoslovakia, 1964) K. Macek and I. M. Hais, Eds.
75. Whatman Paper Chromedia Technical Bulletin C.4 (1965), available from H. Reeve Angel & Co. Ltd., 14, New Bridge Street, London, E.C. 4.
76. A. A. Boulton and I. E. Bush, *Biochem. J.*, *92*, 11P (1964).
77. A. A. Boulton, *2nd Intern. Neurochem. Conf., Oxford*, 10 (1965).
78. W. R. Gray and B. S. Hartley, *Biochem. J.*, *89*, 59P (1963).
79. Z. Deyl and J. Rosmus, *J. Chromatog.*, *20*, 514 (1965).
80. B. Mesrob and V. Holeyšovský, *J. Chromatog.*, *21*, 135 (1966).
81. M. Cole, J. C. Fletcher, and A. Robson, *J. Chromatog.*, *20*, 616 (1965).
82. G. Pataki and E. Strasky, *Chimia*, *20*, 361 (1966).
83. D. Morse and B. L. Horecker, *Anal. Biochem.*, *14*, 429 (1966).
84. H. Rinderknecht, *Nature*, *193*, 167 (1962).
85. A. A. Galoyan, B. K. Mesrob, and V. Holeyšovský, *J. Chromatog.*, *24*, 440 (1966).
86. K. Crowshaw, S. J. Jessup, and P. W. Ramwell, *Biochem. J.*, *103*, 79 (1967).
87. N. Seiler and M. Wiechmann, *J. Chromatog.*, *28*, 351 (1967).
88. N. Seiler, *Hoppe-Seylers Z. Physiol. Chem.*, *348*, 601 (1967).
89. N. Seiler and M. Weichmann, *Z. Anal. Chem.*, *220*, 109 (1966).
90. N. Seiler and J. Weichmann, *Experientia*, *20*, 559 (1964).
91. N. Seiler and M. Wiechmann, *Experientia*, *21*, 203 (1965).
92. I. Durko and A. A. Boulton, *Intern. Neurochem. Conf. Strasbourg, 3rd* (1967).
93. D. Beale, in *Chromatographic and Electrophoretic Techniques*, Vol. 1, I. Smith, Ed., W. Heineman Medical Books, 1968.
94. W. L. Porter and E. A. Talley, *Anal. Chem.*, *36*, 1692 (1964).
95. J. Tempé, *J. Chromatog.*, *24*, 169 (1966).
96. H. J. Kupferberg, A. Burkhalter, and E. L. Way, *J. Chromatog.*, *16*, 558 (1964).
97. W. B. Shelley and L. Juhlin, *J. Chromatog.*, *22*, 130 (1966).
98. J. Bergerman, *Clin. Chem.*, *12*, 797 (1966).
99. A. L. Latner and D. C. Park, *Clin. Chim. Acta*, *11*, 538 (1965).
100. N. M. Papadopoulos and J. A. Kintzios, *Am. J. Clin. Pathol.*, *47*, 96 (1967).
101. G. Pataki and A. Kunz, *J. Chromatog.*, *23*, 465 (1966).
102. N. Seiler, G. Werner, and M. Wiechmann, *Naturwiss.*, *50*, 643 (1963).
103. M. M. Frodyma and R. W. Frei, *J. Chromatog.*, *15*, 501 (1964).
104. E. Sawicki, T. W. Stanley, and W. C. Elbert, *J. Chromatog.*, *20*, 348 (1965).
105. N. M. Neskovic, *J. Chromatog.*, *27*, 488 (1967).
106. W. M. Connors and W. K. Boak, *J. Chromatog.*, *16*, 243 (1964).
107. N. Zöllner, G. Wolfram, and G. Amin, *Klin. Woch.*, *40* (1962).

108. D. A. Forss, P. R. Edwards, B. J. Sutherland, and R. Birtwistle, *J. Chromatog.*, *16*, 460 (1964).
109. K. Genest, *J. Chromatog.*, *19*, 531 (1965).
110. G. Sememuk and W. T. Beher, *J. Chromatog.*, *21*, 27 (1966).
111. R. D. Spencer and B. H. Beggs, *J. Chromatog.*, *21*, 52 (1966).
112. W. M. Lamkin, D. N. Ward, and E. F. Walborg, *Anal. Biochem.*, *17*, 485 (1966).
113. H. Rasmussen, *J. Chromatog.*, *27*, 142 (1967).
114. R. F. McGregor, M. Khan, D. Marrack, and M. P. Sullivan, *Am. J. Clin. Pathol.*, *46*, 163 (1966).
115. A. A. Boulton and C. Felton, *Nature*, *211*, 1404 (1966).
116. K. T. Wang, personal communication.
117. S. G. Waley, *Biochem. J.*, *96*, 722 (1965).
118. J. Boyer and P. Talalay, *J. Biol. Chem.*, *241*, 180 (1966).
119. W. R. Gray and B. S. Hartley, *Biochem. J.*, *89*, 379 (1963).
120. D. J. Neadle and R. J. Pollitt, *Biochem. J.*, *97*, 607 (1965).
121. E. F. Legg, M.Sc. thesis, University of Birmingham, 1967.
122. A. J. Friedhoff and E. van Winkle, *Nature*, *194*, 897 (1962).
123. T. L. Perry, K. N. F. Shaw, D. Walker, and D. Redlich, *Pediatrics*, *30*, 576 (1962).
124. A. A. Boulton, R. J. Pollitt, and J. R. Majer, *Nature*, *215*, 132 (1967).
125. Y. Kakimoto and M. D. Armstrong, *J. Biol. Chem.*, *237*, 208 (1962).
126. A. J. Friedhoff and E. van Winkle, *J. Chromatog.*, *11*, 272 (1963).
127. C. E. Bell and A. R. Somerville, *Biochem. J.*, *98*, 1C (1966).
128. I. Smith, in *Chromatographic and Electrophoretic Techniques*, Vol. 1, Wm. Heinemann Medical Books, 1960, p. 97.
129. S. Udenfriend, in *Fluorescence Assay in Biological Medicine*, Academic Press, 1962, p. 134.
130. A. A. Boulton, Congress of Neurogenetics and Neuro-ophthalmology, in *Excerpta Med.*, *154*, 35 (Sept. 1967).

AUTHOR INDEX

Numbers in parentheses are reference numbers and indicate that the author's work is referred to although his name in not mentioned in the text. Numbers in *italics* show the pages on which the complete references are listed.

A

B

C

M

N

O

T

U

V

W

SUBJECT INDEX

F

H

M

T

Methods of Biochemical Analysis

CUMULATIVE INDEX, VOLUMES 1–16

Author Index

Subject Index